History of Physics

The Springer book series *History of Physics* publishes scholarly yet widely accessible books on all aspects of the history of physics. These cover the history and evolution of ideas and techniques, pioneers and their contributions, institutional history, as well as the interactions between physics research and society. Also included in the scope of the series are key historical works that are published or translated for the first time, or republished with annotation and analysis.

As a whole, the series helps to demonstrate the key role of physics in shaping the modern world, as well as revealing the often meandering path that led to our current understanding of physics and the cosmos. It upholds the notion expressed by Gerald Holton that "science should treasure its history, that historical scholarship should treasure science, and that the full understanding of each is deficient without the other." The series welcomes equally works by historians of science and contributions from practicing physicists.

These books are aimed primarily at researchers and students in the sciences, history of science, and science studies; but they also provide stimulating reading for philosophers, sociologists and a broader public eager to discover how physics research – and the laws of physics themselves – came to be what they are today.

All publications in the series are peer reviewed. Titles are published as both print- and eBooks. Proposals for publication should be submitted to Dr. Angela Lahee (angela.lahee@springer.com) or one of the series editors.

Tijmen Jan Moser · Enders Anthony Robinson

Walking with Christiaan Huygens

From Archimedes' Influence to Unsung Contributions in Modern Science

Springer

Tijmen Jan Moser
The Hague, The Netherlands

Enders Anthony Robinson (Deceased)
Venice, FL, USA

ISSN 2730-7549 ISSN 2730-7557 (electronic)
History of Physics
ISBN 978-3-031-46157-6 ISBN 978-3-031-46158-3 (eBook)
https://doi.org/10.1007/978-3-031-46158-3

This Springer imprint is published by the registered company Springer Nature Switzerland AG
The registered company address is: Gewerbestrasse 11, 6330 Cham, Switzerland

Paper in this product is recyclable.

I dedicate this book to the memory of my co-author Prof. Enders Anthony Robinson, who sadly passed away just before completing the book.

For many generations of geophysicists Enders Robinson has been the everlasting standard. He was one of the founders of modern seismic prospecting and the father of deconvolution. From the start of his studies in geophysics in the 1950s until his passing in 2022, Enders devoted his life to the field, leaving an indelible mark as a researcher and teacher. As a student and later professional in geophysics, I was always aware of Enders' work and his many publications, journal articles and books. I first came in personal contact with Enders when the Society of Exploration Geophysicists asked me to review his book Basic Geophysics. Somehow he must have liked my suggestions, because he then proposed that I write the Introduction to the book. This was already a great honor. Then he invited me as a co-author for his next book, Basic Wave Analysis. Our cooperation culminated in the current book on Christiaan

Huygens. We exchanged many emails when working on this project. We shared a passion for writing and publishing, for history of science, philosophy, art and poetry. I learned many new ideas from him in all these areas. Above all, I cherished his inspiration in the field of didactics and how to teach and explain the basics of applied mathematics, physics, and geophysics to a wider public. Christiaan Huygens was our hero and using Huygens' principle for reconstructing unknown subsurface geology from seismic data was our central credo.

It was such a pleasure and honor to work with him. He was always encouraging and full of spirit. He had a great sense of humor, as is evidenced by his publications, lectures and courses. He was the personification of the joy of doing Earth sciences and using physics in exploring the Earth. He was still full of plans and ideas. His passing is a great loss for the scientific community. It is our solemn duty to continue in his spirit.

Tijmen Jan Moser

Preface

In the 16th and 17th centuries, Dutch ships became global freight carriers for European nations. The Dutch navy was formidable, and Dutch merchants were astute in politics and economics. They established and protected their own colonies in the Caribbean, South America, South Africa, and the East Indies. Their cautious, resource-driven expansion brought immense wealth to the Dutch Republic, making Amsterdam a European financial center. Despite their small size and population, the Dutch understood that they had world enough and time, provided that they made good use of their resources and did not waste time.

This period of great wealth and cultural achievement was called the Dutch Golden Age. Constantijn Huygens (1596–1687), Lord of Zuilichem, was a Dutch Golden Age statesman, poet, and composer. He served as the secretary of two Princes of Orange and was the father of scientist Christiaan Huygens (1629–1695). Another representative of this period was the Dutch artist Rembrandt van Rijn (1606–1669), one of the greatest painters in history. His works represent a wide range including portraits, landscapes, allegorical scenes, historical scenes, and biblical scenes. Rembrandt was productive and innovative. In fact, it was Constantijn Huygens who, in 1629, first discovered Rembrandt's talent and who secured commissions for him from the court of The Hague. Constantijn later described Rembrandt as:

"A Dutchman, a beardless miller, could bring together so much in one human figure and express what is universal. All honor to thee, my Rembrandt! To have carried Illium, indeed all Asia, to Italy is a lesser achievement than to have brought the laurels of Greece and Italy to Holland, the achievement of a Dutchman who has seldom ventured outside the walls of his native city."

With this background, Christiaan Huygens knew what he had to do in order to live up to the expectations of his country and especially of his father. He had to show that he had world enough and time to solve many outstanding problems in timekeeping, mechanics, astronomy, and navigation. Huygens did not dabble in scientific enterprises that had no practical value.

Some of Huygens' inventions are:

The **pendulum clock**, designed and built in 1659, remained the most precise timekeeping device until the 1930s.

Centrifugal governors, which were designed by Huygens to regulate the speed of machinery, in particular windmills. Their adaptation to control steam engines in 1788 by James Watt marked a turning point in the Industrial Revolution.

The **cycloidal pendulum**. The conventional pendulum was not quite isochronous, meaning that its period increased somewhat with its amplitude. Huygens determined which curve an object must follow to descend by gravity to the same point in the same time interval, regardless of starting point: the so-called tautochrone curve. He showed this curve was not the circular arc of a pendulum, but a cycloid. By doing so he entered in the field which we now call differential geometry.

The **compound pendulum**. Huygens solved the problem of how to calculate the period of an arbitrarily shaped pendulum (called a compound pendulum), discovering the center of oscillation of a moving object.

Gunpowder engine, also known as explosion engine. Huygens harnessed the explosive energy for mechanical work. This laid the groundwork for developments in combustion science and internal combustion engines.

The **Huygens eyepiece**, a multi-lens compound eyepiece, which improved the clarity and magnifications of telescopic observations and revolutionized astronomy.

Huygens' lantern, or magic lantern (Laterna Magica). This device formed enlarged images from slides and projected them onto a screen, predating the modern projector.

Balance spring. Invented by Huygens in 1657, the balance spring marked a monumental leap in the accuracy of timekeepers, in particular watches, to within 10 minutes per day. The innovation was disputed by Robert Hooke; until today this dispute is unresolved.

Huygens' telescope. This invention eliminated the need for a traditional tube by mounting the objective lens on a pole, allowing the observer to view celestial objects from the ground while holding the eyepiece. The telescope could therefore be much longer allowing larger magnification.

31 tone system. By dividing the octave into 31 equal parts, Huygens achieved an approximate equal temperament system for musical instruments (1691). For this purpose, he entered in the field of continued fractions (which he already used to construct his planetarium in 1680).

For most of these inventions, Huygens pioneered the underlying physics and mathematics as well. By the turn of the 17th century, mathematics and physics were still largely in the same state as in antiquity. Ancient philosophy placed great value on moderation and considered it a great virtue. Ancient mathematics was mainly concerned with finite numbers and was dominated by the geometry of bodies at rest; it lacked calculus to describe moving bodies or propagating waves, like sound, water, or light waves. Since the Renaissance, the known geographical world had expanded dramatically. Mathematics and physics stretched to infinity. Microscopy made possible the study of infinitesimal creatures, the telescope made it possible to study remote celestial bodies, in both cases invisible to the human eye. This expansion of human knowledge and aspiration to explore the infinite is sometimes associated with the legendary figure of Faust, reflecting the unbounded curiosity and ambition of the era.

As we show in this book, Christiaan Huygens played a key role in bridging ancient and modern science. However, history has often undervalued his contributions for several reasons. Firstly, the Dutch Republic, though influential at the time, was just a small nation, compared to its neighbors England and France. Secondly, the Dutch Golden Age, while splendid, had already waned by 1700, leading to a perception of Dutch stagnation. In contrast, England and France continued to expand their geographic and scientific pursuits. Isaac Newton assumed a prominent role in the scientific realm in the Anglo-Saxon world. Thirdly, within the Dutch-speaking world, the name 'Huygens' is often associated with Constantijn, Christiaan's father, rather than with Christiaan. Fourthly, the lack of historical focus on Huygens may be related to Christiaan's character. While he displayed confidence and a certain degree of bravado in his scientific correspondence, he remained modest in promoting his work.

Our primary motivation to write this book is to increase the recognition of Huygens in the development of natural science among a general readership. All of the traditional books about Huygens are, in effect, languishing in the 19th century. None of these books contain any of the stunning results Huygens achieved in relation to 20th-century physics, such as quantum mechanics and Einstein's relativity. General knowledge of the spectacular results of Huygens would put Huygens in the same category as Leonardo da Vinci.

Huygens' collected works are available in digital format and represent an almost inexhaustible source of information, much of which has not been explored yet. One of the goals of the book is to convey the excitement of delving into these original scientific texts, discovering that, so many centuries ago, they dealt with problems similar to ours and experiencing the satisfaction of solving scientific puzzles.

Two examples: In September 1646, the French mathematician Marin Mersenne (1588–1648) wrote to Constantijn Huygens and presented an intriguing problem to his children: "Is the difference of two squares, one of which is the sum and the other the difference of two squares, itself a square?". On December 28, 1661, Father Caspar Schott (1608–1666), a Jesuit professor in Würzburg, corresponded with Christiaan Huygens about the problem "Constructing a quadrangle when its area and four sides are given." Huygens subsequently solved the problems. Part of the charm of these challenges is of course their deceptively simple formulation.

In 1990, an International Symposium was held in The Hague/Scheveningen, titled "Huygens' principle 1690–1990, Theory and Application," commemorating the tricentennial of the publication of Huygens' famous treatise "*Traité de la Lumière*" in 1690. One of the contributors, J. J. Duistermaat, noted in the proceedings: "the theory of oscillatory integrals and Fourier integral operators has several elements which are actually closer to what Huygens did than the results of (20th-century mathematicians) Hadamard and M. Riesz. In order to explain this, I will give a sort of guided tour of this theory, shouting 'Huygens!' whenever I think it is appropriate." This spirit is exactly what motivated us to write this book.

While browsing Huygens' collected works, one discovers that there is a lot of walking, no doubt to discuss with contemporaries problems such as those above. This inspired the title of our book, "Walking with Christiaan Huygens." Huygens walked with the ancients. In particular, he walked with Archimedes, to whom he was often compared and for whom he had immense respect. As we show, modern science walks with Huygens.

In this sense, Huygens walked with Leonardo da Vinci, more than with any of his contemporaries. Leonardo is the father of non-mathematical technology. Huygens is the father of mathematical technology (that is, high technology). Huygens can be named the first mathematical physicist. Our book puts all of this in a readable form by showing how Huygens obtained his greatest results. We believe that it is time that Huygens got due credit for all of what he has done.

Several recent publications highlight the historical undervaluing of Huygens and we believe that there are many more to follow. Some publications approach this Huygens revival from a historical and biographical perspective. Other publications are advanced specialist texts in various fields where Huygens' name and role resurface time and again. Our book is meant to approach Huygens' role from the viewpoint of the generalist/non-specialist interested in natural sciences. We took a lot of inspiration from C. D. Andriesse's biography "*Huygens: The Man Behind the Principle*" (2005), H. Aldersey-Williams' "*Dutch Light*" (2021) and V. Icke's book on Huygens' principle "*De principes van Huygens*" (2013, in Dutch). J. G. Yoder's "*Unrolling Time, Christiaan Huygens and Mathematization of Nature*" presents valuable insights into Huygens' mathematical work with a focus on the mathematics related to the pendulum clock (involutes and evolutes). J. Aarts' book "*Christiaan Huygens, Horologium Oscillatorium*" (2015, in Dutch) presents a translation and very detailed commentary to Huygens' book on the pendulum clock. Working one's way through all examples in Aarts' book takes patience, but it is very well rewarded in the spirit of puzzle-solving.

A case in point is the physics of light. The nature of light has been one of the main problems in physics throughout time. In 1900, the American physicist Henry Crew (1859–1953) published memoirs by Christiaan Huygens (an English translation of the first three chapters of Huygens' *Traité de la Lumière*, or *Treatise on Light*), Thomas Young (1773–1829) and Augustin Jean Fresnel (1788–1827). In an often-quoted preface to these memoirs, Crew made an inventory of the state of affairs in the physics of light in the 17th century:

"At the time when Huygens and Newton began their work on light, the following phenomena were demanding explanation:

1. The existence of rays and shadows, known from the earliest times.
2. The phenomenon of reflection, known from the earliest times.
3. The phenomenon of refraction, as described by Snell's law.
4. The rainbow and the production of color by the prism.
5. The colors of thin plates—Newton's rings.
6. Diffraction bands outside the geometrical shadow, described by Grimaldi, 1665.

To these might be added the two following phenomena which were discovered before the final publication of Newton's *Opticks* (1704) and Huygens' *Traité de la Lumière* (1690).

7. The polarization of light by crystals (Bartholinus 1670).
8. The finite speed of light (Römer, 1675)."

Crew continued by noting that among these questions, Huygens had provided explanations of the second (reflection), third (refraction), and eighth (finite speed of light) phenomenon, based on his assumptions that light propagates as a wave, that it travels at a constant speed through a luminiferous (light transmitting) ether, and that in different media it travels at a speed inversely proportional to the refraction index of those media. Newton gave a correct explanation of the fourth phenomenon (rainbows) and an as yet incorrect one of the fifth (colors of thin plates). By the early 19th century, Young gave a correct explanation for the colors of thin plates, and Fresnel correctly explained the existence of rays.

Our book covers a comprehensive range of topics of Huygens' work and for most of them makes the relation to modern science. This applies in particular to the special theory of relativity which was mathematically anticipated by Huygens. Huygens' work on the centrifugal force alludes to the equality of gravitational and inertial mass and hence heralded the general theory of relativity. Huygens' formulation of his (that is, Huygens') principle for the propagation of light waves is shown to be fundamental in all wave propagation phenomena (from light to sound and seismic to quantum mechanics). Huygens the instrument maker (of the pendulum clock, telescope, planetarium) is presented to show his practical inventiveness in relation to initiating and developing applied science (for instance continued fractions for the construction of the planetarium). There is even a link between Huygens and the fictional character Sherlock Holmes. We have extensively studied Huygens' scientific correspondence and found some new viewpoints illustrated by little-known quotations. These quotations show that Huygens could be impatient but also had a refreshing sense of humor.

In addition, we point out that many of the phenomena encountered in geophysics have already been described by Huygens, especially in seismology and seismic exploration. In various applications, Huygens already anticipated modern remote sensing. We demonstrate this by a number of examples—even from the modern petroleum industry—and apparently this is an aspect of Huygens' contributions which is largely unknown to the general public. Huygens is very common in geophysical literature. Open any textbook on geophysical prospecting and you will find Huygens ubiquitously present.

We go to some level into the role of Huygens in the natural philosophical debate of the 17th century. Huygens was embedded in mainstream philosophy (Cartesian rationalism) and religious belief (Protestantism), but in his scientific endeavors, he took a very modern and undogmatic approach. He valued scientific theory, but the physical experiment was his final judge. We believe that Huygens' attitude to science is exemplary even for modern scientists and science enthusiasts.

One particular topic in Crew's list is diffraction, the phenomenon of wave propagation around obstacles. One of the novel aspects in this book is what Huygens might have said on the topic of diffraction. We argue that one of the major mysteries of the history of science is why Huygens did not consider diffraction, even though it was a subject of intense interest and debate among his contemporaries. Diffraction was described for the first time by the Italian Jesuit priest Francesco Maria Grimaldi (1618–1663). Newton spent time trying to understand diffraction but lacked a proper physical explanation. Huygens, on the other hand, could have described and explained diffraction straightforwardly using his (Huygens') principle for wave propagation, but for reasons not yet entirely clear he refrained from doing so. This stalemate seems to have held up development for more than a century until the arrival of Young and Fresnel in the 19th century.

The particular place of diffraction is also the reason why we devote a separate chapter to it (Chap. 11), separately from the two chapters on Huygens' principle. While Huygens' principle, introduced in Chap. 4 and further developed in the 19th century by Fresnel and Kirchhoff (Chap. 7), offers a comprehensive and, under certain conditions, exact framework for solving wave propagation problems, diffraction presents a distinct challenge. It typically involves the application of approximations due to its intricate nature. The history of diffraction is a fascinating journey marked by efforts to generalize these necessary approximations. In this book, we allocate space to recount this historical narrative, staying true to the spirit of Huygens' quest for understanding wave phenomena.

Huygens played a pivotal role in bridging ancient and modern science. He still had one foot firmly planted in antique mathematics, where his skills were unequaled (as acknowledged by his contemporaries Newton and Leibniz). We show how Huygens was instrumental in the use of so-called proto-calculus (that is, the earliest form of calculus). It is a form of calculus which predates even Archimedes. For example, Huygens was able to find the second derivative in terms of proto-calculus. Except in a few isolated instances, Newton did not use his own form of modern calculus in his masterwork *Principia*. The reason is unknown. It might be argued that Newton's own calculus was still too immature to be easily used in formulating Newton's laws of classical mechanics. Instead, Newton's *Principia* makes use of proto-calculus which was well understood at the time. Expressing *Principia* in terms of modern calculus would be the work of the next generation of mathematicians. We walk with Huygens during a few of his ingenious derivations.

Finally, we pay due attention to Huygens' remarkable last book, the "*Cosmotheoros*", in which he not only presented some daring speculations about extraterrestrial life but also promoted the concept of remote sensing and the cosmological principle.

The Hague, The Netherlands — Tijmen Jan Moser
Venice, USA — Enders Anthony Robinson

Acknowledgements

Compiling this book dedicated to our scientific hero, Christiaan Huygens, has been a rewarding experience made possible with the invaluable support of many.

We extend our sincere gratitude to the following organizations and contact persons for granting permission to reproduce figures in the book:

Rijksmuseum Amsterdam (Maria Smit),
Leiden University Library (Mart van Duijn),
Society of Exploration Geophysicists (Jennifer Cobb),
Optica Publishing Group (Hannah Greenwood),
Rijksmuseum Boerhaave Leiden (Mara Scheelings),
Nationaal Archief (René Janssen),
Taylor & Francis Group (Annabel Flude),
Huygensmuseum Hofwijck Voorburg (Wang Ch. Choy),
Académie des Sciences (Julien Pomart),
Treasure Mountain Mining (Eric Greene),
Museum Zaanse Tijd Zaandam (Pier van Leeuwen),
Paraselene (Eva Seidenfaden).

We are most grateful to Arkady Aizenberg, Evgeny Landa, Shmoriahu Keydar, Kamill Klem-Musatov, and Jan Pajchel for raising the interest in diffraction, engaging in many compelling discussions, collaborating on projects, and preparing publications.

The first author of the book is especially indebted to Chris de Koster, for the introduction to the world of Christiaan Huygens, several decades ago.

We are very grateful to Richard van den Brink for generous support over so many years.

Sincere thanks go to Michael Pelissier, Mihai Popovici, and Ioan Sturzu, for helping raise the industrial interest in seismic diffraction.

Special thanks go to Henning Hoeber for generously sharing material on the history of diffraction.

The work of the reviewers has been indispensable. We are very grateful to the following individuals who took their time to go through the book and make many valuable comments:

Leon Thomsen, especially for his insightful remarks on Chap. 8 on Huygens and Special Relativity and for generously sharing his photo of Jupiter and its moons.

Sven Treitel, a long-time friend of Enders Robinson, who played a crucial role in ensuring that, after his passing, Enders' spirit was optimally preserved within the pages of the book.

Alexey Stovas, for his clarifying comments on anisotropy.

Aldo Vesnaver, who skillfully put his countryman Galileo into focus.

Rob Memel, for his detailed review of Chap. 6 on Huygens and the Clock, and adding professional knowledge on clock making.

Last but not least, Bernardijn de Koster-Schneider, whose thorough proofreading saved the book from quite a few faux-pas and pitfalls.

Finally, we thank the Springer staff and especially Annelies Kersbergen and Madanagopal Deenadayalan, for their support and advice.

Contents

About the Authors

Tijmen Jan Moser received his Ph.D. from the University of Utrecht, concentrating on the shortest-path method for seismic ray tracing. He has worked as a geophysical consultant for a number of companies and institutes, including Amoco, Institut Français du Pétrole, Karlsruhe University, Bergen University, Statoil/Hydro, Geophysical Institute of Israel, Fugro-Jason, and Horizon Energy Partners. He was an Alexander-von-Humboldt stipendiat from 1996 to 1997. Since 2005, he has been working independently with SGS-Horizon, Seismik, and many other associations. Based in The Hague, The Netherlands, he is close to the former estate of Christiaan Huygens, whom he considers the greatest geophysicist of all time. Moser's main interests include seismic imaging, asymptotic methods, seismic reservoir characterization, diffraction, and geothermal exploration. He has authored many influential papers on ray theory and ray methods, Born inversion and modeling, macro-model independent imaging, and diffraction imaging, several of which have received best paper awards, including an Honorary mention in 2005 from SEG and the Eötvös Award in 2007 and 2009 from EAGE. He has co-edited two reprint volumes for SEG, Classical and Modern Diffraction Theory and Seismic Diffraction, and teaches courses on diffraction for EAGE and SEG. He co-organizes the Active and Passive Seismics in Laterally Inhomogeneous Media (APSLIM) workshops in the Czech Republic (2015, 2021). He is a member of SEG and the Mathematical Association of America (MAA), an honorary member of EAGE, and Editor-in-Chief of Geophysical Prospecting.

Enders A. Robinson (1930–December 2022) was professor emeritus of geophysics at Columbia University in the Maurice Ewing and J. Lamar Worzel Chair. He received a B.S. in mathematics in 1950, an M.S. in economics in 1952, and a Ph.D. in geophysics in 1954, all from Massachusetts Institute of Technology. As a research assistant in the mathematics department at MIT in 1950, he was assigned to seismic research. Paper-and-pencil mathematics on the analytic solution of differential equations was expected. Instead, he digitized the seismic records and processed them on the MIT Whirlwind digital computer. The success of digital signal processing led to the formation of the MIT Geophysical Analysis Group in 1952 with Robinson as director. Almost the entire geophysical exploration industry participated in this digital enterprise. In fact, this effort later would be recognized in the Boston Globe newspaper, which published a special magazine (May 15, 2011) recognizing the 150 most valuable contributions MIT has made in science and technology– #32 was the Geophysical Analysis Group for spurring the "digital revolution" in oil prospecting. In 1965, Robinson and six colleagues formed Digicon, one of the first companies to do commercial digital seismic processing. In 1996, Digicon and Veritas combined to form VeritasDGC, which combined with CGG in 2007. With Sven Treitel, Robinson received the SEG Award for best paper in geophysics in 1964, the SEG Reginald Fessenden Award in 1969, and the Conrad Schlumberger Award from the European Association of Exploration Geophysicists (EAGE), also in 1969. In 1983, he was made an honorary member of SEG. In 1984, he received the Donald G. Fink Prize Award from the Institute of Electrical and Electronic Engineers (IEEE). In 1988, he was elected to membership in the National Academy of Engineering. He received the SEG Maurice Ewing Medal and the SEG Award for best paper in geophysics in 2001, the Blaise Pascal Medal for Science and Technology from the European Academy of Sciences in 2003, and the Desiderius Erasmus Award from EAGE in 2010. He is the author of 20 books and the co-author of 13.

List of Figures

Chapter 1
Historical Overview of the Life of Christiaan Huygens

There are many degrees of Probable, some nearer Truth than others, in the determining of which lies the chief exercise of our Judgment.
— *Christiaan Huygens Cosmotheoros*

Constantijn Huygens

It is told, perhaps apocryphally, that Christiaan Huygens (1629–1695) dropped a stone into the canal next to his house when he was still a boy. He paid close attention to the circular wavelet spreading across the water's surface. Of course, the wavelets that he supposedly saw were never perfectly circular, but he had a clear mental image of a perfect circle, and it can be argued that, in the manner of Plato, he dedicated his life to revealing the crucial role that geometry plays in science.

It may be said that the three greatest contributions of Christiaan Huygens are Huygens' principle, the first astronomical telescope, and the first accurate clock. An astronomical telescope is an optical instrument which is used to see the magnified image of distant heavenly bodies like stars, planets, satellites and galaxies. The pendulum clock, invented in 1656 by Christiaan Huygens, was until the 1930s the world's most precise timekeeper, accounting for its widespread use.

According to the words of the poet William Blake:

To see a World in a Grain of Sand
And a Heaven in a Wild Flower
Hold Infinity in the palm of your hand
And Eternity in an hour.

T. J. Moser and E. A. Robinson, *Walking with Christiaan Huygens*, History of Physics,
https://doi.org/10.1007/978-3-031-46158-3_1

Using the above as a model, we can write:

To see every point on a wavefront as a source of wavelets,
And the new wavefront as the envelope of all these wavelets.
Hold Infinity in the lens of a telescope.
And Eternity in the hands of a clock.

In the above, we recognize the three most important contributions of Christiaan Huygens. The first two lines state Huygens' principle, the third line the telescope, and the fourth line the clock.

Constantijn Huygens (1596–1687), the father of Christiaan Huygens, was broadly educated in languages, law, and social protocol. As a diplomat of the Dutch government, Constantijn Huygens was regarded as an unmatchable precious entity. Constantijn Huygens had other talents as well and his accomplishments were varied and remarkable. He was an art connoisseur and advised the Prince of Orange in this area. Rembrandt van Rijn is generally considered one of the greatest visual artists in the history of art. Unlike most masters of the seventeenth century, Rembrandt's works depict a wide range of style and subject matter. Constantijn Huygens was the first to bring the young Rembrandt van Rijn to the attention of the elite. Anthonie van Leeuwenhoek, a leading scientist in the Golden Age of Dutch science and technology, is commonly known as "the Father of Microbiology". Constantijn Huygens introduced Anthonie van Leeuwenhoek to the Royal Society in London. Constantijn Huygens made visits to England on seven occasions. In 1622, while Constantijn served as a diplomat in England, he was knighted by King James I. In 1632, King Louis XIII of France appointed Constantijn as Knight of the Order of Saint Michel.

Constantijn Huygens contributed to the transformation of European culture in the early-modern period by influencing the cultural, scholarly, and political life of his time. He corresponded with virtually every European contemporary of importance, including King Charles I of England, Descartes, Rembrandt, Corneille, Francis Bacon, Van Leeuwenhoek, Van Dyck, Rubens, and John Donne. Constantijn wrote a significant body of poetry, primarily in Dutch, but also in Latin, and occasionally in French and English. His poetry demonstrates the breadth of his interests and his pivotal role in European culture. In addition to writing poetry, Constantijn composed and performed music; was secretary to two Princes of Orange, Frederick Henry and William II (Frederik Hendrik and Willem II); and became a friend to John Donne, Rembrandt, Descartes, and many other notable people of his time. He is considered by some the discoverer of the talent of Rembrandt (and Lievens), who at the time were working in nearby Leiden. He planned for his sons a private education in mathematics, languages, literature, and music.

The following poem to Anna Roemers Visscher (1583–1651) by Constantijn is an English translation of the Dutch original. Anna Roemers was the daughter of the Renaissance man of letters, Roemer Pieterszoon Visscher. She and her sister Tesselschade were close friends of Huygens throughout their lives; but this poem was written at a time when his acquaintance with them was relatively recent. They were the acknowledged beauties at the center of the circle of the poet P.C. Hooft, called the Muiderkring, which regularly convened at the Muiden Castle. Constantijn Huygens, at this time, was a young man of twenty-two, and still on its periphery. The last two lines of the poem refer to Icarus, son of Daedalus, who escaped from Crete with his father on wings held together with wax, but Icarus was tempted to fly too near the Sun, melting the wax, and plunging Icarus to his death. The poem is:

To Anna Roemers

As my misfortune makes my trespass clear,
I come before your court to make my plea,
In worthless rhyme if I've spelled out your name.
Then punishment well-earned I can't evade.
I see it now: you joy in your revenge,
Devising laughing torments for your friend:
Your hand denying what your mouth allows,
Leave him half-pardoned, half the way to joy.
But hear me, Anna, write your judgement sweet,
That I may ponder it and meditate
The endless difference of your soul and mine.
You make me taste your honor and my shame.
What sun my waxen wing has flown too near:
Thus double your revenge and double thus my joy.

As a return gift for Constantijn Huygens, Anna Roemers Visscher engraved one of her own poems on a wine glass (see Fig. 1.1). Her pen has run dry, and her mind has turned to rust, so the verses are both a lament and a request. She asks that Huygens bring her some water from Mount Helicon, the Muses' home, so that her ink can flow freely and she can write poems again.

Susanna van Baerle was born in 1599 in Amsterdam. She was the daughter of Jan van Baerle and Jacomientje Hoon. Susanna's father passed away in 1605 and her mother in 1617. Described as beautiful, Susanna married Constantijn Huygens in 1627 in Amsterdam (see Fig. 1.2). They had four boys and one girl. The five children were: first son Constantijn Jr in 1628; Christiaan in 1629; Lodewijk in 1631; Philips in 1633; their daughter Susanna in 1637. The mother Susanna died in 1637 shortly after giving birth to their daughter Susanna. Constantijn never remarried. His wife Susanna is remembered today by the long poem Dagh-werck. Constantijn wrote this poem apparently in close collaboration with his wife Susanna.

As these circumstances of his life show, Christiaan Huygens was born into an ideal social and intellectual environment that provided him with the resources, education, and encouragement to pursue a career in science. His family's wealth, influence, and interest in the arts and sciences created an optimal environment to thrive and make significant contributions to science. In addition to being a scientist, Christiaan was also a gifted draftsman and painter of technical devices, landscapes, buildings and human figures. He learned to play the lute and viola da gamba. At the age of 9 he

Fig. 1.1 Wine glass with poem by Anna Roemers Visscher dedicated to Constantijn Huygens (front and backside) [with permission from the Rijksmuseum, Amsterdam]

was fluent in Latin with his family and a year later he made progress in Greek. There is a beautiful drawing by his hand of his father Constantijn (see Fig. 1.3).

René Descartes

Young Christiaan was influenced by the French mathematician-philosopher René Descartes (1596–1650, Latin *Cartesius*, see Fig. 1.4), an older friend of the Huygens family and frequent visitor to their home. Constantijn Huygens senior knew Descartes and was instrumental in the publication of his *Discours de la méthode.* They most likely first met in May 1642 at Hofwijck castle, Huygens' summer residence in Voorburg, just outside of The Hague. Descartes wrote about Constantijn Huygens. "The honor to know him I cherish as one of the most happy things that happened to me." Huygens wrote about Descartes: "Truly, he is a man superior to all esteem that one would wish to render him."

Descartes can be considered the father of modern philosophy. His approach to acquiring knowledge was to start with a radical skepticism and question everything that can be questioned. His conclusion was that even if in this state of extreme doubt,

Fig. 1.2 Suzanna van Baerle (1599–1637), and her husband Constantijn Huygens (1596–1687), painted by Jacob van Campen around 1635

we can at least be sure of the very act of doubting itself. He summarized this in the statement "Cogito ergo sum" (I think therefore I exist). All knowledge would have to be *clare et distincte* (clear and distinct) and built up from the foundations like solid architecture. These thoughts are considered to be the start of the seventeenth century movement of rationalism. Descartes spent 20 years of his life in Holland where he wrote many of his works including *Discours de la Méthode* published in 1637. This book contained three appendices: *La Dioptrique*, *Les Météores*, and *La Géométrie*. In *La Dioptrique*, he introduced his version of Snell's law, in *Les Météores* he offered an optical explanation for the formation of rainbows. The last of these, *La Géométrie*, is one of the most important works in the history of mathematics. Until then, the dominant geometry had been Euclidean, which used ruler and compass to solve problems. The novelty of Descartes' approach was that it allowed geometric problems to be solved through the exclusive manipulation of algebraic expressions. This geometry introduced by Descartes is called analytic geometry. The points of the plane are identified with pairs of numbers (x, y), called Cartesian coordinates. He introduced the notation for the exponent x^y and the square root $\sqrt{x}$ (with the horizontal overbar or "vinculum").

True to the spirit of his rationalism, Descartes sought to develop science through reasoning and speculation, while placing less value on experiment. Christiaan Huygens studied mathematics at Leiden University with Frans van Schooten who was a dedicated Cartesian (follower of Descartes). From Descartes, Huygens learned of

Fig. 1.3 Constantijn Huygens drawing by his son Christiaan

the mechanistic philosophy of nature with the idea that all natural phenomena would one day be explained by reason. Still, Huygens was a sufficiently independent genius not to ignore the value of the physical experiment and during his life he corrected several of Descartes' misconceptions in physics (such as on refraction).

In religious matters, Descartes was a devout Catholic and Huygens a Protestant. Descartes believed in free will, but only for humans, not for animals which he regarded as machines. Huygens, on the other hand, believed in determinism for all living creatures, but all conscious and sensitive to joy and pain.

A vivid illustration of the impact and controversy that Descartes created is given by the correspondence between Huygens and Gottfried Wilhelm Leibniz (1646–1716), who was a true rationalist himself. On March 2 1691 Leibniz, amazed by the lack of

Fig. 1.4 René Descartes, painting by Frans Hals

theory in medicine and referring to Antonie van Leeuwenhoek (1632–1723, Dutch microbiologist and the inventor of the microscope), wrote to Huygens.

> Is there nobody who considers philosophy in medicine? The Cartesians are too much occupied by their hypotheses. I would rather have a Leeuwenhoek say what he sees than a Descartes say what he thinks. Nevertheless, it is still necessary to add reasoning to observations.

To which Huygens answered on March 26 the same year:

> We have doctors who pretend to follow Descartes' philosophy, but they are the last ones I would consult if I would need them.

Joking aside and on a more serious level, Huygens' view of natural science is adequately formulated in his letter to Pierre Perrault in 1673, which was in turn quoted by Nobel laureate Pieter Zeeman in 1929 during a speech on the occasion of Huygens' 300th birthday (Crommelin, Plantenga and Vollgraf, 1929):

> In no matter is our knowledge absolutely certain, in everything only probable. But there are degrees of probability which are very unequal, and some from 100,000 to one, as in the geometric proofs, which may be false, but have been tried so many times and so long that there is hardly any reason to doubt their correctness, especially of those which are short. In matters of physics, there is no other proof than in the deciphering of a message in secret code, where one begins by making assumptions on loose conjecture. If, then, these prove to be correct in so far as they lead to the finding of some well-connected words, then a very

great certainty is ascribed to these suppositions, although otherwise there is no proof for them, and it is not impossible that others may be found, which match better the truth.

Huygens hits the nail on the head with these statements. While the ancients (like Socrates) were right in saying that we know nothing for sure, there is a great (infinite) variation in the probabilities of the truth. In mathematics, proofs are typically accepted based on logical rigor and the consensus of the mathematical community. In physics, we construct theories to explain our data, but we should never rule out the possibility of finding other theories that better match the truth. For Huygens, our physical hypotheses are never more than conjectures. During his scientific life, Huygens stuck to Descartes' ideal of *clare et distincte* but stayed clear of Cartesian dogmatism. Huygens' views on epistemology, the acquisition of human knowledge, positioned between empiricism and rationalism, may be considered as anticipating some aspects of the critical philosophy of the German philosopher Immanuel Kant (1724-1804). However, unlike Descartes and Leibniz, who can be considered figureheads of Western philosophy, Huygens had no major philosophical aspirations other than pragmatism. In a sense we can say that he admired Archimedes more than Aristotle, and valued practice more than speculation. Indeed in many letters, his father Constantijn referred to Christiaan as the Dutch Archimedes. Whether or not Christiaan liked this, in a letter to J. Golius of 28 December 1651 he states that Archimedes is to be compared to no one.

Scientist, Genius, Polymath

Can we call Huygens a scientist? From our modern perspective most certainly. Until the late 19th and early twentieth century scientists were still referred to as "natural philosophers" or "men of science". In 1833, the English philosopher and science historian William Whewell (1794–1866) coined the term scientist, to refer to a person who studies science systematically, and it first appeared in print in his anonymous review of Mary Somerville's "*On the Connexion of the Physical Sciences*" published in the Quarterly Review (Whewell, 1834). In light of the above Huygens can be called a scientist even if in his time the word had not been adopted yet.

Can we call Huygens a genius? Genius has no precise definition and its meaning has evolved over time, culminating in Romanticism in the nineteenth century. According to a well-known quote by the German philosopher Schopenhauer talent hits a target no one else can hit; genius hits a target no one else can see. Huygens' exceptional creativity and imagination placed him in the second category. Practical circumstances also helped. Apart from a degree in law which he and his brother Constantijn purchased (!) from the University of Angers and never used, he had no other academic degree nor an academic position. Thanks to the wealth of Constantijn senior, Christiaan could devote all his time to science with sufficient support for daily tasks and scientific needs; he never held office. Huygens was certainly talented, but his achievements reach far beyond. His visionary wave theory of light qualifies him as a genius.

Can we call Huygens a polymath or a homo universalis, a person who is universally versed in all fields of science, like Leonardo da Vinci? Huygens was versatile and had a broad range of interests, covering mathematics, physics, astronomy and engineering. Yet he cared less for botany, anatomy (except for the operation of the eye), chemistry, law (despite his degree), theoretical logic and, as we have seen, for speculative philosophy and metaphysics. In the seventeenth century, chemistry was not yet an established science and was still rooted in alchemy. The practical importance of knowledge of all commodities traded, of prescriptions, apothecaries, hairdressers, and traveling quacks was of course evident. Chemistry was also needed for the practice of seventeenth century painting. Nevertheless, referring to the *Satires* of the Roman poet Horace, Huygens called chemists *ciniflones* (ashblowers, slaves who heated the irons necessary for curling their masters' hair in glowing ashes) (Huygens, Complete Works, Biography,1888–1950). He was most interested in mathematics and physics and is regarded as one of the first mathematical physicists. In this respect he held the middle ground between being a specialist and a generalist. Therefore we may be hesitant to call him a polymath. Huygens would modestly agree.

In summary, Huygens was unquestionably a scientist and a genius, but less so an academic or a homo universalis. Incidentally, Einstein is also usually referred to as a genius but not a polymath.

Scientific Correspondence

Progress of science cannot take place without communicating discoveries to contemporary colleagues. In the seventeenth century this mainly occurred through one-to-one correspondence. The quickest way to publish a result was by sending a letter to announce it to a number of close colleagues. Large volumes of letters were written in this way between Huygens, Newton, Leibniz and many others. Many of these letters have been preserved for posterity. We are fortunate to be able to consult Huygens' correspondence (incoming and outgoing) in his Complete Works (Huygens, 1888–1950). This is a rich and almost inexhaustible source of information, much of which has not been fully studied even today. The letters were usually written in Latin or French, Huygens occasionally used English or Dutch, depending on the recipient of the letter.

Correspondence by letter was fast. Sending a letter from The Hague to London took about one week; Huygens wrote to John Wallis on 13 June 1655 and received his answer on 1 July 1655. However, there was no copyright and the letters carried the risk that scientific inventions and discoveries would be claimed by someone else. Priority disputes were not uncommon. A well-known priority dispute of the seventeenth century is the one between Newton and Leibniz regarding the invention of infinitesimal calculus (Sonar, 2018). Huygens was also involved in some priority disputes related to the invention of the pendulum clock. Patenting technical inventions was a practice that was still in its infancy in the seventeenth century (Huygens did obtain a number of patents), but the granting of patents at that time was not yet a standardized practice and the modern system of patent law did not emerge until the eighteenth century.

A relatively safe and quick way to claim priority for a scientific discovery was to send a letter with a general description of the discovery but to encrypt the important (today we would say classified) details in an anagram. Huygens made extensive use of anagrams, as he was careful to disclose his results in open publications and often waited several years before he found the appropriate time to publish them in books. Having the truth for himself was more important to him than demonstrating his knowledge.

The anagram usually stated the scientific claim in Latin, which was encrypted by rearranging the letters, sometimes into a quote from a well-known poet or author. At a later stage, the anagram would be proof that the author had made the claim and would ensure priority. Reasons for encrypting the claim could be scientific, commercial or even political.

Let us give some examples. Huygens announced his discovery of Saturn's moon Titan with the following anagram (letter to John Wallis, 13 June 1655):

ADMOVERE OCVLIS DISTANTIA SIDERA NOSTRIS, VVVVVVVC CCRRHNBQX

The first phrase of which was taken from Roman poet Ovidius' poem Fasti: *They brought the distant stars closer to our eyes.* This was one of Huygens' favorite quotes. Ovidius suggested that if the giants of mythology could not approach the stars, an alternative method would be to bring the stars closer to us. This was only accomplished in the seventeenth century through the invention of the telescope. The anagram would read (taking the letters U and V as identical):

SATURNO LVNA SVA CIRCUNDUCITUR DIEBUS SEXDECIM HORIS QUATUOR

Which translates into: "*Saturn is orbited by its moon in sixteen days and four hours.*"

His formula for the oscillation period of the pendulum clock $T = 2\pi\sqrt{l/g}$ (see Chap. 6) was communicated on 4 September 1669 as an anagram to Henry Oldenburg, secretary of the Royal Society in London. Oldenburg himself was addressed with the anagram "Grubendol", to prevent the letter from being withheld by the postal service for political reasons.

On 30 January 1675 Huygens wrote to Oldenburg about his invention of the spiral spring (which revolutionized time measuring) by the following anagram:

a b c e f i l m n o r s t u x

4 1 3 5 3 7 3 1 2 3 4 3 2 4 2

Which reads as "*Axis circuli mobilis affixus in centro volutae ferreae*" ("*The axis of a movable circle fixed in the center of an iron coil*"; *the numbers indicate the frequency of the corresponding letters*). Here he mentioned the risk of disclosing the commercial secret of the mechanism of the balance spring as the specific reason for using the anagram.

The composite of lenses to approximate the hyperbolic lens was encrypted as:

a b c d e h i l m n o p r s t u y

5 2 2 1 4 1 2 3 3 1 3 2 2 3 2 4 1

Which reads as "*Lens e duabus composita hyperbolicam aemulatur* " ("*A hyperbolic lens is made of two compounds*") (letter to Henry Oldenburg of 6 February 1669).

The list goes on for quite a while. The anagrams give a nice and lively overview of what Huygens considered important. The use of anagrams would sometimes be parodied: on 24 February 1658 Hendrik van Heuraet, Dutch mathematician and student of Frans van Schooten, would send Huygens a list of seemingly meaningless Latin phrases.

Anagrams could also cause confusion. In 1610 Galileo sent an anagram to Johannes Kepler that codified his observation of peculiar spots on the sides of Saturn which later turned out to be its ring. One month later he sent an anagram on the phases of Venus. In both cases, Kepler misinterpreted them and took them for messages on Mars and Jupiter, respectively.

Yet another way to publish results was through scientific journals. During the seventeenth century they were still in their infancy. The oldest scientific journal was the French *Journal des sçavans*, founded in January 1665 and still existing today. Three months later Henry Oldenburg founded the *Philosophical Transactions* in London, followed in 1682 by the *Acta Eruditorum* in Leipzig by the German scientists Otto Mencke and Gottfried Wilhelm Leibniz. Huygens and his contemporary colleagues Hooke, Newton, Cassini, Roemer and many others would publish their discoveries in these journals.

Finally, the most enduring way to disseminate scientific results was by publishing books. In the Netherlands the publishing house Elsevier had been established in 1592, which published works by Galileo, Snellius, Hugo Grotius and Huygens' father Constantijn. Many of Christiaan Huygens' books were self-published.

Health

Throughout his life, Christiaan Huygens suffered from poor health. He was plagued by headaches and migraines for long periods of time (Koehler, 2015). This imposed certain restrictions on his activities. It affected his traveling and stays abroad, mainly in France and England. He placed great value on a balanced lifestyle and devotion to duty, in particular scientific duty. It is known that Huygens also suffered from melancholy, but it is not known whether this was innate or caused by periods of health crisis, although they will certainly have contributed to it. His health issues made him wonder if the human mind, which is so dependent on the body, is really something immaterial.

Fig. 1.5 Portrait of Christiaan Huygens and Albrecht Dürer's engraving "Melencolia"

In some sense, melancholy is related to genius. Aristotle wrote "Why is it that all men who have become outstanding in philosophy, statesmanship, poetry or the arts are melancholic?" The link between melancholy and genius became a favorite subject in the Renaissance, as evidenced by the engraving "Melencolia" by Albrecht Dürer, for which Huygens almost could have been the model (See Fig. 1.5).

Huygens had his ups and downs. The 1650s were his first period of scientific glory. However on 13 August 1652 he mentioned his headaches for the first time in a letter to his teacher Frans van Schooten. The headaches would cause him to delay correspondence and occasionally he would have to excuse himself for not fulfilling duties. 1656–1660 were again years of great activity, without traces of illness.

In 1676 he was forced to interrupt his stay in Paris for health reasons and return to The Hague. He wrote to his brother Constantijn junior that he did not expect to return to Paris because of "unpleasant experiences" there, and his father wrote to the physician Menjot in Paris that "his face shows in a way what he has been through" (Huygens, Complete Works, Biography,1888–1950).

During his last stay in Paris (1678–1681) Huygens started in good health, but his family decided that he needed a housekeeper, Mrs la Cour, a former beguine (Christian laywoman active in charity) from Leiden, to look after him. In 1680 he fell ill again and was brought back to The Hague by his family. His hopes of continuing his work for the Académie Royale des Sciences remotely were dashed by the passing away of his patron the French minister Colbert.

We may say that some of the portraits of Christiaan Huygens bear witness to his melancholy side. He suffered a great deal and yet he never gave up, strong in will to strive, to seek, to find, and not to yield. Science was his consolation.

Study and First Discoveries—Saturn's Ring and Largest Moon Titan

Huygens left his home in 1645 to study law and mathematics at the University of Leiden and, in 1647, continued with them at the College of Orange in Breda. In 1649 he returned to The Hague, where he remained until 1666. While there Huygens performed important research. By developing a new method for grinding lenses of unparalleled clarity, Christiaan Huygens transformed the rudimentary telescope of the early 1600's into the powerful instrument that revolutionized astronomy. Using a telescope that he built with his brother, he discovered in 1656 the rings of Saturn. Huygens began to chart the sky in earnest.

During the period from 1655 to 1660, Huygens made a number of landmark astronomical discoveries, including Saturn's rings and the planet's largest moon, Titan. He was the first to observe the surface features and rotation about its axis of Mars, as well as the Great Nebula in the constellation Orion. His search for celestial bodies was relentless and at times he could be impatient. In his letter to Ismaël Boulliau of 20 November 1659, he expresses his disappointment and complains that astronomers in the Dutch university city of Leiden are careless in the observation of eclipses: "I don't find any change in the star Cygnus. As for the gentlemen of Leiden I don't know if they are paying attention. It is a shame that there is nobody there who makes observations neither of any eclipse nor of anything else".

Titan is larger than the planet Mercury and is the second largest moon in our Solar System (Jupiter's moon Ganymede is about 2% larger). Titan is the only moon known to have a dense atmosphere, and the only known body in space, other than Earth,

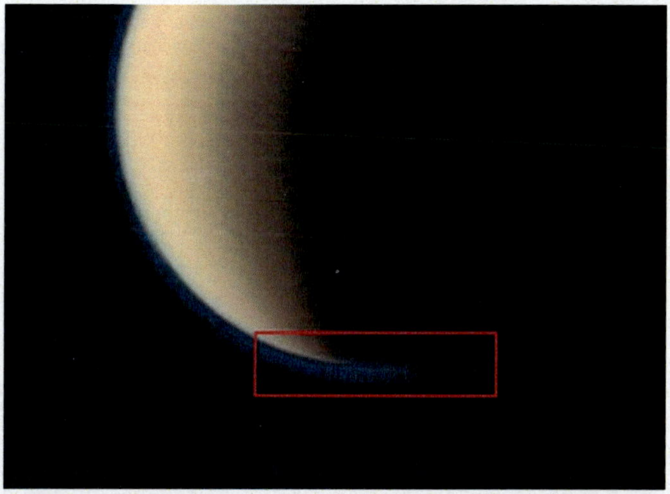

Fig. 1.6 The south polar region of Saturn's largest moon, Titan, showing a depression within the moon's orange and blue haze layers near the South Pole (indicated by the box)

Fig. 1.7 Saturn's rings and four of its moons

where clear evidence of stable bodies of surface liquids has been found. Titan's atmosphere is made mostly of nitrogen, like the Earth's, but with a surface atmospheric pressure 50% higher than the Earth's (Fig. 1.6).

In order to facilitate his astronomical observations, Huygens designed several new pieces of astronomical equipment. Chief among these were his lenses, which provided far greater resolution and magnification than any earlier ones. After several years he perfected an achromatic lens that corrected the "false color" fringes often associated with inferior lens systems and reduced chromatic aberration. This lens, called the Huygenian eyepiece, is still used in many telescopes today. A variety of optical instruments, including telescopes and microscopes, are equipped with eyepieces, also known as ocular lenses. They are called eyepieces because, when someone looks through the device, it is frequently the lens that is closest to the eye. The objective lens, or mirror gathers light and concentrates it to create an image. To enlarge this image, the eyepiece is placed close to the objective's focal point. The focal length of the eyepiece determines the degree of magnification. The Huygenian eyepiece is an ocular having two plano-convex lenses that are formed from similar glass and that are separated by a space equal to half the sum of their focal lengths. This eyepiece is free of lateral chromatic aberration, but is not suitable for use with crosshair lines because the image plane falls between the two lenses.

There are many different length scales that were used in Europe in the past. The standard ancient Roman foot (pes) was about 29.57 cm. In the Roman provinces, the pes Drusianus was used, with a length of about 33.4 cm, named after Nero Claudius Drusus, a Roman general who lived 38 BC–9 BC and was involved in changing Dutch waterways for transporting troops (the river IJssel is thought to have become a distributary of the Rhine through his intervention). The Dutch voet (foot) was of the same order of magnitude as the English foot (30.48 cm), but its exact size varied from city to city and from province to province. The Rijnland (Rhineland) foot in use since 1621 was the most commonly used voet in both the Netherlands and in parts of the German lands. In 1807, the Rijnland foot in the Leiden observatory was measured as 31.39465 cm, approximately taken as 31.4 cm.

In 1655 Christiaan Huygens and his older brother, Constantijn Huygens Jr, constructed a telescope, initially designed to improve upon the original design of Galileo's telescope. The telescope's objective lens had a focal length of 10 Rijnland feet. The brothers equipped the telescope with a single-lens eyepiece with a focal length of 3 Rijnland inches (equivalent to 0.25 Rijnland foot), which yielded a magnification of about 43x. Huygens described the new telescope as having a length of 12 Rijnland feet. The practice at that time was to consider the telescope tube's total length by including the eyepiece. The lens was finished and ready for use in February 1655 and stunning results, as mentioned already, were quick in arriving. At around 8 p.m. on March 25, 1655, Christiaan Huygens pointed his newly constructed telescope at Saturn, the sixth planet in our Solar System. He noticed a companion that turned out to be the first satellite ever found in the Saturnian system, as later observations showed. He called this satellite *Saturni Luna* (Moon of Saturn), it was renamed to Titan by Herschel in 1847; Giovanni Domenico Cassini would be a champion in discovering four more Saturnian moons before 1684. In addition, Huygens was also struck by the curious extension of Saturn, which had intrigued astronomers ever since Galileo first observed it in 1610. Huygens gave the solution (in the form of an anagram, see later in Chap. 2): "Saturn is encircled by a ring, thin and flat, and nowhere touching, inclined to the ecliptic." (Fig. 1.7).

The only component of Huygens' 1655 telescope that survives until today is the objective lens. A replica was built in the early twentieth century and can be seen on display at the Museum Boerhaave in Leiden, Netherlands. Huygens further developed the design of his telescopes to provide a clearer view of the sky. It was well known that longer telescopes with longer focal lengths could magnify objects more. Huygens went above and beyond with this idea, building telescopes up to twenty-three feet (7.0 m) in length. Huygens was dissatisfied even though these amazing instruments provided a magnificent view of the planets. Since the traditional telescope design relied on metal tubes that would bend if they were too long, he thought it was too constrained. His response was an "aerial telescope", a tubeless telescope. The larger objective lens and the smaller eyepiece could be as far apart as practical lens construction would allow because there was no tube to connect them. These aerial telescopes ranged in size, with the largest measuring more than 100 feet (30 m). In addition to developing optical tools, Huygens also created a micrometer in 1658 that allowed him to measure angular separations of objects with an accuracy of a few seconds of an arc, such as the apparent distance between Saturn and its moon. Huygens published his discoveries in 1659 in a book entitled *Systema Saturnium*.

This publication also contains his first description of the workings of a micrometer in an astronomical telescope. In this telescope, the objective lens creates a true image which is viewed through the ocular lens. By fitting a distance gauge onto the true image, the position of one heavenly body in relation to another could be determined with great accuracy.

The Pendulum Clock

Huygens' early enthusiasm for astronomy led him to tackle the problem of an accurate clock. Making use of Galileo's realization that a swinging pendulum keeps a regular rhythm, Huygens built the first pendulum clock. The pendulum clock was so accurate that its precision was not challenged until the introduction of electronic clocks. The significance of Huygens' invention in the history of science is that it marked the new interest taken in time as a dimension. Because of Huygens, the observation of astronomical bodies could be precisely made and the timing of astronomical events could be accurately measured. All of this was a result of his profound optical research.

Because of his invention of the first powerful telescope and the first accurate clock, Huygens was courted by both England and France. The Royal Society of London was created in 1660 and one of their first acts was to elect Huygens as a member. In 1664 the French offered Huygens membership in a similar academy to be founded in Paris. However, it was not until 1666 that the Académie Royale des Sciences was created. As its most prominent member, Huygens received a generous stipend from King Louis XIV and an apartment in the Bibliothèque Royale. Thus Huygens began a stay in Paris that lasted from 1666 to 1681. As mentioned before, because of ill health, his residence in Paris was interrupted by two periods which he spent in his native Holland. One period was from 1670 to 1671. The year 1672 was catastrophic ("rampjaar", English "catastrophic year") because four foreign powers tried to invade Holland and wipe it off the map (France, England, Münster and Cologne). The Dutch were "radeloos, reddeloos, redeloos" (without hope, without help, without reason). Still the seventeenth century is the Dutch Golden Century, not the least because of Huygens (who actually was back in Paris during the 'rampjaar'—apparently he was not distracted by political events and the French did not regard him as a possible collaborator with the enemy). His other stay back in Holland was from 1676 until 1678. In 1681 Huygens returned to Holland for good. In 1689, albeit in poor health, Huygens visited England in order to see Newton and other old friends among the English men of science.

During those days, the central problem of navigation was the determination of longitude. In contrast, latitude is much easier to determine by the use of a sextant to measure the altitude of the Sun at noon or the altitude of the North Star Polaris at night. Accurate longitude is required both for the location of a ship at sea and for mapping the globe. Because the Earth rotates around its axis every day, measurement of longitude becomes a matter of time. If one knows the difference in local time at two points, one knows the longitudinal distance between them. In seventeenth-century exploration,

Fig. 1.8 Old church at Scheveningen, drawing by Christiaan Huygens in 1658

improving the accuracy of the clock was at the center of scientific investigations. The existing mechanical clocks, which were accurate to only about 15 min a day, were unsuitable for navigation.

Huygens' invention of the pendulum clock was a breakthrough in timekeeping. It used an escapement, a mechanism that counts the swings, while a driving weight provides the push. In effect, the escapement is a feedback regulator that controls the speed of a mechanical clock. The escapement is used in typewriters as well. Huygens produced his first pendulum clock in December 1656; it was accurate to 10 s a day. His pendulum clocks were accurate enough to allow measurement of the orbital periods of the satellites of Jupiter over the span of a year. In fact, pendulum clocks were the most accurate clocks in the world for the next 300 years.

However, it was a challenge to use pendulum clocks during stormy seas. This problem was solved by John Harrison (1693–1776). Huygens obtained a patent on the pendulum clock and performed experiments in the Oude Kerk (Old Church) at Scheveningen. There is a drawing by Huygens of this church still (Fig. 1.8).

Galileo Galilei (1564–1642; see Fig. 1.9) believed that a pendulum is isochronous, meaning that its period is independent of the magnitude of its swing. However, Huygens discovered through intricate mathematics that a pendulum swinging through the arc of the circle is not isochronous. It only appears isochronous if the length of the arc is relatively small compared to the length of the pendulum. Due to this property, a clock with a long pendulum has an advantage over a clock with a short pendulum. To reduce the amount of energy needed to keep them moving, the pendulums of early clocks were kept short and light. Hence, due to their relatively large pendulum swings, early pendulum clocks had poorer accuracy.

Huygens went to great efforts to make a pendulum-like mechanism that would be isochronic. In such a case, the pendulum could swing through a large angle. In doing so, Huygens originated the discipline today known as differential geometry.

The circle and the cycloid are closely connected. The cycloid is the location of a point on a circle's rim rolling on a straight line. A cycloidal clock is an improved pendulum clock that Huygens describes in detail mathematically in his famous work *Horologium Oscillatorium* (1673). The pendulum of the cycloidal clock is forced to swing along an arc that resembles a cycloid, making it isochronous.

Unfortunately, there was too much friction from the pendulum's motion on the metal cheeks. The anchor escapement for a pendulum clock was created by Robert Hooke (1635–1703), an English physicist famous for his discovery of Hooke's law, which links stress and strain. The angle of swing required by the escapements of the early pendulum clocks was greater than that required by the anchor escapement. The cycloidal clock was abandoned as a result of pendulum clocks' improved accuracy and less frictional loss. Although Huygens' cycloidal clock failed to withstand the test of time, another innovation he made did. In place of a pendulum, he created a chronometer in 1675 that utilized a balance wheel and a spiral spring. Before the development of the quartz crystal oscillator in the twentieth century, practically all watches used balance wheels and spiral springs.

The cycloidal clock is significant historically because it was the first complicated device to be successfully designed using higher mathematics. The crucial idea in this instance was Huygens' additional discovery of the evolute (the locus of the centers of the osculating circles of a curve). Others, like Heron of Alexandria and Leonardo da Vinci, used mechanical concepts rather than much mathematics beyond Euclidean geometry to design their innovations. The introduction of the use of higher mathematics to accomplish mechanical design gives Huygens a strong claim as the one of the founding fathers of modern technology.

Fig. 1.9 Galileo Galilei, Isaac Newton, Gottfried Wilhelm Leibniz

First Book on Probability Theory

Chevalier de Méré, a wealthy French nobleman, wanted to find the reason why he consistently lost money in a certain game of dice. Until then, gamblers would blame losses on bad luck. Not so with the Chevalier de Méré. He posed the question to Blaise Pascal, who solved the problem with mathematics. Pascal started a correspondence with Pierre de Fermat (1607–1665, mathematician and councillor of Toulouse) on the subject. Because of their correspondence, they are recognized as the originators of the theory of probability.

In Paris, Christiaan Huygens was told of their correspondence, which was in an unorganized form. In 1657, Huygens wrote a treatise that gave the first comprehensive treatment of probability theory. In 1662 Sir Robert Moray sent to Christiaan Huygens the "life expectancy table" which had been compiled by John Graunt (1620–1674, English pioneer in demography). From the life table, Christiaan Huygens and his brother Lodewijk Huygens originated the concept of "life expectancy," which was the beginning of the discipline of mathematical statistics.

Pascal was the first to use mathematics to solve probabilistic problems. With "mathematical intelligence," mankind could now use the nuances of symbols and equations to express probabilistic ideas. It was a breakthrough that was essential to the development of the modern world. Pascal was also the first to invent (1642) a mechanical digital calculating machine that worked. Called the "Pascaline", it was manufactured and sold on the open market. This invention, a result of mathematical intelligence, led to the modern digital computer (for more details on the Pascaline, see Chap. 6).

Centrifugal Force

Huygens derived the formula for the centrifugal force exerted by an object undergoing a circular motion. The velocity of a moving object is characterized by both its magnitude and direction. By comparing the change in direction of the velocity of an object in uniform circular motion with the change in magnitude of the velocity of the same object falling under a uniform gravitational force, he was able to derive the well-known formula $a = v^2/r$ where a is the acceleration, v the circular velocity and r the radius of the circle. This was first published posthumously in 1703 in his book *De Vi Centrifuga* ("*About the centrifugal force*").

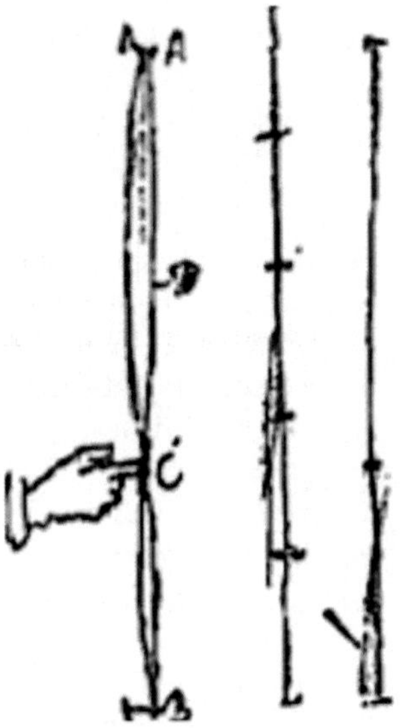

Fig. 1.10 Monochord (Huygens, Complete Works, Vol XIX)

Tone System

The Italian music theorist and composer Nicola Vincentino (1511–1576) worked on an improvement of the musical theory of Pythagoras and proposed the 31-tone system, which divided the octave in 31 equal steps. The 31-tone system was rediscovered and further developed by Christiaan Huygens. He took the meantone temperament as a starting point, and transformed it into a proportional tonal distribution. This has the advantage that it can be modulated arbitrarily, without sounding false or out of tune. While deriving the formula for the intervals, Huygens found himself having to find a rational approximation to the fifth in the quarter-comma meantone, which is the irrational number $\frac{1}{4}\log_2 5$. It is probable that he achieved approximations by the method of continued fractions. The number $\frac{1}{4}\log_2 5$ would be represented as a sequence of nested fractions as follows:

$$\frac{1}{4}\log_2 5 = \cfrac{1}{1 + \cfrac{1}{1+\cfrac{1}{2+\cfrac{1}{1+\cfrac{1}{1+\cfrac{1}{1+\cfrac{1}{1+\cfrac{1}{5+\cfrac{1}{1+\dots}}}}}}}}}.$$

By truncating the continued fraction after a finite number of steps, this would lead to the simple approximations 1/2 = 0.5, 3/5 = 0.6, 4/7 = 0.5714.., 7/12 = 0.5833.., 11/19 = 0.5789.., 18/31 = 0.58064.., 101/174 = 0.58045… The 18/31 convergent in these approximations appeared to Huygens as the most useful approximation, sufficiently accurate but with a limited number of intervals (31) still manageable for musical instrument design.

Continued fractions had been first used by Euclid in an algorithm for the greatest common divisor. During the Renaissance, they became a widely used tool for the approximation of real numbers by rationals. Logarithms had been proposed by John Napier in 1614. Here we see Huygens using the contemporary mathematical tools to achieve his goals. Another application of continued fractions by Huygens was the construction of the Planetarium (see Chap. 6).

In his letter to French Huguenot H. Basnage de Beauval of October 1691, Huygens wrote "Now if we divide the octave into 31 equal intervals we will find in the tones produced by these different lengths a system which comes so close to the meantone temperament that it is entirely impossible that the most delicate ear finds a difference."

Huygens experimented with the 31-tone system using a monochord, an instrument consisting of one string, see Fig. 1.10. In 1950, the Dutch physicist Adriaan Fokker built a 31-tone equal-tempered organ and regular performances still take place. However, in spite of the beautiful mathematics, it is fair to say that the music produced by these instruments is quite unconventional for listeners used to the normal 12-tone system.

Celestial Worlds

The light from a star radiates out in all directions, and is largely unimpeded in its path from the star to us. Why do we not see more stars? Olbers' paradox, named after the German astronomer Heinrich Wilhelm Olbers (1758–1840) is also called the "dark night sky paradox." It is the argument that the darkness of the night sky conflicts with the assumption of an infinite and eternal static universe. The magnitude of a star decreases quadratically with distance, but the number of stars at a certain distance increases quadratically with distance, therefore the sky should be equally bright as the Sun in all directions, which is clearly not the case. Edgar Allan Poe (1809–1849) was an American author and poet, who wrote the poem:

I have reached these lands but newly

From an ultimate dim Thule—

From a wild weird clime that lieth, sublime

Out of SPACE—Out of TIME.

The explanation of the "dark night sky paradox" was given by Edgar Allen Poe in his lecture in New York, one year before he died at age 40. The content of his lecture is included in his work, "Eureka: A Prose Poem," which comprises nearly 40,000 words (1848). Poe writes, referring to Blaise Pascal, "The Universe of stars has always been considered as coincident with the Universe proper. Thus, at any given point in

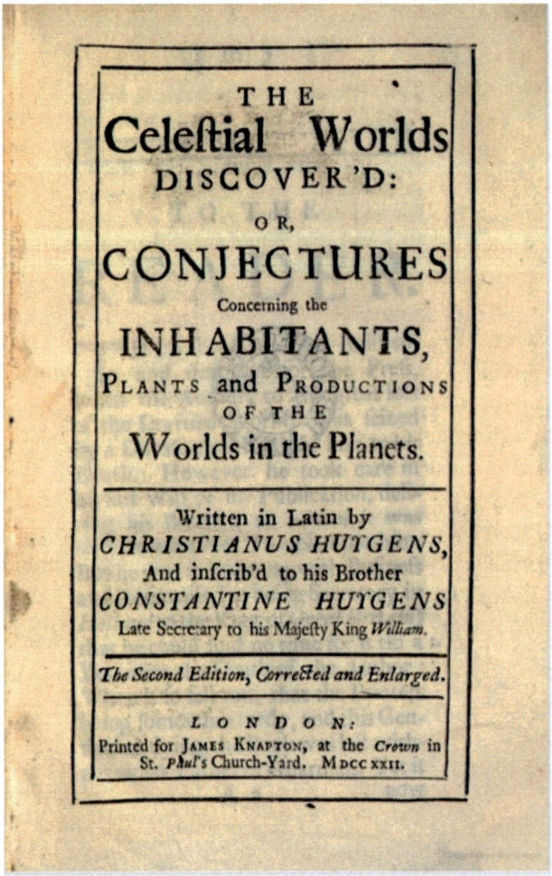
THE
Celeſtial Worlds
DISCOVER'D:
OR,
CONJECTURES
Concerning the
INHABITANTS,
PLANTS and PRODUCTIONS
OF THE
Worlds in the Planets.

Written in Latin by
CHRISTIANUS HUYGENS,
And inſcrib'd to his Brother
CONSTANTINE HUYGENS
Late Secretary to his Majeſty King *William*.

The Second Edition, Corrected and Enlarged.

LONDON:
Printed for JAMES KNAPTON, at the *Crown* in
St. *Paul's* Church-Yard. MDCCXXII.

Fig. 1.11 Frontpage of the English translation of the *Cosmotheoros. The celestial worlds discovered: or, conjectures concerning the inhabitants, plants and productions of the worlds in the planets*

space, we should still find, on all sides of us, an interminable succession of stars. This was the untenable idea of Pascal when making perhaps the most successful attempt ever made, at paraphrasing the conception for which we struggle in the word "Universe". 'It is a sphere,' Pascal said, 'of which the centre is everywhere, the circumference, nowhere.'" Poe points out that the finite size of the observable universe resolves the apparent paradox. Because the universe is of finite age and the speed of light is finite, only finitely many stars can be observed from Earth (although the whole universe can be infinite in space). Any line of sight from Earth is unlikely to reach a star due to the low star density in this finite volume.

Since Poe's time a great many other conjectures have been made regarding the cosmos. Poe's discussion still omitted the role of the expansion of the universe, which stretches the wavelengths of light into the far infrared. The bright night-sky expected by Olbers is, in fact, present, as the cosmic background radiation in the infrared. At present the most puzzling conundrum is the question of things we cannot see at all; namely, dark matter and dark energy. Dark matter is an undetected form of mass that emits little or no light but whose existence we infer from its gravitational influence. Dark energy is an unknown form of energy that seems to be the source of a repulsive force causing the expansion of the universe to accelerate. Dark matter and dark energy do opposite things. Dark matter is an unseen source of gravity that pulls things together. Dark matter dominates on the scale of galaxies, galaxy clusters, and galaxy superclusters. Dark energy is an unseen source of anti-gravity that pushes things apart. Dark energy dominates on the scale of the whole universe. The energy of the universe (including mass energy) consists of 68 % dark energy, 27 % dark matter, 4 % hydrogen, 1% helium, and traces of everything else. About 90% of the hydrogen and helium exists outside of stars. This breakdown confirms Poe's reasoning that the Universe of stars (which we can see) is not coincident with the Universe proper which also contains dark matter and dark energy (which we cannot see).

In the early seventeenth century, all thoughts were on the Solar System. However, Huygens looked beyond the Solar System. Christiaan Huygens was one of the first to realize that stars were like the Sun, and each star could have its own Solar System. An exoplanet or extrasolar planet is a planet outside our Solar System. The first scientific detection of an exoplanet was in 1988. By mid 2021, we knew of 4758 exoplanets. Only a small fraction of these are located in the vicinity of the Solar System. In particular, there are 97 known exoplanets within a distance of 32.6 light-years.

Shortly before his death in 1695, Huygens completed his book *Cosmotheoros*, which was published posthumously in 1698. In this book (see Fig. 1.11), Christiaan Huygens gave the first systematic, but speculative, account of extraterrestrial life, which he imagined would be similar to that of the Earth. *Cosmotheoros* (literally "theorist of the cosmos") was a word coined by Huygens, possibly inspired by *Cosmotheoria* (1528) by the French physician, mathematician and astronomer Jean Fernel, who estimated the Earth's circumference to an accuracy of 1%. In *Cosmotheoros* Huygens writes, "What a wonderful and amazing scheme have we here of the magnificent vastness of the universe! So many suns, so many Earths, and every one of them stocked with so many herbs, trees and animals, and adorned with so many seas and mountains! And how must our wonder and admiration be increased when we consider the prodigious distance and multitude of the stars?" Huygens took the analogy further and speculated that some of these planets could have rational creatures that study astronomy, geometry, music, have hands and feet, walk upright, and live in a society.

Huygens went into great detail, for example, by stating that the availability of water in liquid form was essential for life and that the properties of water must vary from planet to planet to conform with the local temperature range. Here Huygens uses the word "water" to mean a liquid necessary for life. He took the observations of bright spots on the surface of Mars to be evidence of water and ice.

According to modern insights life depends on three conditions: the availability of energy, organic molecules and liquids. There is water on Mars, as we now know, but it is frozen under the north and south polar ice caps. Initially it was thought that the polar ice caps consisted of only dry ice (frozen carbon dioxide); now we know that there is water ice beneath a top layer of dry ice.

Saturn's moon Titan is even more a case in point of Huygens' speculations on extraterrestrial life. Discovered by Christiaan Huygens in 1655, it was visited in 2005 by the space probe appropriately called Huygens. The Huygens landing was the most distant from the Earth so far, the first landing in the outer Solar System and the first on a moon other than our own.

Christiaan Huygens' and our modern interest in Titan is more than justified. Titan's physical and chemical composition is exceptional, which makes it a kind of astrobiological laboratory. It is the only moon in the Solar System known to us with a significant atmosphere. Titan is the only place apart from the Earth known to have rivers, lakes, clouds and seas. Its seas have similar dimensions, depth and extension, like our terrestrial seas. Titan has a hydrological cycle similar to the Earth with liquid hydrocarbons like methane and ethane playing the role of water. Below its surface there appears to be water rather than methane.

Life could exist in both environments: in Titan's subsurface water or in its atmospheric system of liquid hydrocarbons, but perhaps in a different kind than ours, using a different chemistry.

It is no coincidence that Huygens discovered the one extraterrestrial body in the Solar System that might support life.

Huygens and Remote Sensing

The method of remote sensing refers to the acquisition of information about an object or phenomenon that is out of reach of the observer. By use of real-time sensing devices, it is possible to collect data from dangerous or inaccessible areas. The instruments are sensors not in physical or intimate contact with the object. One example is geophysics, the study of the interior of the Earth, by means of seismic waves and Huygens' principle. As we know there are two kinds of remote sensing, passive and active. In *passive* remote sensing, the observer detects signals that are either emitted or reflected by the object or surrounding area being observed. For example, we see the Moon because of the reflected light from the Sun. In fact, reflected sunlight is the most common source of radiation measured by passive sensors. Examples of passive remote sensing include photography (without flash), charge-coupled devices, radiometers, and of course the human eye. In *active* remote sensing, the observer emits signals that travel to the remote object. The object then reflects or backscatters the signals, which are then detected by the observer. Radar is a prime example of active remote sensing. Another example is photography with flash.

Remote sensing has its roots in antiquity, as the method was incorporated in various forms of hunting and gathering. Ancient humans turned their eyes to the inaccessible sky. The presence of stone circles, such as at Stonehenge, shows the importance of astronomy in the prehistoric period. Astronomy played a considerable part in fixing the dates of festivals and determining the hours of the night. Astronomy was central in all early civilizations. A demonstration of the high degree of technical skill in watching the heavens is afforded by the precise orientation of the Egyptian pyramids. Celestial spheres were fundamental entities of the celestial mechanics of the ancient Greeks and this system even survived in the cosmology of Copernicus. In his heliocentric celestial system, the stars and planets are carried by ethereal spheres. With the Sun as center, the orbits of the planets increased in radius in the order of Mercury, Venus, Earth, Mars, Jupiter and Saturn. The outermost sphere, which was called the celestial sphere, held all of the fixed stars.

The telescope, as an instrument of (passive) remote sensing, appeared in the seventeenth century. All at once, astronomy, which was more or less at a standstill since the time of the ancient Greeks, spurted forth a multitude of new discoveries. A disadvantage of the early astronomical telescopes was that higher magnification was accompanied by more spherical and chromatic aberration. The result was geometric distortion and false colors. Lens-grinding and polishing techniques improved gradually during the seventeenth century. The quality of telescopes improved but they became longer. The telescope made by Christiaan Huygens and his brother Constantijn in 1656 was 23 feet (7 m) long. It had a large field of view and a magnification of 100.

Huygens was the first person to distinguish between active and passive remote sensing. He described a method of *active remote sensing* that would give the distance of the Moon from the Earth. The method which Huygens described was first used in earnest in the twentieth century when radar signals were sent from the Earth to the Moon and reflected back to the Earth. Huygens also described an effective way of *passive remote sensing*, which allowed him to make some remarkable discoveries. He chose the Earth-Jupiter system. He did not have the instruments that we have today; but he did have the mathematics, and his mathematics is correct.

Because the Earth moves in its orbit so much faster than Jupiter moves in its orbit, we may consider the two planets as moving away from each other for half the year and moving toward each other for the other half year. In this sense we have a round-trip journey, a great advantage for remote sensing. Of course, we have our clocks on Earth. One such clock is our Moon. In early myths, the Moon carried images of both eternity and time. Eternity was seen in the ever-returning cycle of the Moon and time in the phases of the Moon. The Egyptian Moon god Thoth was the scribe of time. Because of the telescope, moons of Jupiter could be seen, and they acted as clocks on that planet. The presence of an observable clock on the remote object is another great advantage for the purposes of remote sensing. The Earth-Jupiter system was an ideal starting point for Christiaan Huygens. We will return to this in Chap. 8.

Wave Theory of Light

Huygens made many contributions that advanced knowledge, but one of the most important is the wave theory of light. Scientists in the seventeenth century were involved in a deliberation about the fundamental nature of light. The question was whether light was a wave or whether it was a stream of particles. Isaac Newton (1642–1727; see Fig. 1.9) advocated the particle nature of light. On the other hand,

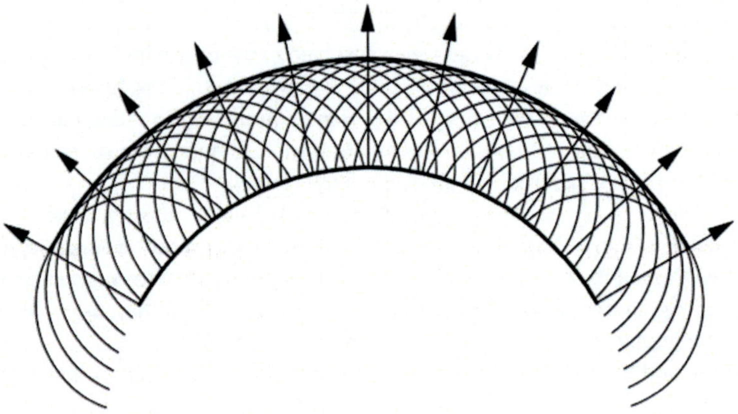

Fig. 1.12 Huygens' principle for propagation of waves. A new wave front is formed by the envelope of wavelets on the previous wavefront

Christiaan Huygens believed that light consisted of waves, much like the waves one sees on a water surface. He supported his theory by devising a principle, or more specifically a construction, that demonstrated the way in which waves propagate. Huygens showed that the wave theory of light explained reflection, refraction and interference. His principle says that the wavefront of a propagating wave at any instant is the envelope of spherical wavelets emanating from every point on the wavefront at the prior instant (see Fig. 1.12).

Let us figuratively describe the Huygens' construction. If you put together a lot of little spherical pearls on a string, you get a beautiful necklace. If you put together a lot of little spherical wavelets on a string, you get a beautiful wavefront. Signaling (the sending and receiving of signals) can be done by light waves and by sound waves. The knowledge of wave propagation is basic in either case. Huygens' all important insight is this: Given a wavefront at a given instant of time, wavelets are emitted by each point on the wavefront. These wavelets are in phase with the original wave and they travel with a speed determined by the medium. For example, the speed of sound in water is greater than the speed of sound in air. The wavelets propagate outward and combine, or interfere constructively, an instant later to form a new wavefront. In the same manner the wavelets from this new wavefront produce yet another wavefront. The process repeats itself again and again. This process of wave propagation is called Huygens' principle.

During the seventeenth century, Isaac Newton said that light is made up of tiny particles. In contrast, Christiaan Huygens looked at the way that light travels. Huygens said that light travels as waves. Every point of a wavefront may be considered the source of spherical wavelets that spread out in all directions with a speed equal to the speed of propagation of the wave. The primary wavefront at some later time is the surface tangent to all of these wavelets, in other words their *envelope.* This concept, known as *Huygens' Principle,* provides the fundamental mechanism for wave propagation. Each point on the wavefront may be regarded as a secondary source of spherical wavelets, which progress with the wave-speed of the medium and whose envelope at later times constitutes the new wavefront. The new wavefront is tangent to each wavelet. In other words, the new wavelets constructively interfere to produce the wavefront at an instant later.

Huygens used the leading edges of the wavelets to describe the propagation of waves. Despite his brilliant insight, Huygens faced formidable obstacles in getting his contemporaries to accept his wavelet theory of wave propagation. Few did and the opponents included very prominent names. Newton claimed that light was not propagated as waves, but as particles, in which case spherical wavelets are impossible. Descartes claimed that the speed of light is infinite, in which case spherical wavelets with finite radius cannot exist. The Danish astronomer Ole Roemer (1644–1710) was the first to establish that the speed of light is finite.

It was not until 1801 when Huygens' wavelet theory of 1678 was finally vindicated by the interference experiments of English physicist and polymath Thomas Young (1773–1829). Young used the side edges of the wavelets in order to describe the diffraction of the light into regions of geometric shadow.

Huygens employed a geometrical method in the seventeenth century, but by the nineteenth century, mathematics had improved to the point that Young, Fresnel, Kirchhoff, Maxwell, and others could establish the mathematical underpinnings of Huygens' principle. Maxwell's equations, first published in 1861 and 1862, can be used to directly derive Huygens' principle. Every point in an electromagnetic wave is said to act as a source of the continuing wave according to Ampere's law and Faraday's law, which fits well with Huygens' explanation. In modern science textbooks, the Huygens wavelet theory is used to describe how electromagnetic and mechanical waves propagate.

Starting in 1672, Leibniz (see Fig. 1.9) spent several years in Paris. With Huygens as his mentor, 17 years his senior, Leibniz began a study of mathematics and physics. In a letter to Henry Oldenburg of the Royal Society in London dated April 16, 1672, Leibniz writes, "The very noble Huygens first proposed the problem [which involved calculus] to me, and I solved it much to the surprise of Huygens himself." Leibniz invented the version of the differential and integral calculus in use today. However, more than a decade before either the publication of Newton's *Principia*, or the first publications of Leibniz on calculus, Huygens in his derivation of centrifugal force had used the second law of motion and twice differentiated a vector valued function. Huygens showed how singularities can arise in the propagation of waves. His investigations of singular points were important in establishing the relationships between waves and rays. In these studies, Huygens made use of the calculus of variations, optimization, and Hamiltonian mechanics all of which were officially discovered by others years later. The insight of Huygens found inspiration in the form of images. In his work Huygens was able to solve the most advanced mathematical problems by means of elementary geometrical constructions.

Before the seventeenth century, it was generally believed that there was no such thing as the "speed of light." In other words, light could travel any distance in no time at all. Later, several attempts were made to measure light-speed. Galileo Galilei tried to determine the speed of light. He concluded that "If not instantaneous, light is extraordinarily rapid." Descartes was so convinced of the instantaneous transmission of light that he unwisely said he would stake all his system of philosophy on its truth. The assumption that light travels with a finite speed is fundamental to Huygens' work. If the speed of light were infinite, his spherical wavelet would have infinite radius, and his wavelet theory would be meaningless because there would be no propagation. However, in mechanical waves his theory was already valid. Huygens used astronomical observations of the nearest moon on Jupiter to determine the speed of light. Since then, more sophisticated techniques have improved the precision of the value obtained by Huygens. Today the use of sophisticated instrumentation allows us to measure the speed of light very accurately.

Chapter 2
Huygens and Spontaneous Order

The sympathy of clocks, an admirable thing, testifies to the accuracy of clocks since it takes so little to keep them in perpetual harmony
— *Christiaan Huygens, letter to Constantijn Huygens Sr, 26 February 1665*

Spontaneous Order in the Universe

Early steam engines were inefficient, converting just a small part of the energy of the coal into work. A great deal of useful energy was dissipated or lost into what seemed like a state of immeasurable randomness. Over the years, physicists investigated this puzzle of lost energy; the result was the concept of entropy. Similarly, in 1859, after reading a paper on the diffusion of molecules by Clausius (1851), James Clerk Maxwell (1831–1879) formulated the Maxwell distribution of molecular velocities, which gave the proportion of molecules having a certain velocity in each specific range. Maxwell's work inspired Ludwig Boltzmann (1844–1906) to interpret entropy in terms of order and disorder. From this perspective, entropy is a measure of disorder. For example, in a company of soldiers, the entropy has its lowest value when all the soldiers are standing at attention. When the soldiers are moving about wildly and disorderly, the entropy has its largest value. The second law of thermodynamics asserts that in an isolated system, entropy never decreases. In other words, the second law of thermodynamics states that nature tends to take things from order to disorder. By comparison, the first law of thermodynamics is simple and states that the total energy in a system remains constant. The first law of thermodynamics prohibits the perpetuum mobile of the first kind—a machine that constantly generates energy without changing its environment. The second law of thermodynamics is a stronger statement and prohibits the perpetuum mobile of the second kind—a machine which is capable of converting heat fully into work without dissipation.

Entropy is one of the few quantities in the physical sciences that require a particular direction for time, sometimes called an arrow of time. As one goes "forward" in

T. J. Moser and E. A. Robinson, *Walking with Christiaan Huygens*, History of Physics,
https://doi.org/10.1007/978-3-031-46158-3_2

time, the second law of thermodynamics states, the entropy of an isolated system can increase, but not decrease. Thus, entropy measurement is a way of distinguishing the past from the future. However, there are many examples of systems where entropy decreases, such as living systems where life forms get more complicated over time. It is said that such systems are not closed, and when the entire closed system is taken into account there is a net increase in entropy. In other words, a local increase in order is only possible at the expense of an entropy increase in the surroundings.

The second law of thermodynamics says that when energy changes from one form to another form, or matter moves freely, entropy (disorder) in a closed system increases. The second law of thermodynamics is about the quality of energy. It states that as energy is transferred or transformed, more and more of it is wasted. The second law also states that there is a natural tendency of any isolated system to degenerate into a more disordered state. Suppose that we went to the trouble of setting two pendulum clocks exactly in phase with each other, so that they both ticked at exactly the same time instants. After a few days we would expect the two clocks to get somewhat out of phase with each other. Such thinking is in line with the fact that an isolated system tends to degenerate into a more disordered state.

However, in recent times instances have been discovered in which an isolated system tends to develop into a more ordered state. Two astronomical bodies in orbit exert a regular, periodic gravitational influence on each other. When the periods of their orbits are related by a ratio of two small integers, resonance can result. The resonance enhances the gravitational effects between the two bodies. As a result, these effects can alter the orbits. When the result is an unstable interaction, the bodies are forced to change their orbits or destruct. The gaps in Saturn's rings are attributable to unstable resonances with the inner moons of Saturn. In some cases, the resonant system is self-correcting with a stable pattern. The three Jovian moons Ganymede, Europa, and Io have a perfect octave resonance of 1:2:4. The resonance between Pluto and Neptune is 2:3. The resonance between Saturn's moons Hyperion and Triton is 3:4.

The universe is full of magnificent structures, such as galaxies, cells, ecosystems, living beings, that have all managed to assemble themselves. The second law of thermodynamics dictates the opposite in saying that nature should inexorably degenerate toward a state of greater disorder, corresponding to greater entropy. What may be considered one of the greatest scientific contributions of Christiaan Huygens is his discovery of the *process of synchronization*. This process is the most plausible one to account for existence of spontaneous order in the universe.

Huygens' Synchronization

Synchronization can be defined as the attempt to have two or more events occur simultaneously, in a coordinated manner, in phase and at the same frequency. Examples of synchronous events are: (1) birds or insects flying in great swarms (2) fish swimming in large schools (3) musicians playing in a symphony orchestra (4) ballet dancers moving in unison to the rhythm of the music (5) athletes in carefully choreographed opening sessions of the Olympic Games (6) soldiers marching—and so on. Synchronization occurs in both living and inanimate nature. Flocks of birds can be remarkable because of their mathematically coordinated organization, which is thought to provide protection against predators. Life requires synchronization. Pacemaker cells, for example, must emit electrical discharges synchronously in order for our heart to beat properly. Circadian rhythms (24 h biological rhythms), the heart, intestinal muscles, insulin-secreting cells in the pancreas, herds of walking elephants, and flying fireflies are all examples of spontaneous synchronization. Synchronization in physical systems is particularly relevant to physicists. Dynamical systems are mathematically described by systems of coupled non-linear differential equations whose solutions can have synchronous properties which are hard to predict beforehand (and chaotic properties as well). Synchronous movement is closely related to resonance. In physics, resonance (from the Latin resonare, "resonate") is the amplified resonating of an oscillating system when it is subject to a time-varying influence. Every system has a natural frequency, or eigenfrequency, at which induced movement is greatest. Examples of resonance are pushing a child on a swing, or an opera singer breaking glass by singing a high, pure note.

Christiaan Huygens was one of the first to observe the phenomenon of synchronization. Let us consider two identical metronomes placed on a wooden bar mounted on two wobbly supports. The result is that the rhythms of these two metronomical devices can synchronize within minutes or even seconds. This strange phenomenon leads to pressing questions: how do the metronomes achieve the common rhythm, and why is this happening? Huygens was the first person to face these questions, and discover and study a real-world example. While working on the longitude problem in 1657, Christiaan Huygens observed this phenomenon in the form of clock synchronization. He referred to it as *sympathy of the clocks.* Huygens was surprised to discover that if he hung two of his pendulum clocks from a common support—a wooden bar supported by two chairs—the devices kept pace relative to each other. The two pendulums always swung at the same frequency albeit in opposite directions to each other. According to Huygens' reports, it took the two pendulums about half an hour to reach this out-of-phase synchronization. And if he distorted the synchronization, it would be automatically restored in the same amount of time.

After Huygens' death there was little progress on what is now known as "Huygens' synchronization" and somehow it fell into oblivion. The phenomenon was rediscovered in 1739 by the English clockmaker and watchmaker John Ellicott (1706–1772), who maintained an observatory at his home in Hackney and is best known for his work on temperature compensated pendulums. In 1740 he submitted two manuscripts to the Royal Society: "An account of the Influence which two Pendulum Clocks were observed to have upon each other" and "Further Observations and Experiments concerning the two Clocks above mentioned". In these papers, (without reference to Huygens) he described an unusual phenomenon: when two pendulum clocks were placed next to each other with their pendulums oscillating in the same plane, one of them always stopped working after about two hours, while the other continued to swing normally. Ellicott concluded that the two pendulum clocks were influencing each other due to their shared structure. In other words, the resonance can be both constructive and destructive. This depends on the ratio of the mass of the pendulum to the mass of the platform. For a small ratio the coupling is weak and the clocks fall into out-of-phase oscillations. For a large ratio the coupling is strong and one or both clocks eventually stop running. This is called "beating death".

Since Ellicott, several scientists, including Euler, Bernoulli, and Poisson, have investigated the phenomenon of sympathy. William Ellis (1828–1916), a British astronomer and meteorologist, observed the phenomenon in 1873 using two pendulum clocks on a wooden stand. He noticed that the times indicated by the clocks were the same for nine days in a row, despite the fact that they were swinging in anti-phase, as in Huygens' experiments. Ellis also tested 9 pendulum clocks and discovered that the previous agreement in the pendulums vanished. Instead, the rate of the clocks showed great variations.

The Dutch mathematician Diederik Korteweg (1848–1941) made the first systematic attempt to explain the synchronized motion in Huygens' pendulum clocks in 1906—incidentally, he also co-edited and delivered the collected works of Huygens (1888–1950). Korteweg derived a linear model (set of coupled linear differential equations) which was simplified by ignoring damping and driving forces. With this model he demonstrated that the anti-phase movement of pendulums is the most dominant mode. This was in agreement with Huygens' observations.

Today, it is fair to say that Huygens correctly understood the basic underlying mechanism. The clocks synchronize by transferring energy to each other in the form of mechanical vibrations via the coupling bar. However, many questions are still unanswered and a complete theory is still missing.

Huygens' clock studies may also shed light on the many similar synchronization effects observed in living organisms. The human body, for example, has numerous oscillating rhythms, such as respiration, heartbeat, neuronal activity, and blood perfusion (flow of blood through body tissues and organs), and when these synchronize, energy is optimally conserved. Synchronization can also go wrong, for instance in the case of epileptic attacks, which are related to the synchronization of millions of neurons. Or in the case of bridges, as vibrating or actually collapsing bridges demonstrate time and again. Bridge builders are therefore very aware of the dangers of destructive resonance.

Saturn's Rings

The rings of Saturn are a case in point of synchronization and Huygens actively contributed to their discovery. Nowadays, we know that the rings of Saturn consist of countless small particles, ranging in size from micrometers to meters, which orbit around Saturn, indeed with remarkable synchronicity. The ring particles consist almost entirely of water ice, with a trace component of rocky material. There is still no consensus as to the mechanism of their formation. Have the rings, in some shape or form, existed since the beginning of the Solar System, 4.6 billion years ago, or are they a more recent addition? In 1610, Galileo Galilei observed Saturn's rings, but could not see them well enough to figure out their true nature. Galilei was troubled by seeing the rings disappear and reappear again and made an allusion to Greek mythology, in which the God Saturn ate his children to avoid the prophecy that they would dethrone him. In 1655, Christiaan Huygens was the first to interpret them as a disk surrounding Saturn. In 1656 he published *De Saturni luna observatio nova* (*New observation of a moon of Saturn*) which contains the following anagram:

aaaaaaacccccdeeeeeghiiiiiiillllmmnnnnnnnnnoooopppqrrstttttuuuuu

acdeghilmnopqrstu

75151174294212155,

Which would read as "*Annulo cingitur tenui, plano, nusquam cohaerente ad eclipticam inclinato*". (*It is encircled by a thin, flat, nowhere coherent ring inclined to the ecliptic.*)

Huygens explained the periodic disappearance and reappearance of the ring system by its inclination as seen from the Earth and Saturn's revolution around the Sun and his assumption that the edge of the disk is dark (see Fig. 2.1). He wrote to great detail about the shape and the size of the ring. He assumed it to be a single solid disk (see Fig. 2.2). In *Systema Saturnium*, Huygens wrote: "As for the width of the space between the ring and the globe of Saturn, observation of its shape by others has taught me—and my own observation later confirmed—that this width is equal to or even greater than that of the ring itself; and that the greatest diameter of the ring is to that of Saturn about as 9 to 4." Both estimates for the inner and outer radius of the main ring system are more or less accurate compared to modern values.

As to the surface of the ring, Huygens prophesied that it cannot be rough, covered with mountains, as the surface of our Moon is for the most part, but must be equal

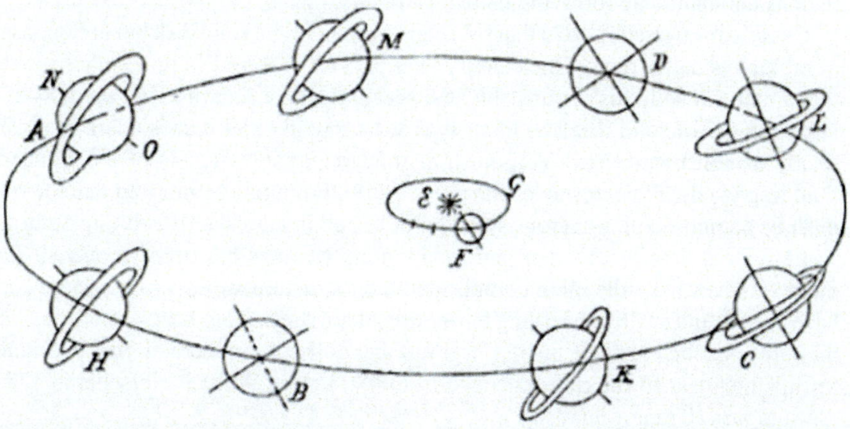

Fig. 2.1 Huygens' explanation of the appearance of Saturn's ring upon evolution around the Sun (from his letter to Jean Chapelain of 28 March 1658)

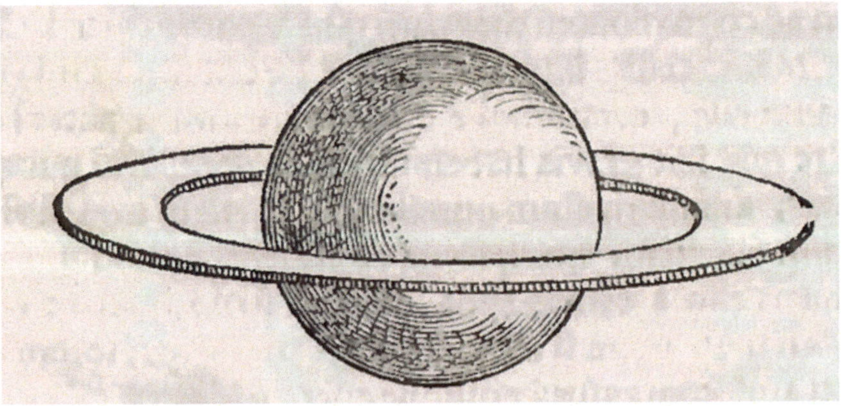

Fig. 2.2 Huygens' proposal for Saturn's ring system as a single solid disk (from Systema Saturnium)

and flat as is the case with those regions of the Moon (the mare) which some had interpreted to be seas.

As to the thickness of the ring, Huygens could not imagine it to be so thin as we know it today. In *Cosmotheoros* he estimated the thickness at 600 German miles (about 4300 km or 1/40 of the ring's width). As a result he found himself forced to explain the invisibility of the ring when viewed from the side. To accomplish this he had to assume that the edge was covered with dark material that would not reflect light.

In *Cosmotheoros*, Huygens gave further details on the ring. There are regions around Saturn's north and south poles (above 54° northern latitude and below 54° southern latitude) where the inhabitants of Saturn cannot see their own planet's ring. That is of course, Huygens dryly added, provided that the cold does not make these regions uninhabitable for living beings in the first place.

Cassini discovered in 1675 that the ring was not a single solid disk but is composed of multiple smaller rings. One century later, in 1787, Pierre-Simon Laplace proved that a uniform solid disk is unstable and speculated that Saturn's ring consists of a large number of solid ringlets. In an awarded essay in 1859, James Clerk Maxwell finally demonstrated that a non-uniform solid ring, solid ringlets or a continuous fluid ring would all be unstable (Maxwell, 1859). From this he inferred that the ring must be composed of numerous small particles, all independently orbiting Saturn.

In modern times much more information on the rings has been recovered. The more we investigate the more complicated their structure appears to be. More gaps have been found and by observing the motion of the rings, it has been established that they are not solid but made up of a large number of small shepherd moons. Estimates for the thickness of the rings changed from 480 km by William Herschel in 1789, down to just 10 m nowadays.

Our perception of the rings has significantly changed as a result of the Pioneer 11, Voyager and Cassini-Huygens missions to Saturn. The now-famous image of the rings taken by the Voyager in backlight from the Sun revealed for the first time that what looked to be the enormous A, B, and C rings was actually made up of millions of tiny ringlets. These tiny ringlets may be caused by gravitational influences of very small moons creating waves in the main rings (spiral density waves, which are rotating periodically in a spiral patterns). Larger gaps (such as the Cassini and Enke divisions) are caused by orbital resonances with some of Saturn's larger moons.

The main dense ring system extends from 67,000 km (the inner edge of the D ring) to 137,000 km (the outer edge of the A ring), so Huygens was not far off with $137/67 \approx 9/4$. Saturn's rings vary in size, with an estimated local thickness of as little as 10 m and as much as 1 km, with an average thickness of 10–20 m. The entire ring system extends up to 282,000 km from the planet, but despite their immense width, they are extremely thin. The pressing question is, why are Saturn's rings so incredibly thin?

This has to do with orbital resonance and synchronization and the key organizing principle is *conservation of angular momentum*. In classical mechanics, the movement of celestial bodies is governed by Newton's laws. The movement of one celestial body around another can be described as an endless repetition of falling, overshooting, slowing down and falling back again, all controlled by gravity and sometimes graciously referred to as a *gravitational dance*.

The two-body problem has been solved exactly. The solution is given by the three laws of Kepler:

1. a planet moves around the Sun in elliptical orbits with the Sun being located at one of their foci;
2. a line segment joining a planet and the Sun sweeps out equal areas in equal intervals of time;
3. the ratio of the squared orbital time and cube of the semi-major elliptical axis is constant for all planets.

As a result, we have a full understanding of the movement of two celestial bodies with given masses around each other (substituting one of them by the Sun), when there is no interaction with third bodies.

The three-body problem is exactly solvable only in special cases. Historically it was used to attempt to solve the movement of the Sun-Earth-Moon system. The Swiss mathematician Leonhard Euler showed in 1760 that the problem can be solved if two of the bodies are assumed to have fixed locations in space, relative to the third body. The Italian-French mathematician Joseph-Louis Lagrange showed in 1772 that a closed solution to the three-body problem exists if one of the three masses is negligible compared to the other two. In such a situation a particular type of resonance occurs where the smaller body is confined to one of five *libration points* or *Lagrange-points*. Libration points are wonderful examples of synchronization which are used to position human-made satellites, such as the James Webb Space Telescope.

The multiple-body problem is generally solvable only by numerical computation. Nevertheless, celestial bodies exert a regular, periodic gravitational pull on each other, and this can lead to remarkable resonance effects. Often, stable orbits of gravitationally interacting objects are related to each other by a ratio of small integers. We have already mentioned the examples of the 1:2:4 resonance of Jupiter's moons Ganymede, Europa and Io, and the 2:3 resonance between Pluto and Neptune. Stable orbits are self-correcting in the sense that minor disturbances (by other bodies) eventually cancel out. Unstable interaction causes bodies to exchange momentum and gradually change orbits until the bodies spiral out of control and resonance is lost. Examples of unstable interaction are some of the interactions between Saturn's innermost moons, leading to gaps in the ring system of Saturn.

This is where the conservation of angular momentum comes into play. Angular momentum is the rotational version of linear momentum. It is central to classical mechanics because it is constant for closed systems of bodies. The angular momentum of an object is the amount of rotational movement and is defined as the product of the moment of inertia (resistance to changes in rotational movement) and angular velocity of the object. Conservation of angular momentum means that the total angular momentum of a system of objects remains the same unless there is some external interaction. Spinning tops and gyroscopes depend on the conservation of angular momentum, figure skaters make use of it, and without it flying helicopters or riding bicycles would be impossible. Closed systems of celestial bodies such as our Solar system are no exception. Starting out as an unorganized (and unsynchronized!) three-dimensional cloud, the bodies rotate and move under the influence of mutual gravity, keeping the total angular momentum constant. As a result, dissident movements up and down are called to order and eventually canceled out until all bodies rotate in a two-dimensional plane. This mechanism applies to most galaxies, to our Solar system and, in the extreme case, to Saturn's ring system.

Saturn's rings are the most prominent ones in our Solar system, but Saturn is not the only planet being encircled by rings (Tiscareno, 2020). Since the telescope was invented and Huygens made his discoveries, rings have been discovered around all giant planets of our solar system, Jupiter, Uranus and Neptune. Saturn's moon Rhea seems to be the first moon known to have a ring itself. The reason why larger planets have rings but the smaller have not is related to tidal forces and the Roche limit. The Earth and Moon periodically attract each other, and the pulling of the Moon causes ocean tides on Earth. If a smaller celestial body gets too close to a larger celestial body (for example, if a moon gets too close to its planet), then the tidal forces can be so strong that the smaller celestial body is torn apart. The critical distance within which this occurs is known as the Roche limit, after French astronomer Edouard Roche (1820–1883). The rings of Saturn are believed to be the result of a moon passing too close to Saturn. Another theory for planetary rings is based on collision of the planet with a moon, or collision between two moons. A possible explanation of the lack of rings for the inner planets is that rings usually consist of ice dust. Close to the Sun it is presumably too hot for ice rings to form.

Saturn's rings are far from static. Small moons orbiting close to or within the rings are constantly interacting. They are referred to as *shepherd moons*. Pan, Daphnis, Atlas, Pandora and Prometheus, are the most important. They measure between 8 and 35 km in diameter. By their gravitational interaction they keep the rings together. The shepherd moons also cause gaps in the rings which are free of particles. They generate density waves which constantly alter the shape of the rings. The rings are also constantly losing ice particles, which rain down on Saturn's surface. Since their formation, the rings may have lost half of their mass. This leads to speculation that in the far future they may disappear. On the other hand, they are constantly fed by the shepherd moonlets that surround them.

The discovery in 2009 of the Phoebe ring, which consists of extremely tenuous dust stretching from 128 to 207 times the radius of Saturn at an angle of 5° to Saturn's orbit and main ring systems, seems to turn all science upside down again. Nevertheless, the extreme flatness of Saturn's main ring system is well understood. The explanation is Huygens' synchronization. Synchronization is the coordination of events to operate a system in unison. For example, the conductor of an orchestra keeps the orchestra synchronized or in time. Systems that operate with all parts in synchrony are said to be synchronous or in sync—and those that are not are asynchronous. Huygens not only discovered the rings of Saturn, but he also discovered the reason for their existence.

Tidal Locking

Another example, closer to home, is the synchronicity of the spin of the Earth's moon and its orbit, which have the same rotation time (one month), so that the same face is always pointed to the Earth and the other side is permanently invisible from the Earth. This effect is referred to as tidal locking. The mutual gravitational pull of the Earth and the Moon not only affects the ocean tides (as mentioned above) but also their orbital mechanics.

Newton had a keen interest in tidal effects and offered an explanation in his *Principia* (1687, Proposition XXXVIII, Book III). Assuming that the Moon is not a rigid body but a homogeneous fluid, Newton reasoned that the Earth's pull on its front and back side would be different and resulting in a stretching into an ellipsoidal (non-spherical) shape in the direction of the Earth. "Hence it is that the same face of the moon always respects the earth; nor can the body of the moon possibly rest in any other position, but would return always by a libratory motion to this situation; but those librations, however, must be exceedingly slow, because of the weakness of the forces which excite them; so that the face of the moon, which should be always obverted to the earth, may, for the reason assigned in Prop. XVI I. be turned towards the other focus of the moon's orbit, without being immediately drawn back, and converted again towards the earth."

Huygens, in his Cosmotheoros (1695), offered a similar explanation, generalizing to other planet-moon interactions: "Who will doubt, given that both in the case of this extreme satellite [of Saturn] and in that of our Moon the same face is always seen from the corresponding primary Planet, that nature has arranged things in the same way in the case of the other satellites of Jupiter and Saturn? As for the efficient cause, it can hardly be anything but this: in all the Moons matter is denser and heavier on the farthest side of the Planet. Indeed, in this way this part will tend with more force to deviate from the center of rotation, while, if such were not the case, the same face should according to the laws of the movement always be directed not towards the Planet but towards the fixed stars."

The key principle of the tidal synchronization is again conservation of angular momentum, now working on celestial bodies that are not totally rigid but deformable. Nowadays, most moons in our Solar system are known to be tidally locked to their planets, with a few exceptions of moons at larger distance from the gas giants (such as Saturn's moon Phoebe). Similar effects are even suggested between planets (Mercury versus the Sun, Venus versus the Earth) and exoplanets.

Would Huygens have realized that the sympathy of clocks and the tidal locking are just two manifestations of the same physical mechanism, namely synchronization?

Two Identical Pendulum Clocks

Historically sailors determined latitude by sighting the height of the Sun above the horizon. However, there was no physical reference for determining longitude. They needed a reference time. Longitude could be obtained from the difference between the local time (as given by the Sun) and the "reference" time such as Greenwich Time. Christiaan Huygens studied the problem of measuring longitude. Huygens felt that his pendulum clock would provide the solution. He combined two identical pendulum clocks for the redundancy necessary to provide reliability and accuracy. For example, if one clock failed, the other clock would still be working.

This situation led Christiaan Huygens to first discover systems that went from disorder to order. He did this work two centuries before the establishment of the second law of thermodynamics. In February 1665, feeling unwell, he stayed in his room for a few days. His two newly made pendulum clocks were hanging on the wall next to one another separated by about two feet (60 cm). Huygens observed a notable effect. The two clocks of the same shape and size were in agreement so exact that their respective pendulums were exactly in cadence with one another. When the pendulum of the left clock reached the top of its swing, so did the pendulum of the right clock. When the pendulum of the left clock reached the bottom of its swing, so did the pendulum of the right clock. Huygens concluded that this behavior occurred through a kind of sympathy. Accordingly, he arbitrarily reset the positions of the pendulums and so destroyed their cadence. He observed that within a half hour the pendulums of the two clocks were again in cadence, and remained so afterwards. There is one further element. When in cadence, the pendulums did not oscillate parallel to one another, but instead they approached and separated in opposite directions. When one swung to the left, the other swung to the right, and vice versa.

Huygens then separated the two clocks, hanging one at the end of the room and the other fifteen feet away. He noticed that the two clocks did not get into cadence. Within a day they suffered a five seconds difference between them. He conveyed these results to his father. In a letter he wrote on February 24, 1665 to his friend René-François de Sluse (1622–1685, a mathematician living in Liège—then an independent state in the Spanish Netherlands, now part of Belgium) Huygens referred to this odd phenomenon at the "miraculous sympathy of two clocks".

Huygens next performed various experiments. He suspected that the clocks might be interacting through tiny vibrations of their common support. He hung each clock from a separate plank. Both planks were lying on top of two chairs positioned about a meter apart, back-to-back. The clocks sympathized within an hour. Again, their pendulums swung back and forth, like a pair of hands clapping. With the clocks in sympathy the chairs were motionless. When he disrupted the sympathy, the chairs began to shake. They were trembling, clattering on the floor. They continued to shake for another half hour until the sympathy restored itself, at which point the chairs fell silent.

Huygens concluded that the swinging of its pendulum imparted slight movement to the clock, which jiggled the planks, which jiggled the chairs. But when the clocks were in sympathy—when their pendulums swung precisely opposite to each other, the equal and opposite forces they exerted on the planks canceled each other out, which allowed the chairs to keep still. Conversely, when he disrupted the sympathy, the opposing forces no longer balanced at all times. A portion of them added up and dragged the planks back and forth from side to aide, shaking the chairs. Once the resonance was achieved the chairs did not move any more. In essence, Huygens had discovered the concept of stabilization by negative feedback, the process whereby a system responds to an input impulse by damping it Fig. 2.3.

Fig. 2.3 The drawing by Christiaan Huygens of his experiment in 1665

Synchronization of Clocks

In summary, Huygens in 1665 discovered that two clocks hanging on the wall had acquired an opposing motion; their pendulums were beating in unison but in opposite direction; that is, 180° out of phase. Today we know that the cause of this phenomenon was:

(1) The two pendulums were affecting each other through slight motions of the supporting mantelpiece.
(2) This process is called entrainment or mode locking and is observed in coupled oscillators.

On February 27, 1665, Huygens wrote a letter to Moray, asking him to convey his observations to the Royal Society.

Huygens had discovered one of the most pervasive rudiments in all of nature. Huygens had discovered inanimate synchronization. An informal word for synchronization is "sync". Entropy is a measure of disorder. The second law of thermodynamics asserts that in an isolated system, entropy never decreases. If we drop an egg, the shell (a well-ordered system) shatters into a random collection of little pieces (a disordered system). The entropy (disorder) increases when the egg breaks. It is impossible to have all the pieces rise up into a pristine egg. If we synchronize the pendulums on two clocks, we would expect them to get out of synchronization over the passage of time. In other words, we would expect the clocks to go from order to disorder. However, in 1665 Huygens discovered the opposite holds under specified conditions. Two clocks loosely connected can go from disorder to order, refuting common sense. It is as if the shattered pieces of egg shell jumped up and formed a perfect egg.

The sympathy of clocks does not contradict the second law of thermodynamics. The capacity for synchronization does not depend upon some universal principle, but only depends upon the laws of classical mechanics. The clue to the apparent contradiction is given by the fact that the second law of thermodynamics refers to isolated systems, without outside interaction. This does not exclude the synchronization of open smaller sub-systems.

Another example of synchronization is the laser, which makes use of a quantum mechanical phenomenon called "Light Amplification by Stimulated Emission of Radiation or LASER. The emission is said to be stimulated (as opposed to spontaneous) because the incoming photons provoke the excited atom into giving out the new photon. The emitted photon is indistinguishable from the one that spawned it. Each photon represents a tiny wave of light, all of which are perfectly synchronized (an example of the ambiguity in quantum mechanics, whereby light has both the characteristics of a particle and of a wave). Their peaks and valleys are perfectly aligned, so all of them comprise light of the same color, in the same direction and with the same phase. The central idea is that the newly created photons are always in synchronization with the ones that made them. Ordinary light consists of a random collection of waves of various colors, directions, and phases. Laser light consists of a collection of waves perfectly synchronized. In going from ordinary light to laser light is going from disorder to order, as first established by Huygens. Here we have a central problem in the study of the scientific work of Christiaan Huygens. He was centuries ahead of his time. It is only now that we can truly appreciate his work.

Chapter 3
Huygens and the Speed of Light

I saw Eternity the other night, Like a great ring of pure and endless light, All calm, as it was bright; And round beneath it, Time in hours, days, years, Driv'n by the spheres, Like a vast shadow mov'd; in which the world, And all her train were hurl'd.
— *Henry Vaughan (1621–1695), The World*

Speed of Light

The seventeenth century saw the rise of modern science, notably with the work of René Descartes, Galileo Galilei, Christiaan Huygens, Isaac Newton and Gottfried Leibniz. Isaac Beeckman (1588–1637) was a Dutch scientist who is not well known today because he did not publish his ideas. However he did have a large influence on many scientists of his time. They included Dutch astronomer and mathematician Willebrord Snell van Royen, Flemish mathematician Simon Stevin, René Descartes, Dutch statesman Johan de Witt, French polymath Marin Mersenne, and French philosopher, astronomer, and mathematician Pierre Gassendi. French physician and natural philosopher Sébastien Basson (around 1573-after 1625) is credited for advancing the concept that matter is composed of atoms. Beeckman independently came upon the same concept. Beeckman was one of the first persons who recognized the physical property of inertia; namely, that an object will remain at rest or in uniform motion in a straight line unless acted upon by an external force. This property eventually became Newton's first law of motion.

Spectacle makers Hans and his son Zacharias Jansen built compound microscopes for the first time between 1590 and 1608 in Middelburg in the Netherlands. Their fellow townsman Hans Lipperhey then invented the telescope in 1608. Galileo improved upon the design of the telescope and used it in his discoveries of the moons of Jupiter (see Fig. 3.1). Dutch astronomer and mathematician Willebrord Snell (1580–1626, latinized as Snellius) in 1621 discovered the mathematical relation (Snell's law) between the angles of incidence and transmission for a light ray refracting through an interface between two media (the same law had already been

T. J. Moser and E. A. Robinson, *Walking with Christiaan Huygens*, History of Physics,
https://doi.org/10.1007/978-3-031-46158-3_3

Fig. 3.1 Photo of Jupiter and its moons, taken with a mobile phone through a small telescope [courtesy Leon Thomsen]

described by the Persian mathematician Ibn Sahl in 984). The first description of what are now known as diffraction effects, in which light passing by an obstacle penetrates into the geometrical shadow, was made by the Jesuit mathematician Francesco Grimaldi (1618–1663) and posthumously published in 1665.

At the beginning of the seventeenth century, the general consensus was that light did not have a speed. Instead, light just appeared instantaneously. In the course of the seventeenth century the idea of instantaneous propagation of light was opposed. It was first challenged by Isaac Beeckman in 1629, who set up a series of mirrors around a gunpowder explosion to see if observers noticed any difference as to when the flashes of light appeared. The results were inconclusive. The problem is that the speed of light is too high to be detected by the human eye this way. Also in the early 1600s Galileo performed a similar experiment to determine the speed of light. Two people with covered lanterns stood a known distance apart. One person uncovered his lantern and as soon as the other person saw the light, he uncovered his own lantern. Galileo attempted to record the time between lantern signals but was unsuccessful because the distance involved was too small and light simply moved too fast to be measured this way. Worse still, Galileo had no accurate clock to measure time. Galileo used his pulse like a clock to measure time intervals by counting the number of heartbeats.

In 1637 Descartes characterized light as a pressure wave transmitted at infinite speed through a pervasive elastic medium, the "luminiferous ether". Although Descartes assumed the speed of light to be infinite, he also assumed in his derivation of Snell's law that the denser the medium, the greater the speed of light. This seems to be a contradiction in Descartes' point of view. His formulation of Snell's law is therefore limited to the (by itself correct) statement that the ratio of the sines of the incident and refracted angles is constant when the incident angle is varied, but does not contain a statement on the ratio of the propagation speeds in the two media. Pierre de Fermat (1607–1665), on the other hand, correctly assumed that the speed of light is finite and his derivation of Snell's law is based on the speed of light being lower in a denser medium (light travels more slowly in water than in air). English physicist Robert Hooke was a proponent of the wave theory of light as developed by

Descartes. In 1665 Hooke studied diffraction effects and thin-film interference and concluded that light is a rapid vibration of any medium through which it propagates.

However, Isaac Newton (1642–1727) imagined light in a different way again. Newton developed the corpuscular theory of light. The theory held that light consisted of tiny particles, which he referred to as corpuscles, in rapid motion. Such corpuscles were single, infinitesimally small, particles which have shape, size, color, and other physical properties. The corpuscles emanated from their source and were scattered in the atmosphere. According to Newton, Grimaldi's diffraction was merely a new form of refraction. He referred to diffraction as "inflexion"; the term "diffraction" was coined by Grimaldi but adopted only in the early nineteenth century. Newton argued that the geometric nature of the laws of reflection and refraction could only be explained if light is made of corpuscles. Newton also assumed, like Descartes, that light travels faster in a denser medium than in a rarer medium. For instance light would travel faster in water than in air. This assumption is incorrect for electromagnetic waves (such as light), if we understand density to be the number of particles in a given volume of the medium. However, it is correct for some types of mechanical waves (such as seismic waves in the Earth), if we understand density to be mass density.

Newton believed a medium could either attract or repel the corpuscles of light. He assumed that reflection of light is due to the repulsion between the corpuscles and reflecting surface, while refraction of light is due to the attraction between the corpuscles and refracting surface. When corpuscles (light particles) approach the refracting surface, they are attracted near the surface. When they enter the denser medium from a rarer medium, their speed increases and hence changes their direction; this phenomenon we now call refraction. Newton's corpuscular theory (Fig. 3.2) was not able to explain diffraction, interference or polarity. Huygens' wave principle could, as we will see, but Huygens mysteriously never studied diffraction; it is not mentioned even once in his collected works.

It was Christiaan Huygens who determined the speed of light using astronomical observations taken at the Paris observatory by the Danish astronomer Ole Roemer. In 1678, Huygens presented his *wave theory of light*, which was published in 1690 in

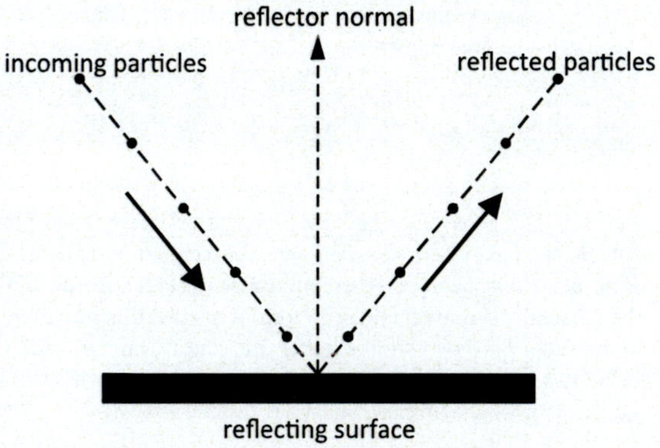

Fig. 3.2 Newton's reflection of light particles

his work *Traité de la Lumière (Treatise on Light).* He held that light was emitted in all directions as a set of waves that moved through a medium that he called luminiferous ether ("ether" comes from the Greek word for "brightness" and is alternatively spelled as "aether"). The idea of the luminiferous ether was motivated by experiments by Evangelista Torricelli (1608–1647) who in 1643 showed that light propagates in vacuum, whereas Robert Boyle showed in 1660 that sound does not propagate in vacuum. Therefore, there had to be a dedicated medium other than air for light to propagate in. If light was a wave phenomenon, something had to vibrate. Blaise Pascal conducted further experiments on pressure and vacuum, which are relevant to the understanding of the properties of air and the concept of vacuum. Descartes did not believe in a vacuum based on rationalistic views dating back to Aristotle that space and matter are essentially the same: *horror vacui* or nature abhors vacuum. Descartes jokingly argued that vacuum exists primarily in Pascal's head.

Huygens' view on light propagation was based on collision theory of particles making up the medium (see Fig. 3.3). He assumed the speed of the waves was reduced when they entered a denser medium. His model proved particularly useful in explaining the laws of reflection and refraction.

Pythagoras first described the wave motion of sound in about 550 BC. It is said that he noticed that anvils of different weights produce different musical notes when they are struck. The work of Pythagoras demonstrated that music can be expressed mathematically. Water surface waves can be seen. Sound waves can be heard. Over the years an analogy between water waves and sound waves was made. In the following paragraph from the first chapter in *Traité de la Lumière*, Huygens introduced light waves by making the analogy with sound waves. Huygens wrote:

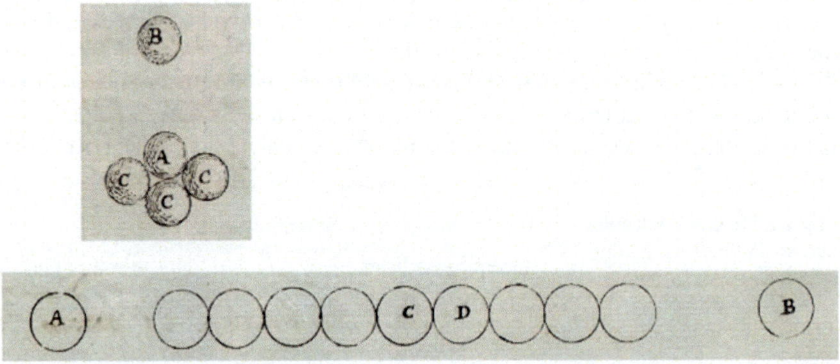

Fig. 3.3 Huygens' view on light propagation through a luminiferous ether: (top) "It is that when a sphere, such as A here, touches several other similar spheres CCC, if it is struck by another sphere B in such a way as to exert an impulse against all spheres CCC which touch it, it transmits to them the whole of its movement". (bottom) if against this row there are pushed from two opposite sides at the same time two similar spheres A and D, one will see each of them rebound with the same velocity which it had in striking, yet the whole row will remain in its place [with permission from Leiden University Library, HUG 31, ff. 255r.-312v.]

> We know that by means of the air, which is an invisible and impalpable body, Sound spreads around the spot where it has been produced, by a movement which is passed on successively from one part of the air to another; and that the spreading of this movement, taking place equally rapidly on all sides, ought to form spherical surfaces ever enlarging and which strike our ears. Now there is no doubt at all that light also comes from the luminous body to our eyes by some movement impressed on the matter which is between the two; since, as we have already seen, it cannot be by the transport of a body which passes from one to the other. If, in addition, light takes time for its passage—which we are now going to examine—it will follow that this movement, impressed on the intervening matter, is successive; and consequently, it spreads, as Sound does, by spherical surfaces and waves: for I call them waves from their resemblance to those which are seen to be formed in water when a stone is thrown into it, and which present a successive spreading as circles, though these arise from another cause, and are only in a flat surface.

Action at distance is a concept in physics that an object can be moved or otherwise changed without actually being physically touched by another object (as in mechanical contact). It is understood to mean the immediate remote interaction between objects that are at a distance from each other in space. Descartes refused to accept the concept of action at a distance, even although, paradoxically, he maintained that light travels at infinite speed. Instead, Descartes introduced his concept of ether in *Principia Philosophiae* (1644). For Descartes, the ether was an intermediate medium consisting of vortices that transmit forces between bodies at a distance (truth be said, the theory of vortices has not survived the test of time).

Newton and Huygens had great mutual respect for each other, but conceptual differences in their approach to science. This was especially apparent in their theory of gravity. In 1687 Isaac Newton published *Philosophiæ Naturalis Principia Mathematica* (often referred to as simply the *Principia*), in Latin. In it, Newton presented his famous three laws of motion as well as his law of universal gravity, which became the foundation of classical mechanics. Two years later in 1689 Christiaan Huygens travelled to London and gave lectures on his own theory of gravitation to the Royal Society. Huygens met with Newton but avoided a direct debate with him. Huygens had a lot of respect for the mathematical brilliance of the *Principia*. Still, his correspondence, especially with Leibniz, makes it clear that he thought a theory of gravitation lacking any mechanical explanation was fundamentally unacceptable. Here we clearly see the clash between the rationalist Huygens and the empiricist Newton. Huygens' own theory, published in 1690 in his *Discours de la cause de la pesanteur* (*Discourse on the Cause of Gravity*), dating back to 1669, rejected action at a distance. Instead, Huygens included a mechanical explanation of gravity based on Cartesian vortices.

Another field where Newton's and Huygens' concepts clashed was the theory of light. Huygens' *Traité de la Lumière* (*Treatise on Light*), already largely completed by 1678, was published in 1690. In *Traité,* Huygens introduced his luminiferous (that is, light-transmitting) ether. This medium in which light propagates penetrates all matter and is even present in the vacuum. Huygens thereby again showed his demand for ultimate mechanical explanations in his discussion of the nature of light. His explanations of reflection and refraction (considered superior to Newton's) were exclusively and consistently based on his (Huygens') principle of secondary wavefronts, which had its mechanical basis in the theory of collision of ether particles (not Newton's corpuscles of light).

In the nineteenth century Young and Fresnel kept the luminiferous ether when they revitalized Huygens' wave theory of light. Maxwell showed in the 1860s that light waves are electromagnetic waves that are polarized transverse (perpendicular) to the direction of the propagation of the waves. Following Young and Fresnel, Maxwell retained the ether of Huygens. In 1878 Maxwell wrote the following in *Encyclopedia Britannica*:

> Ethers were invented for the planets to swim in, to constitute electric atmospheres and magnetic effluvia, to convey sensations from one part of our bodies to another, and so on, until all space had been filled three or four times over with ethers. ... The only ether which has survived is that which was invented by Huygens to explain the propagation of light.

By the end of the nineteenth century physicists regarded the ether as a quasi-rigid solid (not completely rigid because it can vibrate), luminiferous (that is, light transmitting) medium that is massless and transparent, at absolute rest, and present everywhere in the universe. The ether was finally made redundant in 1905 by Einstein's theory of special relativity In which Einstein replaced the ether by vacuum. In his theory of general relativity, Einstein suggested that the shape of spacetime is what gives rise to the force we experience as gravity. Space becomes curved in the vicinity of accumulations of mass such as planets and stars, much like water flows around obstacles in a river. Nowadays, the concept of ether does no longer play a role in physics. Modern physics does not reject, in principle, the idea of an ether, but there is simply no need for it. In effect, the theory of transmission of light was right back where it started from. In the seventeenth century Torricelli performed an experiment that demonstrated that light travels in a vacuum and from astronomical observations Huygens calculated the numerical value of the constant speed of light. In the twentieth century Einstein postulated light travels in a vacuum at a constant speed.

Huygens and Remote Sensing

The method of remote sensing refers to the acquisition of information about an object that is out of reach of the observer. The first significant instrument of remote sensing was the telescope. As has been described in Chapter 1, the telescope appeared in the seventeenth century. The science of astronomy had been at a standstill in Europe since the time of the ancient Greeks, whose legacy was guarded and continued by Arab scientists. With the invention of the telescope in the Renaissance, astronomy burst forth with a multitude of new discoveries. A shortcoming of the early astronomical telescopes was geometric distortion and false colors. These defects occurred because high magnification was accompanied by spherical and chromatic aberration.

Like we said, lens grinding and polishing techniques improved gradually during the seventeenth century. The quality of telescopes improved but they became longer.

A telescope made in 1656 by Christiaan Huygens and his brother Constantijn was 23 feet (7 m) long. See Fig. 3.4. It had a large field of view and a magnification of 100. More on the telescope in Chap. 5.

In 1659 Christiaan Huygens wrote (in *Systema Saturnium*): See Fig. 3.5

> In the sword of Orion are three stars quite close together. In 1656, I chanced to be viewing the middle of one of these with a telescope, instead of a single star twelve showed themselves (a

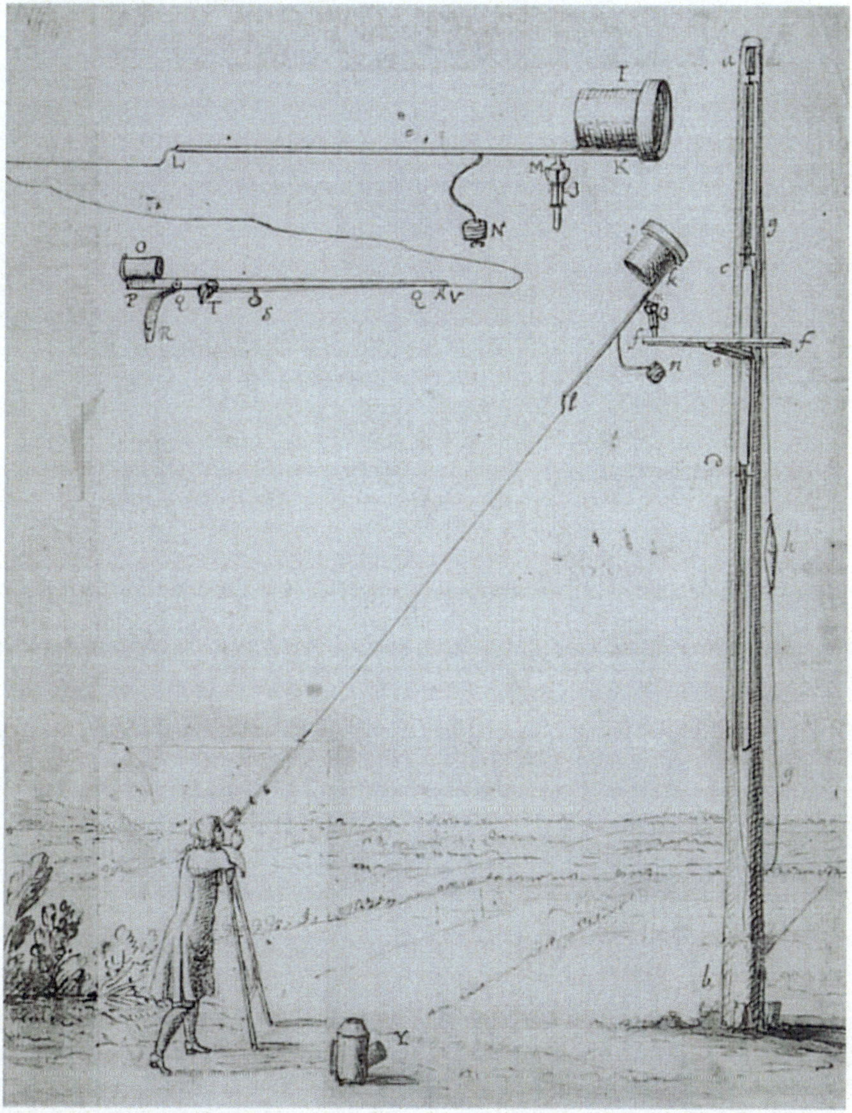

Fig. 3.4 Huygens' aerial telescope

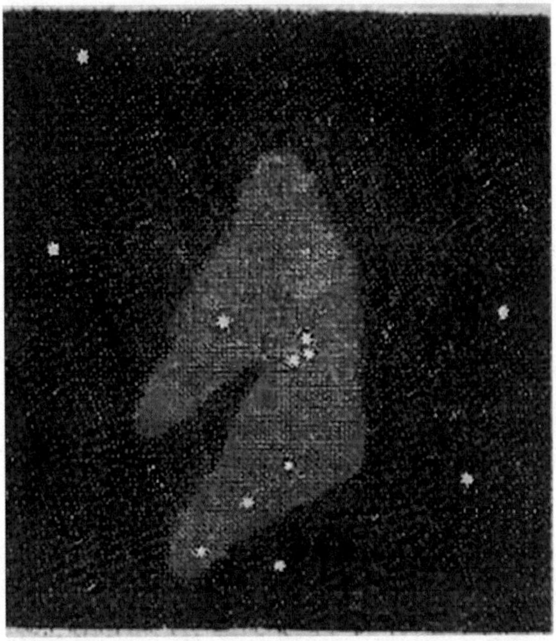

Fig. 3.5 Display of the sword of Orion from Huygens' *Systema Saturnium*

> not uncommon occurrence). Three of these almost touched each other, and with four others shone through the nebula, so that the space around them seemed far brighter than the rest of the heavens, which was entirely clear and appeared quite black, the effect being that of an opening in the sky through which a brighter region was visible.

This suggested already a departure from the heliocentric Copernican system which held that the Sun was the center of the universe and a surrounding celestial sphere contained all of the "fixed" stars. Huygens went even further when he wrote in his *Cosmotheoros* (1698).

> That all these Stars are not in the same [celestial] Sphere, as that the Sun, which is one of them, cannot be brought to this Rule [the Copernican system]. But it is more likely they are scattered and dispersed all over the immense spaces of the Heaven, and are as far distant perhaps from one another, as the nearest of them are from the Sun.

Huygens concluded,

> I must give My vote to have the Sun of the same nature with the fixed Stars. For then why may not every one of these Stars or Suns have as great a Retinue as our Sun, of Planets, with their Moons, to wait upon them?

In other words, Huygens said that the stars are spread out all over the universe and the Sun is just another star. Little was made of this perceptive statement of Huygens. With the observations of William Herschel, Friedrich Bessel, and other astronomers in the nineteenth century, it was realized that the Sun, while near the barycenter of the Solar System, was not at any center of the universe. The Cosmological Principle assumes that the Universe is homogeneous (the same observational evidence is available to observers at different locations in the Universe) and isotropic (the same observational evidence is available by looking in any direction in the Universe). A homogeneous, isotropic, infinite Universe does not have a center. The statement of Huygens is more in line with the Cosmological Principle. The Copernican system was shattered; the Sun is not at the center of the universe. The celestial sphere of stars becomes nothing more than a mathematical construction. The Copernican heliocentric system was rather short-lived, introduced by Copernicus in 1534 in *De revolutionibus orbium coelestium* ("On the revolution of heavenly spheres") and superseded already by Huygens in 1698 (in the "Cosmotheoros").

Remote sensing is obtaining information about a distant object or phenomenon without making physical contact with the object, in contrast to local in-contact observation. Remote sensing can be subdivided into "active" and "passive" remote sensing. In *active remote sensing*, a signal is sent from a source to a distant object and the reflection of the signal is detected by a receiver. In *passive remote sensing*, a receiver sits and waits until it receives a desired signal from afar. In his masterpiece *Traité de la Lumière*, Huygens distinguished between active remote sensing and passive remote sensing. He was the first person to do so. Huygens described a method of *active remote sensing* that would find the distance of the Moon from the Earth, see Fig. 3.6. Unfortunately, the only signals he had available were astronomical shadows, which were entirely unsatisfactory. However, the active method of remote sensing described by Huygens came to fruition in the last century. On 10 January 1946, a team at Fort Monmouth, New Jersey, reflected radar signals off the Moon. The signals took 2.5 s to travel to the Moon and back to the Earth. This achievement marked the beginning of radar astronomy and space communication.

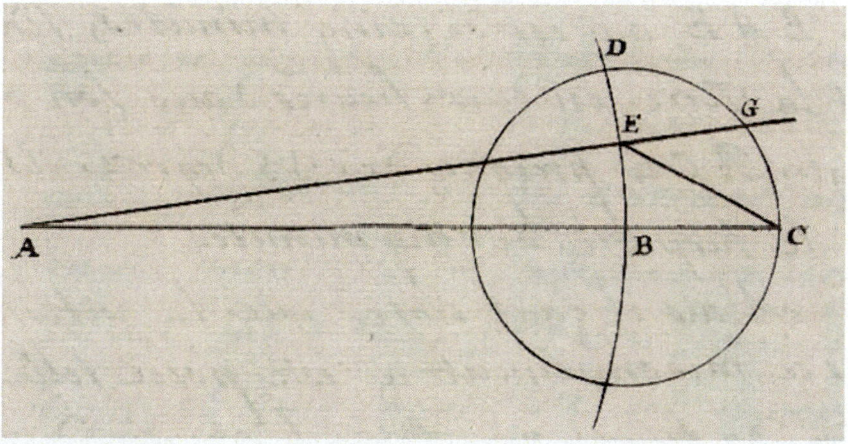

Fig. 3.6 Huygens' drawing for active remote sensing of a lunar eclipse (C) by the Earth moving along the orbit B-E-D illuminated by the Sun at A. Because of the finite speed of light the eclipse is seen at C when the Earth has already moved to E and the Moon to G. (from manuscript of Traité de la Lumière) [with permission from Leiden University Library, HUG 31, ff. 255r.-312v.]

Huygens also described an effective method of passive remote sensing, with which he made some startling discoveries. He chose the Earth-Jupiter system and used observational results of Ole Roemer. The two planets move apart for half of the year and move together for the other half year. In this sense we have a round-trip journey, which is the essence of active remote sensing. We will treat this state of affairs in more detail when we discuss Roemer's work in coming sections.

In particular, Huygens used Roemer's astronomical observations to find the numerical value of the speed of light. Because of the recent discovery of original reports of the Paris observatory, the common understanding that Roemer gave an estimate for the speed of light no longer holds. The attainment of the numerical value of speed of light was a major breakthrough in the history of science. Before the seventeenth century, it was generally believed that there was no such thing as the speed of light. In other words, light could travel any distance in no time at all. Later, several attempts were made to measure light-speed. Galileo concluded that "if not instantaneous, light is extraordinarily rapid." Descartes was so convinced of the instantaneous transmission of light that he never gave up this belief despite the fact that he rejected action at distance.

The assumption that light travels with a finite speed is fundamental to Huygens' work. Specifically, the fact that light has a finite speed is essential to the veracity of his best-known contribution, namely, *Huygens' principle* which is also known as *Huygens' construction.* If the speed of light were infinite, his spherical wavelet would have infinite radius, his construction using propagation (at finite speed) and envelopes would fall apart, and his wavelet theory would be meaningless. As we will see in Chap. 8, Huygens used astronomical observations of the nearest Galilean moon of Jupiter to determine the speed of light. Since then, more sophisticated techniques have improved the precision of the value obtained by Huygens (212,000 km/s). Today the speed of light is very accurately known (299,792.458 km/s).

The brilliant insight of Huygens led to magnificent scientific advances. Huygens' principle is based on the assumption that light is not instantaneous, but travels in waves from one place to another at the finite speed. Unfortunately, Huygens faced formidable obstacles in getting people to accept his wavelet theory of wave propagation. Few people supported Huygens and the opponents included very prominent names. Newton claimed that light was not propagated as waves, but as particles ("corpuscles"), in which case spherical wavelets are impossible. In the 1700s in later editions of *Principia*, Newton did include the speed of light, not the original 1678 value of Huygens but a value obtained by using more recent estimates of astronomical quantities.

Throughout the eighteenth century, Newton's corpuscular theory held sway. It was not until 1801 when Huygens' wavelet theory of 1678 was finally vindicated by the interference experiments of Thomas Young. Huygens used the leading edges of the wavelets to describe the propagation of waves and used the trailing edges of the wavelets to describe the reflection of waves. Young used the side edges of the wavelets in order to describe the diffraction of the light into regions of geometric shadow. Young's experiments as the confirmation of wave theory were done in England, where most scientists still adhered to Newton's invalid corpuscular theory.

Meanwhile in France, Fresnel gave the mathematical foundation of wave theory. The diffraction experiment of Arago (1786–1853) designed to refute Fresnel's wave theory, and the subsequent acceptance by Poisson (1781–1840) of the spot named after him are traditionally seen as the decisive victory of wave theory. Huygens paradoxically never studied diffraction (more on diffraction in Chap. 11). Whereas Huygens used a geometrical approach in the seventeenth century, subsequent advances in mathematics allowed such nineteenth century scientists as Young, Fresnel, Kirchhoff, and Maxwell to establish the mathematical foundations of Huygens' Principle (see more on Huygens' principle in Chaps 4 and 7). Huygens' principle can be obtained directly from Maxwell's equations, which are wave equations for electromagnetic waves. Ampere's law and Faraday's law predict that every point in an electromagnetic wave acts as a source of the continuing wave, which fits perfectly with Huygens' analysis. Today every science book uses the Huygens wavelet theory to explain the propagation of electromagnetic waves and mechanical waves as well.

James Clerk Maxwell

James Clerk Maxwell (1831–1879, see Fig. 3.7) was the first person to fully appreciate the fundamental implications of the work of Huygens with respect to speed of light and relativity theory. Maxwell was insatiably curious as a youth. At age 14 he was writing scientific papers, and by age 25, he was a university professor. The importance of Maxwell's theory of electromagnetic radiation rivals that of the discoveries of Newton and Einstein. Albert Einstein said, "The special theory of relativity owes its origins to Maxwell's equations of the electromagnetic field." In 1861 and 1862 Maxwell published the early version of the equations, which form the foundation of electromagnetism, optics, electronics, and radio technologies. From these equations Maxwell established the existence of electromagnetic waves that travel in vacuum at the speed

$$c = \frac{1}{\sqrt{\mu_0 \varepsilon_0}} = 299792458\mathrm{m/s}\,, \tag{3.1}$$

where ε_0 denotes the electric constant (formally: the vacuum permittivity) and μ_0 denotes the magnetic constant (formally: the vacuum permeability). The speed c is thus approximately 300,000 km/s or 186,000 mi/s (the highly precise value shown above was determined in 1983). Maxwell's equations explain specifically how electromagnetic waves physically propagate through space. As soon as Maxwell calculated the numerical value expression for c, he saw immediately that it was essentially the same as the speed of light in vacuum. Maxwell concluded that light is an electromagnetic wave, even though no other type of electromagnetic wave had been identified at the time. In 1886 Heinrich Rudolf Hertz established the existence of radio waves. These are electromagnetic waves as predicted by Maxwell. Maxwell's

Fig. 3.7 James Clerk Maxwell

equations not only unify the theories of electricity and of magnetism, but of optics as well. In other words, electricity, magnetism, and light can all be understood as aspects of a single object: the electromagnetic field. Light (in vacuum) travels with the same constant speed c, in all directions, in each and every reference frame. In brief, Maxwell gave physical confirmation of the Huygens theory of the wave propagation of light.

According to special relativity, c is the upper limit for the speed at which conventional matter and information can travel. Electromagnetic waves and massless particles (for instance, photons) travel at speed c regardless of the motion of the source or the inertial reference frame of the observer (an *inertial frame* is a frame that is not undergoing any acceleration). Particles with nonzero rest mass (such as muons) can approach c, but can never actually reach it.

When Einstein visited Cambridge in 1920 and was asked if he stood on the shoulders of Newton, he replied: "No, on the shoulders of Maxwell." Einstein put a picture of Maxwell at the best place on the wall of his room. Einstein (1940) wrote, with obvious admiration: "The precise formulation of the time–space laws was the work of Maxwell. Imagine his feelings when the differential equations he had formulated proved to him that electromagnetic fields spread in the form of polarized waves, and at the speed of light! To few men in the world has such an experience been vouchsafed. It took physicists some decades to grasp the full significance of Maxwell's discovery, so bold was the leap that his genius forced upon the conceptions of his fellow workers."

Maxwell had successfully computed the speed of light from the electromagnetic constants but still, as Einstein said, Maxwell had to convince others. Nineteenth-century physicists expected that, if they took the Earth as the frame of reference and measured the speed of light, they would obtain varying values of the speed. The reason for their conclusion was that the Earth in its motion around the Sun is constantly changing its velocity with respect to a reference frame called the ether. The luminiferous ether (also known as the light-transmitting ether) was the term Huygens had used to describe a medium for the propagation of light.

In the eyes of Maxwell, it was only the approach of Huygens that remained standing (see his contribution to the *Encyclopedia Britannica* already quoted above). Huygens did not have the modern terminology to describe special relativity but he did have the correct equations. Einstein maintained that "James Clerk Maxwell's [work is the] most profound and the most fruitful." In 1879, Maxwell decided to measure the velocity of light in various reference frames. The wisdom of Maxwell is not in doubt. He looked back into the work of Huygens and used the method of passive remote sensing which Huygens had described in *Traité de la Lumière* two centuries earlier. In Greek mythology Io was a Greek princess and one of the mortal companions of Zeus. The German astronomer (and student of Tycho Brahe) Simon Marius named a moon of Jupiter after Io in 1614. Huygens' method, using signals provided by the moon Io, held promise because the signals traveled over astronomical distances with relatively long travel times. Unfortunately, Maxwell was hampered by astronomical data that were not sufficiently accurate to obtain the desired result.

As a result, in 1879 Maxwell again turned to Huygens and used the method of active remote sensing written down by Huygens in *Traité de la Lumière*. Instead of using two-way signals to the Moon as Huygens had suggested, Maxwell proposed using two-way signals over small distances on the Earth. Such signals, which have short travel times, could be used because of great advances in instrumentation. In 1879 Maxwell proposed a two-way experiment using a beam of light. Specifically, the source and receiver are at the same point, and the signal is returned by reflection. However, Maxwell died of cancer that same year at the age of 48 so he could never carry out the experiment.

In 1887, A.A. Michelson and E.W. Morley famously performed Maxwell's experiment and the results were revolutionary. Their experiment was designed to determine the absolute motion of the Earth through the then privileged inertial frame, called the ether, which almost all scientists assumed existed. However, no absolute motion could be detected. This null result of the experiment could not be explained in terms of classical physics. As a result, the ether was deposed from its position as a privileged frame. The ether frame, if it exists, is unobservable. In other words, the experiment showed that no privileged frame exists at all. Thus, for light, all inertial frames are of equal status. Light, now without any privileged frame, finds refuge in no frame at all, namely the vacuum. Since it belongs to no frame, light has the same velocity in all inertial frames and this is the cornerstone of relativity.

There is no reason at all to assume the existence of the unobservable material medium known as the ether, because it actually confuses the essential difference between electromagnetic waves and mechanical waves. The propagation of electromagnetic waves (such as radio waves, microwaves or light waves) is essentially different from the propagation of mechanical waves (such as sound waves, seismic waves, or water waves). Matter (like air, rocks, or water) is required for the propagation of mechanical waves. Matter is not required for the propagation of electromagnetic waves. Electromagnetic waves can only propagate in vacuum. Light is a stream of particle-like concentrations of electromagnetic energy. These particles are called photons. Photons can exist only in vacuum going at the speed c of light. Furthermore, photons propagate not in the fashion of material objects such as tennis balls, but instead photons propagate in a wavelike fashion such as ocean waves. However, light waves are not at all like ocean waves, because ocean waves propagate only in a material medium (sea water), whereas light waves propagate only in vacuum. Let us elaborate. Absorption of electromagnetic radiation explains how matter (typically electrons bound in atoms) takes up a photon's energy, and so transforms electromagnetic energy into internal energy of the absorber (for example, thermal energy). As light travels through a transparent or semi-transparent medium (such as glass), its photons are absorbed by electrons bound in atoms along the way. After a slight delay, new photons are reemitted and proceed at speed c. In other words, the photons travel at speed c only in the vacuum between atoms. The photons cease to exist once they are absorbed, in an opaque medium where light does not penetrate. The delay due to absorption reduces the speed of light through matter to a value less than c.

Zeno and Proper Time

In 1900, William Thomson, 1st Baron Kelvin (1824–1907), addressed the British Association for the Advancement of Science with the following words: "There is nothing new to be discovered in physics now. All that remains is more and more precise measurement." Contrary to this pronouncement, the entirely new discipline known as modern physics was established with the introduction of quantum mechanics and Einstein's theory of relativity. Today there are many things yet to be discovered and a plethora of perplexing questions yet to be answered. For example, gravity pulls inward yet the cosmos keeps expanding outward faster and faster. To account for this, astrophysicists have proposed an invisible agent called dark energy. It counteracts the inward pull of gravity by pushing outward. Dark energy is an unknown form of energy that affects the universe on the largest scales. The first observational evidence for its existence came from supernovae measurements, which showed that the universe does not expand at a constant rate; instead, the expansion of the universe is accelerating. Without introducing a new form of energy (called dark energy), there seems to be no way to explain how an accelerating universe could be measured.

To give some philosophical context (also for later chapters), let us take some steps back go back to ancient days, 2500 years ago. Even then, there were many perplexing questions to be answered. Elea is a town in Italy, originally founded by

Fig. 3.8 Zeno of Elea (circa 490-430 BC), Greek philosopher and mathematician

the Greeks around 535 BC. It is the home of the philosophers Parmenides and Zeno. Parmenides believed that everything which exists must exist permanently, which meant to him that change was an optical illusion of some kind. Parmenides taught that change is logically not possible. In other words, there can be no motion. In this respect he notably differed from other Greek philosophers, such as Heraclitus who stated that *everything flows* ("panta rhei") and that one cannot step into the same river twice. Zeno (see Fig. 3.8) was the favorite disciple of Parmenides. Zeno vigorously defended his teacher Parmenides with ingenious arguments about space and time.

The concept of infinity has tantalized and sometimes troubled mankind. In measuring and counting, we meet infinity in the form of distances that are too great to be measured and numbers too large to be counted. The book of *Genesis* in the *Bible* says, "I will surely do you good, and make your offspring as the sand of the sea, which cannot be counted because of their number." In this metaphor the sand forms a pile so large that the number of grains is uncountable. Archimedes refuted the uncountability claim, but first, we should review the ancient Greek numbering system for large numbers. The key is the myriad, which is the name for the number ten thousand (10,000). The Greeks had names and symbols for all numbers up to a myriad. Myriad-myriad would be ten thousand times ten thousand. In other words myriad-myriad is one hundred million (100,000,000). The book of *Revelation* in the Greek translation of the *Bible* says, "Then I looked, and I heard the voices of many angels and living creatures and elders encircling the throne, and their number was myriads of myriads", implying they were truly uncountable. However, in his book *The Sand Reckoner* ("Psammites" in Greek), Archimedes writes: "There are some who think that the number of sand grains is infinite in multitude." Archimedes proposed a new number system, one that used powers of a myriad-myriad. Such a system was capable of giving names and symbols to all numbers up to the number 10^{63}. Archimedes then argued that this finite number was large enough to count the number of grains of sand that would fill the entire universe.

The Greek philosophers were always writing about infinity. Zeno rejected the epistemological validity of sense experience, and instead took logical standards of clarity and necessity to be the criteria of truth. In this sense he can be regarded as a radical rationalist. Zeno is remembered most of all for his paradoxes of motion. These paradoxes open up deep questions about the nature of time and space. Unfortunately, the paradoxes also led to some misconceptions about infinity.

Zeno's most famous paradox states that the quickest runner can never overtake the slowest. Achilles is the quickest; he runs at 1 mile per minute. Tortoise is the slowest; he runs at 1/2 mile per minute. In other words, Achilles runs twice as fast as Tortoise. Jupiter sits in his seat on Mount Olympus. At time 0, both Achilles and Tortoise are with Jupiter. Tortoise immediately starts running. At time 1, Jupiter sends Achilles to catch Tortoise. We will now give the reasoning propounded by Zeno. In the first minute from time 0 to 1, Tortoise runs 1/2 mile, while Achilles waits. In other words, Tortoise has a head start of 1/2 mile. While Achilles runs this 1/2 mile, Tortoise runs 1/4 mile and is still ahead. While Achilles runs this 1/4 mile, Tortoise runs another 1/8 mile. While Achilles runs this 1/8-mile, Tortoise runs another 1/16 mile. And so on, this process continues ad infinitum. By this reasoning Achilles can never catch Tortoise. In other words, whenever Achilles reaches somewhere the Tortoise has been, Achilles still has farther to go. Therefore, because there are an infinite number of points Achilles must reach where the Tortoise has already been—he can never overtake the tortoise.

The ancient Greeks made many important mathematical discoveries. One has to do with the summation of sequences of numbers. They found that there are sequences for which one can add arbitrarily many numbers together without the sum surpassing a given bound. Let us use this method with the numbers provided by Zeno. We have the table:

Step	Partial sum of times	Partial sum of distances
1	1 =1	$\frac{1}{2} = 0.5$
2	$1 + \frac{1}{2} = 1.5$	$\frac{1}{2} + \frac{1}{4} = 0.75$
3	$1 + \frac{1}{2} + \frac{1}{4} = 1.75$	$\frac{1}{2} + \frac{1}{4} + \frac{1}{8} = 0.875$
4	$1 + \frac{1}{2} + \frac{1}{4} + \frac{1}{8} = 1.875$	$\frac{1}{2} + \frac{1}{4} + \frac{1}{8} + \frac{1}{16} = 0.9375$

If we continue adding more steps, the partial sums for time will approach to the number 2. The partial sums for distance will approach to the number 1. Both statements can be quickly seen from $s = 1 + \frac{1}{2} + \frac{1}{4} + \frac{1}{8} + \cdots = 1 + \frac{1}{2}(1 + \frac{1}{2} + \frac{1}{4} + \frac{1}{8} + \ldots) = 1 + \frac{1}{2}s$, so $s = 2$. Therefore, we say that the infinite sum for time will approach the number 2. The infinite sum for distance will approach the number 1. As a result, we can write:

$$1 + \frac{1}{2} + \frac{1}{4} + \frac{1}{8} \ldots = 2 \text{ for time,} \tag{3.2}$$

$$\frac{1}{2} + \frac{1}{4} + \frac{1}{8} + \frac{1}{16} + \ldots = 1 \text{ for distance.}$$

This result, which was first noted by the Flemish Jesuit and mathematician Grégoire de Saint-Vincent (1584–1667), states that Achilles will catch Tortoise at time 2 min and distance 1 mile. This conclusion represents one of the generally accepted explanations of Zeno's paradox. Over the years, many philosophers have thought and written about the paradox. Some have not accepted this reasoning, but by and large the matter is settled. Philosophers generally consider distance, not time. Here we have nailed down both time and distance, adding to the argument's persuasiveness.

Zeno abolished motion. Here is an example used by Zeno that demonstrates that there can be no motion. For motion to be occurring, an object must change the position which it occupies. Take an arrow in flight and consider any one instant of time. For the arrow to be moving, it must either move to where it is, or it must move to where it is not. The arrow cannot move to where it is because it is already there. The arrow cannot move to where it is not, because it is not there. If the arrow cannot move in a single instant, then the arrow cannot move in any instant. Therefore, any motion is impossible. In conclusion: Whatever moves, it moves neither in the place where it is, nor in a place where it is not. The arrow is flying through the air. Since time has been assumed to be discrete, we may "freeze" the arrow's motion at an indivisible instant of time. For it to move during this instant, time would have to pass, but this would mean that the instant contains still smaller units of time, contradicting the indivisibility of the instant. So at this instant of time the arrow is at rest; since the instant chosen was arbitrary, the arrow is at rest at any instant. In other words, the arrow is always at rest, and so motion does not occur.

Whether Zeno took his own arguments literally is hard to say. Presumably after all the thinking, he walked home for lunch, so he must have moved. It is said that Diogenes of Sinope (404–323 BC), Greek philosopher and founder of the school of thought known as Cynicism, simply stood up and walked around silently when asked about Zeno. Could it be that Zeno did not reject all motion, only the relative motion of Achilles and the Tortoise? Or, did he reject altogether our common-sense perception of reality? However it may be, philosophers of all times have tried to propose solutions to Zeno's paradoxes and have taken their implications very seriously.

Electromagnetic Waves and Mechanical Waves

The simplest picture of light given by classical physics is of an electromagnetic wave in a vacuum. Maxwell's equations predict that these waves travel at a specific speed. This specific speed of light in vacuum is usually denoted by a lowercase c. The postulate of the constancy of the speed c of light in all inertial reference frames lies at the heart of special relativity and has given rise to a popular notion that the "speed of light is always the same". However, in many situations light is more than a disturbance in the electromagnetic field.

Light slows as it travels through a medium other than vacuum (such as air, glass or water). This slowing of light is not because of scattering or absorption. Rather it is because the light itself causes electrons within the material, to oscillate. The oscillating electrons emit their own electromagnetic waves which interact with the original light. The resulting "combined" wave has wave packets that pass an observer at a rate slower than c. In other words, the light has effectively been slowed. When light returns to a vacuum and there are no electrons nearby, this slowing effect ends and its speed returns to c.

As we have just observed, light propagates through a physical material at a speed less than c. The velocity of such light is called the phase velocity v. The ratio of c over phase velocity v is called the refractive index n of the material. That is, $n = c/v$, where c is the speed of light in vacuum and v is the phase velocity of light in the medium. For example, the refractive index of water is 1.333, meaning that light travels 1.333 times slower in water than in a vacuum. Increasing the refractive index corresponds to decreasing the speed of light in the material.

The waves encountered in ordinary experience can be divided into two classes; electromagnetic waves and mechanical waves. Electromagnetic waves are waves that require no medium in which to travel whereas mechanical waves need a medium for their transmission. In other words, electromagnetic waves can travel in a vacuum whereas mechanical waves cannot. Mechanical waves travel in a medium such as air, water, metal, or rock. Sound waves, water waves, and seismic waves are examples of mechanical waves. Light and radio signals are examples of electromagnetic waves. All electromagnetic waves travel at the speed of light in vacuum. All mechanical waves travel at a speed less than the speed of light in vacuum. When light travels through a medium such as glass its velocity is reduced to a value less that. As a result, such light is treated as a mechanical wave.

Maxwell recognized the difference between light waves and sound waves. However, Huygens had also recognized the difference. Huygens writes in Chap. 1 of the *Traité de la Lumière*:

> Now if one examines what this matter may be in which the movement coming from the luminous body is propagated, which I call Ethereal matter, one will see that it is not the same that serves for the propagation of Sound. For one finds that the latter is really that which we feel and which we breathe, and which being removed from any place still leaves there the other kind of matter that serves to convey Light. This may be proven by shutting up a sounding body in a glass vessel from which the air is withdrawn by the machine which Mr. Boyle has given us, and with which he has performed so many beautiful experiments. But

> in doing this of which I speak, care must be taken to place the sounding body on cotton or on feathers, in such a way that it cannot communicate its tremors either to the glass vessel which encloses it, or to the machine; a precaution which has hitherto been neglected. For then after having exhausted all the air one hears no Sound from the metal, though it is struck. One sees here not only that our air, which does not penetrate through glass, is the matter by which Sound spreads; but also that it is not the same air but another kind of matter in which Light spreads; since if the air is removed from the vessel the Light does not cease to traverse it as before.

At this point Huygens could have said that light travels in vacuum. However, Huygens had no idea of an electromagnetic wave. He could only visualize mechanical waves. Water waves are formed by vibrations in water and sound waves are formed by vibrations in air. The water waves with which we are most familiar are visible on the surface of the water, but water also transmits sound waves through the interior of the water, just as air transmits sound waves through the air. (Whales communicate with such sound waves in water.) These mechanical waves travel through water and air respectively by causing the particles to bump into each other. Huygens believed that light is a mechanical wave. A mechanical wave requires a medium. As a result Huygens could not say that light travels in vacuum. Instead, Huygens said that light travels in ethereal matter, also known as (luminiferous) ether. Compare this to the ancient Greeks, who believed that the Earth was made of Earth, air, fire, and water, but that the heavens and their inhabitants were made of a purer, less tangible substance known as "ether" or "quintessence."

Chapter 4
Huygens' Principle

> Glory is like a circle in the water, Which never ceaseth to enlarge itself, Till by broad spreading it disperse to nought.
> —*William Shakespeare, Henry VI, Part I, Act I, Scene II*

Huygens' Principle in the Words of Christiaan Huygens

Christiaan Huygens is best known for devising the principle that bears his name, namely Huygens' principle (or Huygens' construction). Huygens' principle appeared in Huygens' classic book *Traité de la Lumière*. The book is the first publication of Huygens' theory of the properties of light. As opposed to the corpuscular theory of light adhered to by Newton's followers, Huygens developed a coherent wave theory of light that explained the laws of reflection and refraction as well as double refraction and the polarization of light. Although Huygens is dealing with waves of light, Huygens' principle applies to all kinds of waves: electromagnetic waves (such as light) but also mechanical waves (such as water waves, sound waves and seismic waves). The application of the principle was extended in the nineteenth century to account for the phenomena of interference and diffraction.

With reference to Fig. 4.1, Huygens explains wave propagation in this way:

> So it arises that around each particle there is made a wave of which that particle is the center. Thus, if DCF is a wave emanating from the luminous point A, which is its center, the particle B, one of those comprised within the sphere DCF, will have made its particular or partial wave KCL, which will touch the wave DCF at C at the same moment that the principal wave emanating from the point A has arrived at DCF. It is clear that it will be only the region C of the wave KCL which will touch the wave DCF, to wit, that which is in the straight line drawn through AB. Similarly, the other particles of the sphere DCF, such as *bb*, *dd*, etc., will each make its own wave. But each of these waves can be infinitely feeble only as compared with the wave DCF, to the composition of which all the others contribute by the part of their surface which is most distant from the center A.

T. J. Moser and E. A. Robinson, *Walking with Christiaan Huygens*, History of Physics,
https://doi.org/10.1007/978-3-031-46158-3_4

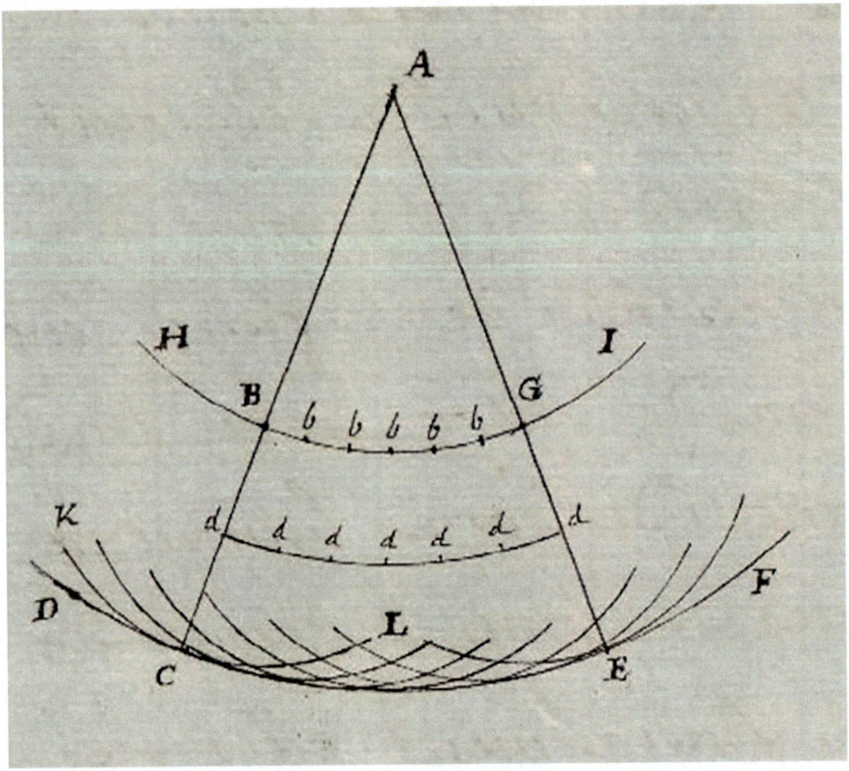

Fig. 4.1 Wave propagation according to Huygens' principle (from manuscript of Traité de la Lumière) [with permission from Leiden University Library, HUG 31, ff. 255r.-312v.]

Spherical Wavelets

Christiaan Huygens was a great observer of nature. Most certainly as a boy he looked at waves formed in water when rain drops were falling into it. Figure 4.2 is a picture of such waves, forming two-dimensional circles on the surface of the water. Huygens imagines that light spreads in three dimensions around the spot where it has been produced, by a movement which is passed on successively from one part of the air to another; and that the spreading of this movement, taking place equally rapidly on all sides, ought to form spherical surfaces ever enlarging. If, in addition, light takes time for its passage, it will follow that this movement, impressed on the intervening matter, is successive; and consequently, it spreads by spherical surfaces and waves. As we see, the finite speed of light is a first requirement for this construction to work for light; the assumption of instantaneous transmission of light has to be ruled out right away.

A wavefront is a surface over which the wave disturbance has a constant "phase". For example, consider a small section of a spherical wavefront coming from a

Fig. 4.2 Waves formed in water when rain drops are falling into it. Any two waves pass through each other without distortion

monochromatic (single frequency/color) point source S in a uniform medium. It is obvious that if the wavefront radius is r at one point in time, it will simply be $r + vt$ after some time t, where v denotes the wave velocity. However, how would that work in general? Suppose that the wave passes through a non-uniform sheet of material and it will meet some obstacles. In such a case, the wavefront will be distorted and deformed from the simple spherical shape. How can we find its new form? What will the wave look like some time later if it is allowed to continue without meeting obstacles? These questions are fundamental for wave propagation.

We recall that *Huygens' principle* or *Huygens' construction* states that *every point on a primary wavefront serves as the source of spherical wavelets such that the primary wavefront at some later time is the envelope of these wavelets*. These wavelets advance with speed and frequency equal to that of the primary wave at each point in space. If the medium is homogeneous, the spherical secondary wavelets may be constructed with constant radii. On the other hand, if the medium is inhomogeneous (meaning that the wave velocity varies from place to place), the wavelets will have radii that depend on the wave velocity of the medium at the respective centers of the wavelets. If, in addition, the medium is anisotropic (meaning the wave velocity varies with the direction of the wave), the wavelets will be ellipsoidal.

In more advanced discussions the envelope prescription is not used. Instead, the superposition of the wavelets is described in detail by means of the Kirchhoff integral (named after the German physicist Gustav Robert Kirchhoff (1824–1887), see Chap. 7). The Kirchhoff approach also gives rise to the inclination (or obliquity) factor which is of considerable interest. In the Huygens' construction the inclination factor was used to as an artificial trick to ensure that waves would propagate only in the forward direction. In the final expressions derived by Kirchhoff the inclination factor appears as a natural by-product. In fact, to explain the rectilinear forward propagation of waves in a homogeneous medium without obstacles was one of the main difficulties for Huygens' wave theory. In Newton's corpuscular theory of light propagation, the straight rays are a direct consequence of the theory.

Huygens' ability to bring the disciplines of mathematics, mechanics and optics to bear on his interest in astronomy enabled him to design, construct and operate a telescope with which he discovered the fourth satellite of Saturn. Eminent as a physicist as well as an astronomer, Huygens established the wave theory of light. It is intriguing that many of the basic principles of optics are predicated on the wave theory of light and yet are completely independent of the exact nature of that wave. It is for this reason that Huygens' principle can serve to describe not only light waves but wave motion in other disciplines as well. In fact, Huygens assumed that light propagates in the same way as sound, namely by pressure (longitudinal) waves in a certain medium. For light he assumed a luminiferous ether as medium. In a sense, in its original form Huygens' principle applies to anything that propagates as a wave and satisfies a wave equation.

Figure 4.3 shows a wavefront, as well as a number of semi-spherical secondary wavelets (hemispheres), which during time Δt have propagated out to a radius of $v\Delta t$. The envelope of all these wavelets is the advanced primary wavefront. For simplicity, we will deal with a homogeneous medium so the spherical wavelets may be constructed with finite radii all of equal magnitude $v\Delta t$, where v is the constant wave velocity and Δt is the time increment between wavefronts. For a homogeneous medium, the envelope is a curve (surface in three dimensions) which is *parallel* or *equidistant* to the primary wavefront. Again for reasons of simplicity, we make use of only two spatial dimensions in our drawings, so we use circles instead of spheres. However, for physical correctness, the reader should make the transition in his mind to three spatial dimensions. As a matter of fact, Huygens' principle in its simplest form only applies to spaces with an odd number of dimensions. It applies to wave propagation in one dimension and three dimensions. Notably, in two dimensions (like in the description of the propagation of surface waves on water) it breaks down in its simplest form and needs a generalization. We will come back to this later in this chapter. Although our drawings and diagrams are two-dimensional, we concentrate here on three-dimensional wave propagation (this is a common feature in textbooks dealing with Huygens' principle). Let us now look at the spherical wavelet. For propagation in the forward direction, we only require a hemisphere (that is, a half sphere). Such Huygens' hemispheres are shown in Fig. 4.3.

As mentioned in Chap. 3, Huygens regarded light as a wave effect in an ether, and in order to explain rectilinear propagation of light waves he used collision theory, a

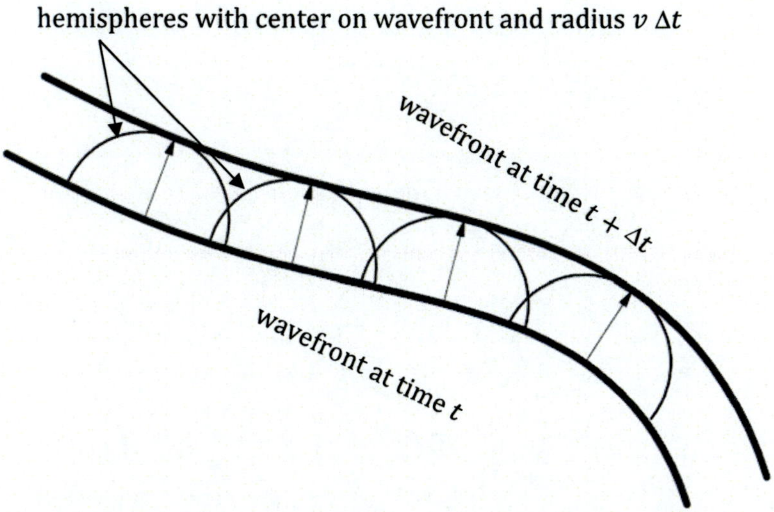

Fig. 4.3 The propagation of a wavefront according to Huygens' principle. The advanced wavefront is the envelope of the hemispheres

topic he previously worked on leading to the discovery of kinetic energy conservation. The idea was to explain the rectilinear propagation of waves by the interaction of the particles filling the ether (Chap. 3, Fig. 3.3). An often-quoted statement from *Traité de la Lumière* (first chapter) reads: "*But although we do not understand the cause of the elasticity, we cannot fail to observe that most bodies possess this property: it is not unnatural, therefore, to suppose that it is a characteristic also of the small, invisible particles of the ether. If, indeed, one looks for some other mode of accounting for the gradual propagation of light, he will have difficulty in finding one better adapted than elasticity to explain the fact of uniform speed.*" Clearly, Huygens did not have the same idea of light wave propagation as we have today. Still his construction was strictly based on modern principles and therefore still valid and useful, with a number of generalizations and refinements.

The first idea embodied in Huygens' principle is that of *secondary point sources*. Huygens explained this by the example of light propagation from a burning candle (Fig. 4.4). Each point in the flame acts as a source of light and what we see is the superposition of all the light points. In his own words (*Traité de la Lumière*, first chapter):

> But we must consider, in greater detail, the origin of these waves and the manner of their propagation from one point to another. And, first, it follows from what has already been said concerning the production of light that each point of a luminous body, such as the Sun, a candle, or a piece of burning carbon, gives rise to its own waves, and is the center of these waves. Thus if A, B and C represent different points in a candle flame, concentric circles described about each of these points will represent the waves to which they give rise. And the same is true for all the points on the surface and within the flame.

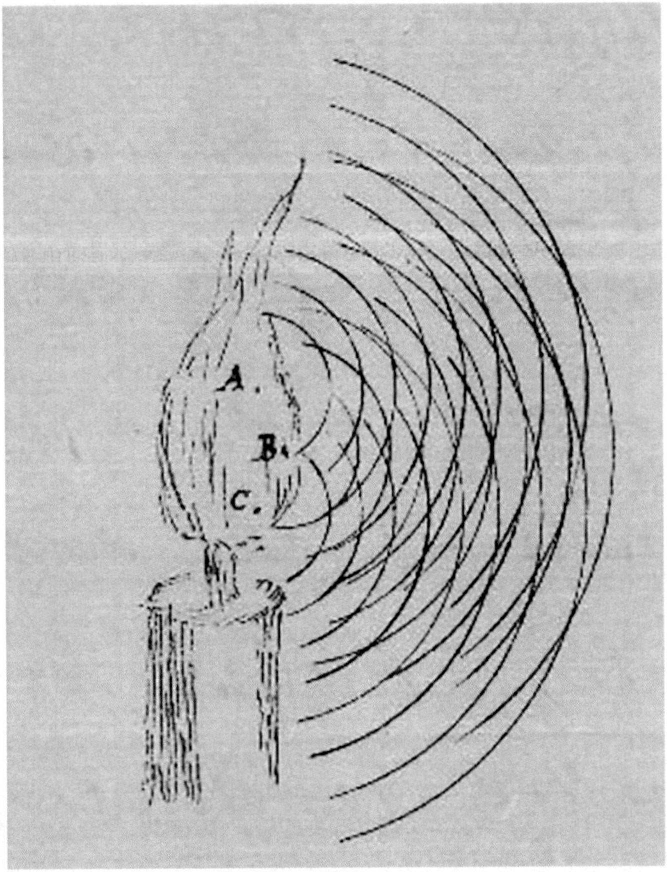

Fig. 4.4 Huygens' principle, burning candle (from manuscript of *Traité de la Lumière*) [with permission from Leiden University Library, HUG 31, ff. 255r.-312v.]

The second idea is that of the *envelope* of the secondary waves. The wavefront observed from the flame is at any place tangent to a particular secondary wave and consists therefore of the envelope of the contributions of all points in the flame. Although interference was still unknown to Huygens, he claimed that this phenomenon is explicitly forbidden between the "*enormous number of waves which are crossing one another without confusion and without disturbing one another*". His accurate assumption was that only the common tangent points of the secondary waves and the new wavefront would contribute.

The formulation of Huygens' principle for the propagation of waves can then be described by the following steps, where we refer to the original diagram in the *Traité de la Lumière* (Fig. 4.1):

1. First the wave is initiated at a primary source point (point A in Fig. 4.1).
2. At each instant during the propagation, all points on the current wavefront (denoted by $B - b - b - b - b - b - G$) act as secondary point sources.
3. A wavefront at a later instant (denoted by the curve $C - E$) is formed by the envelope of the propagation from secondary sources.
4. And so on for all instants ...

Again, Huygens' principle and its later improvements and generalizations can be shown to be valid for any physical type of wave propagation: acoustic waves (such as sound), electromagnetic waves (such as light), elastodynamic waves (such as seismic waves). In fact, with appropriate generalizations, it is mathematically valid for any phenomenon obeying the wave equation.

Huygens' principle has a constructive, or even algorithmic, character. It almost reads as a computer program:

Step 1 is the initialization,
Step 2 can be seen as a "for each ... do ..." statement,
Step 3 is the statement carrying out the actual construction,
Step 4 can be read as "goto step 2" and represents the iteration loop.

Such algorithmic steps form the basis of many modern software programs for the modeling of wave propagation, notably in geophysics. The method applies to waves of any geometrical shape; in particular, to plane and spherical waves. The initiation as a point source is non-essential; we could start the program at step 2 with a wave of any shape. It just propagates a wave from any time t_1 to any other later time t_2.

Reflection and Refraction

Let us now look at reflection and refraction as given by Huygens' principle. Look at Figs. 4.5 and 4.6, which are original figures taken from Huygens' book *Traité de la Lumière*. Each circular curve represents a wavelet originating from the wavefront under consideration. The envelope of these wavelets is the new wavefront.

Figure 4.5 shows Huygens' construction for the reflection of waves. An incident plane wavefront $A - H - H - H - C$ travels downward and strikes the surface $A - K - K - K - B$ of a mirror. The situation is like a wave hitting a straight beach at an angle. The breakers travel along the beach as the wavefront hits successive points on the beachfront. With points K on the horizontal line AB as centers, draw a number of circles which represent the secondary wavelets, starting at instances when the incident wave hits them. Huygens' principle states that the envelope of the circles

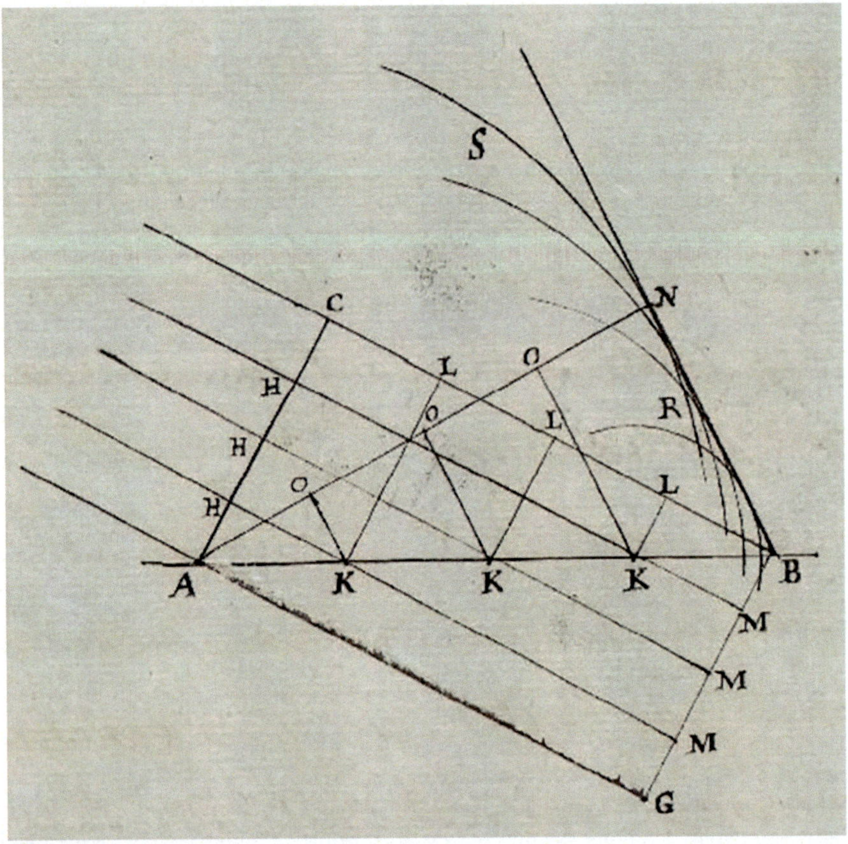

Fig. 4.5 Reflection of waves (from manuscript of *Traité de la Lumière*) [with permission from Leiden University Library, HUG 31, ff. 255r.-312v.]

gives the new wavefront. The envelope BN is the straight line that is tangent to all the circles. The reflected wavefront BN travels upward from the horizontal line.

According to Huygens' principle, the process of the reflection of a wave is indeed instantaneous. While the wave approaches the interface, only the front edges of the wavelets reinforce each other to form the new wavefront. At the moment the wavefront hits the interface, the front edges become moot and the back edges become operative. As the reflected wave leaves the interface in reflection, the former back edges become the reflected wavefront. There is no deceleration followed by acceleration, as Newton had to assume in his corpuscular theory.

In Fig. 4.6 line AB is the refracting surface and line AC is the incident wavefront. The refracted wavefront NB propagates in the lower part of the medium which has a different refractive index than the upper part. As a result, the wave propagates at a different velocity. The envelope of the circles constructed at the points $A - K - K - K - B$ gives refracted wavefront NB.

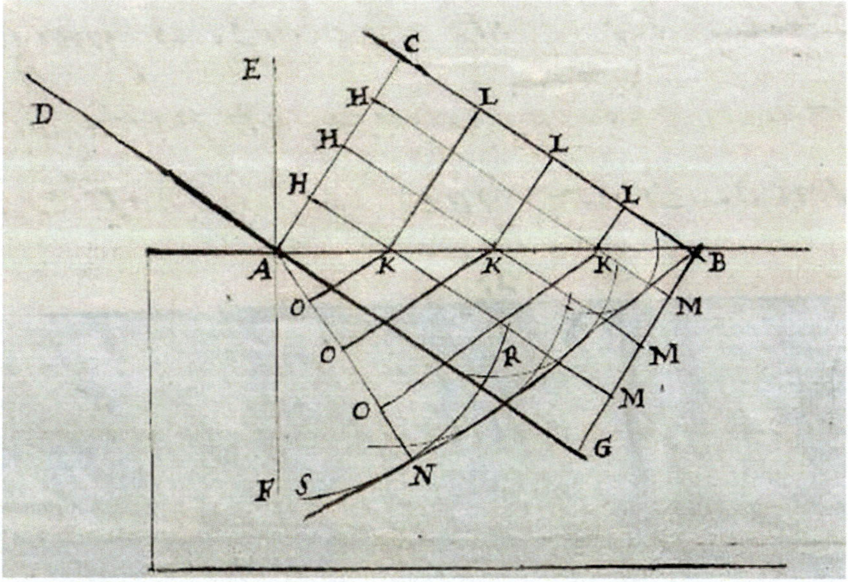

Fig. 4.6 Refraction of waves (from manuscript of *Traité de la Lumière*) [with permission from Leiden University Library, HUG 31, ff. 255r.-312v.]

If we put a stick in a water pool, the light rays from the underwater part are refracted by the water surface. See Fig. 4.7. This gives the illusion that the stick is broken.

By a simple argument Huygens showed that his principle implies Snell's law (see Fig. 4.6): the paths CB and AN are propagated in the same time, so if the media above and below the interface have velocities v_1 and v_2, respectively, then

Fig. 4.7 Illusion of broken stick in pool

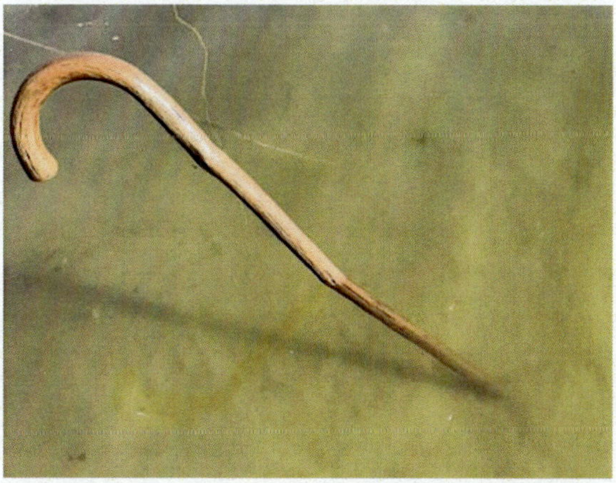

$$\frac{CB}{v_1} = \frac{AN}{v_2}\,. \tag{4.1}$$

The paths CB and AN are expressed in terms of the incident and refracted angle as

$$CB = AB\ \sin\sphericalangle EAD \quad AN = AB\ \sin\sphericalangle FAN. \tag{4.2}$$

Therefore,

$$\frac{\sin\sphericalangle EAD}{v_1} = \frac{\sin\sphericalangle FAN}{v_2}, \tag{4.3}$$

which is Snell's law. Note that Fig. 4.6 shows the case where the wave velocity in the lower space is higher than in the upper space ($v_1 < v_2$); for the opposite case ($v_1 > v_2$) the construction is similar.

By a similar argument, Huygens showed that Fermat's principle follows from Snell's law. Fermat's principle, also known as the principle of least time, states in its original form that the ray path between two points has the shortest travel time among all paths connecting the same pair of points.

In presenting his proof of Fermat's principle, Huygens noted in passing that his proof is simpler than the proof given by Descartes. In fact, Descartes, although he still believed that the speed of light is infinite, in deriving Snell's law he paradoxically (and incorrectly) assumed that light travels faster in a denser medium.

Huygens' derivation runs as follows. In Fig. 4.8 the ray ABC follows Snell's law and HB and FG are wavefronts before and after passing the interface, so the paths HF and BG have the same travel time:

$$T(HF) = T(BG). \tag{4.4}$$

Since

$$T(AB) = T(OH) \text{ and } T(GC) < T(FC), \tag{4.5}$$

it follows that

$$T(ABC) < T(OFC) < T(AFC). \tag{4.6}$$

In words, the travel time along ABC is shorter than along OFC, and by force shorter than along AFC. The same applies to AKC:

$$T(LB) = T(KM) \text{ and } T(BC) = T(MN), \tag{4.7}$$

$$T(LBC) = T(KMN)\,,$$

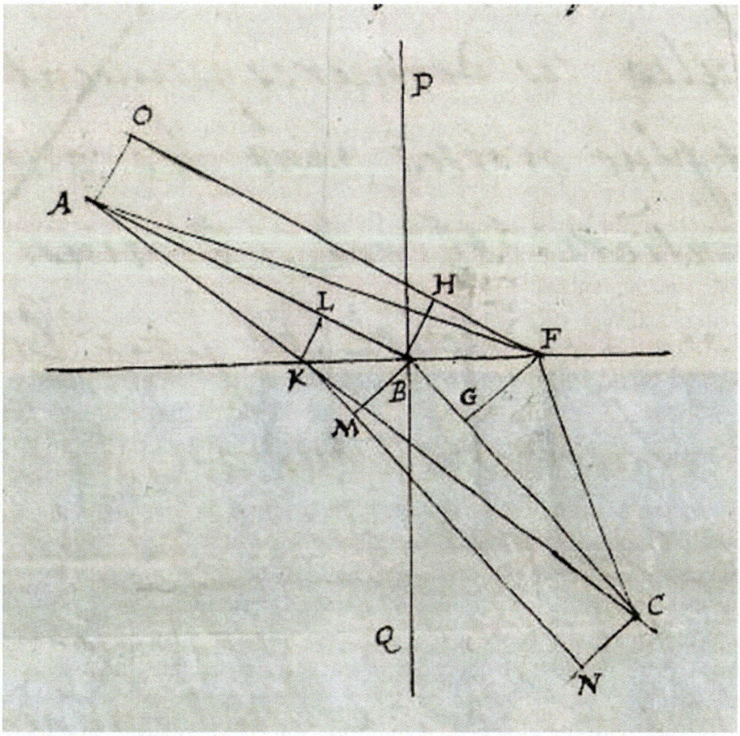

Fig. 4.8 Huygens' principle and Fermat's principle (from manuscript of *Traité de la Lumière*) [with permission from Leiden University Library, HUG 31, ff. 255r.-312v.]

$$T(AKC) > T(AKN) > T(AL) + T(KN) = T(AL) + T(LBC) = T(ABC).$$

Therefore, the ray satisfying Snell's law has minimal travel time among all rays connecting A and C, which is the essence of Fermat's principle.

Note that with this simple argument Huygens makes no use of differential calculus, which was about at the same time developed independently by Newton and Leibniz. In Chap. 10 we take a close look at Huygens' view towards the newly developed infinitesimal calculus. Basically, he was of the opinion that it solves problems which can be equally well solved by conventional methods of his time, in which he happened to be extraordinarily skilled.

Also note that in a more general formulation, Fermat's principle says that the travel time between two points is stationary, meaning that it changes only to second order when the ray path changes to first order (keeping the end points fixed). This formulation includes the principle of least time as a special case. The formulation in terms of stationary travel time implies that the travel time has a smooth minimum (or maximum or extremum) along the ray path, so that it lends itself to differentiation with respect to the ray path. As we point out in Chap. 10, Fermat's principle plays a role in the variational calculus developed after Huygens.

Caustics

In optics, a caustic is the envelope of light rays reflected or refracted by a curved surface or object, or the projection of that envelope of rays on another surface. The caustic is a curve or surface to which each of the light rays is tangent, defining a boundary of an envelope of rays as a curve of concentrated light.

Figure 4.9 illustrates the fact that Huygens' principle can be used to explain caustics (where the wavefront folds over itself). A plane wave AD incident on a spherical transparent body (for instance a raindrop) generates secondary waves at the points of incidence (for instance F, G, H, A). The envelope of the secondary waves from these points is the wave transmitted in the rain drop (EVK). After further propagation the wavefront folds over itself (the ray from A travels to C, the ray from E to Y), leaving a caustic (ENC). In Chap. 10, we return to the caustic to discuss its geometrical relation to the evolute and involute, another topic of great interest to Huygens.

The occurrence of a caustic has some implications for the assumption that light travels along rays. This assumption, which is basically due to Newton, is based on the geometrical optics or high-frequency approximation for wave propagation. If the frequency is considered infinite, the waves are fully described by rays and wavefronts. In the geometrical optics approximation energy propagates forward only along rays, not across them. There is no energy diffusion between neighboring rays. Rays propagate perpendicularly to the wavefronts (in an isotropic medium, where the velocity does not vary with the direction of the ray). Since at the caustic neighboring rays become tangent to each other the ray amplitude becomes infinite at the caustic, hence its name (caustic comes from Greek for "burning"). The ray field has a singularity (the wavefront curvature makes a jump). A connection can be made between Huygens' principle and modern singularity/catastrophe theory. In contrast Leibniz stated that nature is continuous: "*Natura non facit saltus*" (Nature does not jump). Indeed, for finite frequencies, when a band-limited wavelet is attached to the wavefront, the wave field will be smooth and the wave amplitude finite. As we will see in Chap. 7, it was Young, Green and Helmholtz who in the nineteenth century combined Huygens' principle with band-limited wavelets.

Anisotropy and Ellipsoidal Wavelets

An important part of *Traité de la Lumière* is devoted to the study of light propagation through "Icelandic crystal" (optical calcite), which Huygens obtained from Erasmus Bartholin from Aarhus University, Denmark. A remarkable effect of light through this crystal is the "double refraction", meaning that a light ray splits in two, depending on the angle of incidence, with each ray having a different polarization. After many experiments and conjecturing, Huygens was

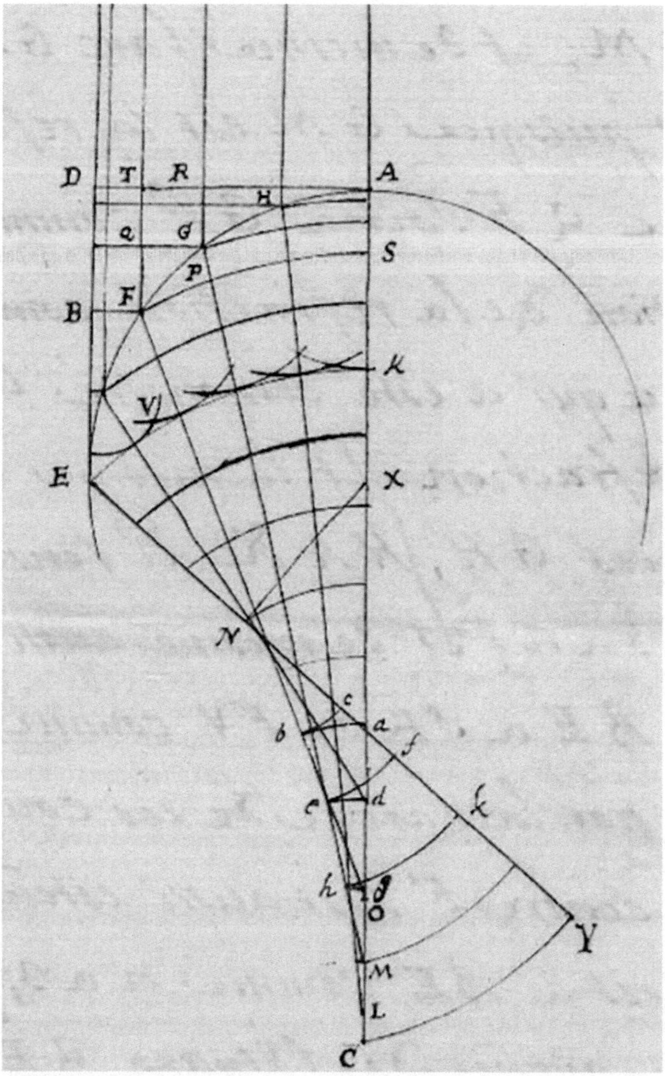

Fig. 4.9 Huygens' principle and the formation of caustics (from manuscript of *Traité de la Lumière*). The caustic is the envelope of the incident ray and is given by the line $ENbehC$. Subsequent wavefronts EVK, cba, fed, ... fold over themselves and develop a cusp [with permission from Leiden University Library, HUG 31, ff. 255r.-312v.]

able to reconcile his wave principle with this effect. He concluded that the crystal is anisotropic, meaning that the speed of light depends on its direction. In this case, the wavefront is no longer spherical, but ellipsoidal in shape. The anisotropy of the crystal could be understood by the special ordering of the particles making up the medium. If the primary wave and secondary waves in

Fig. 4.10 Double image produced by a crystal of Iceland spar (calcite). Look straight down on a calcite crystal and an object beneath it is seen double [with permission from Treasure Mountain Mining]

Huygens' principle are allowed to be ellipsoidal, the same construction applies. The spherical shape is not essential in the construction. The explanation of the light effect in Icelandic crystal in terms of his wave principle was seen as a considerable accomplishment for Huygens and his wave theory of light (Fig. 4.10).

In 1669 Erasmus Bartholin (also known by the Latinized Bartholinus), doctor of medicine and professor of mathematics at the University of Copenhagen, received a piece of Iceland spar and noted the remarkable optical phenomenon which he called double refraction. In his essay entitled *Experimenta Crystalli Islandici disdiaclastici, quibus mira et insolita refractio detegitur* (*Experiments with a dysdiaclastic Icelandic crystal, by which a strange and unusual refraction is discovered*, 1669) he wrote:

> Greatly prized by all men is the diamond, but he who prefers the knowledge of unusual phenomena will have no less joy in a new sort of body, recently brought to us from Iceland, which is perhaps one of the greatest wonders that nature has produced. As my investigation of this crystal proceeded there showed itself a wonderful and extraordinary phenomenon: objects which are looked at through a crystal do not show, as in the case of other transparent bodies, a single refracted image but they appear double.

Iceland spar (calcite) is 'birefringent' or 'double refracting'. Bartholin's double image is shown in Fig. 4.11.

Huygens' explanation of these peculiar phenomena was in terms of waves. The waves that behave as if they merely passed through a plate of glass were called the ordinary (or *o*) waves. The waves that behave unusually, were known as the extraordinary (or *e*) waves. The *o* waves correspond to the case where the medium is isotropic and therefore the secondary Huygens wavelets were spheres. But Iceland spar (calcite) is *anisotropic* and in such a medium Huygens reasoned there are two secondary wavelets for each point source: one is spherical (corresponding to the *o* waves) and the other is an ellipsoid (corresponding to the *e* waves).

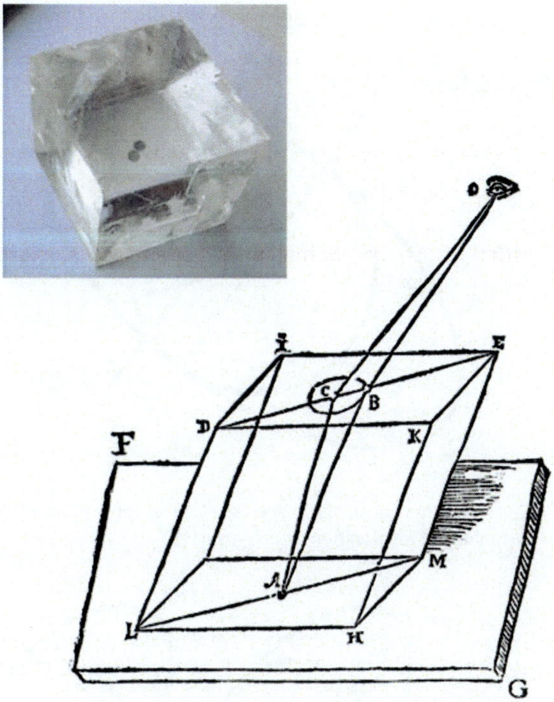

Fig. 4.11 Double refraction. An incident ray is split in two, one obeying (isotropic) Snell's law and the other not. The dot *A* underneath a spar crystal thus appears as two fainter dots. If the crystal is rotated while on the table, the dot representing the "extraordinary" ray *OBA* is rotated with it. Diagram from R. Bartholin's essay of 1669

The spherical wavelet spreads with constant velocity in all directions; the ellipsoidal wavelet spreads with a velocity which varies with the direction of the ray. In the case of calcite, the velocities of both types of secondary wavelets are identical in the direction of the "optical axis" (line of rotational symmetry) but in other directions, the velocity of the ellipsoidal wavelets depend upon their angles of propagation. The ellipsoidal wavelets spread at their greatest velocity when propagating in a plane perpendicular to the optical axis. In other directions, the velocity of the ellipsoidal wavelets range between the lower limit of its velocity along the optical axis and the upper limit of its velocity normal to the optic axis.

A ray represents an energy flow. As seen in Fig. 4.12, a ray for the ordinary wavelet goes from the origin to a point on the sphere (which represents the wavefront of the *o* wavelet). A ray for the extraordinary wavelet goes from the origin to a point on the ellipsoid (which represents the wavefront of the *e* wavelet). The length of a ray is proportional to the velocity in the ray's direction. Obviously, ray velocity is the same in all directions for the spherical wavelet; but for the ellipsoidal wavelet, the velocity is minimum for a ray along the optic axis and maximum for one perpendicular to the optic axis.

In 1690 (although the work had largely been completed in 1678), Huygens published his *Traité de la Lumière* and attempted to explain the double refraction of Iceland spar. His argument is diagrammed in Fig. 4.13.

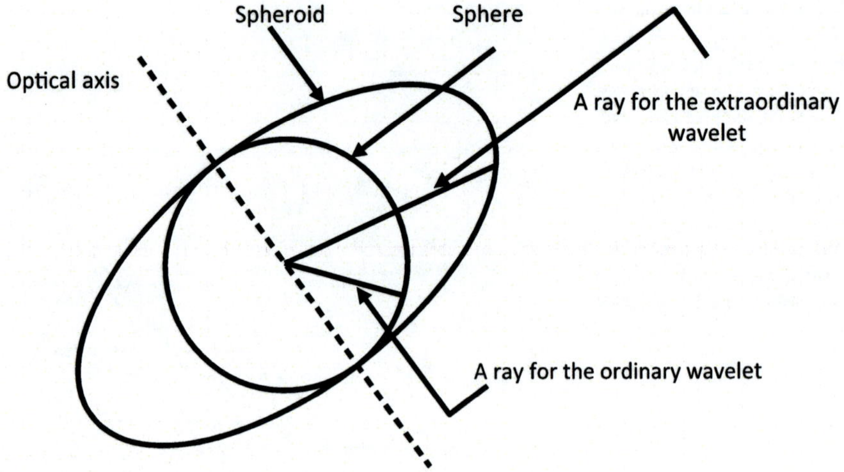

Fig. 4.12 Rays for the ordinary wavelet and for the extraordinary wavelet [with permission from the Society of Exploration Geophysicists]

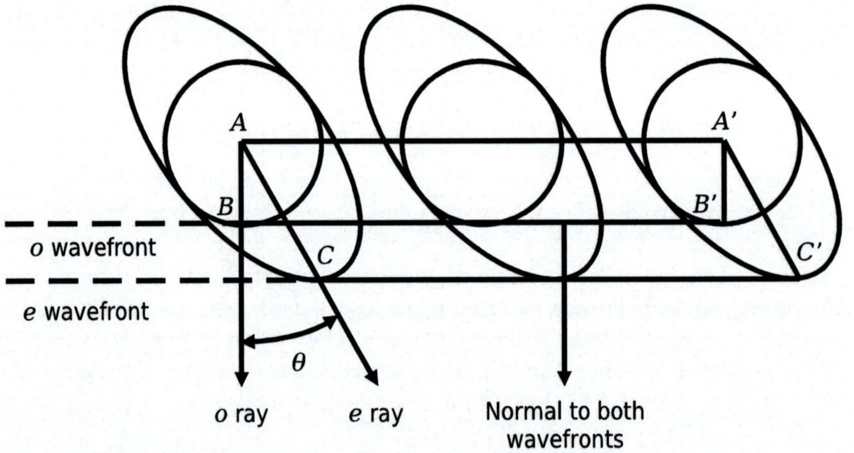

Fig. 4.13 Huygens explanation of double refraction [with permission from the Society of Exploration Geophysicists]

Suppose that line AA' represents the primary wavefront at a given time. Each point on AA' acts as a source of secondary wavelets. The envelope BB' of the spherical wavelets locates the wavefront of the *o* wave at a later time. The envelope CC' of the ellipsoidal wavelets gives the wavefront of the *e* wave at the same later time. Energy travels along rays, so the energy of the *o* wave is moving vertically down. However, as can be seen in Fig. 4.13, the energy of the *e* wave is moving at angle θ to the vertical.

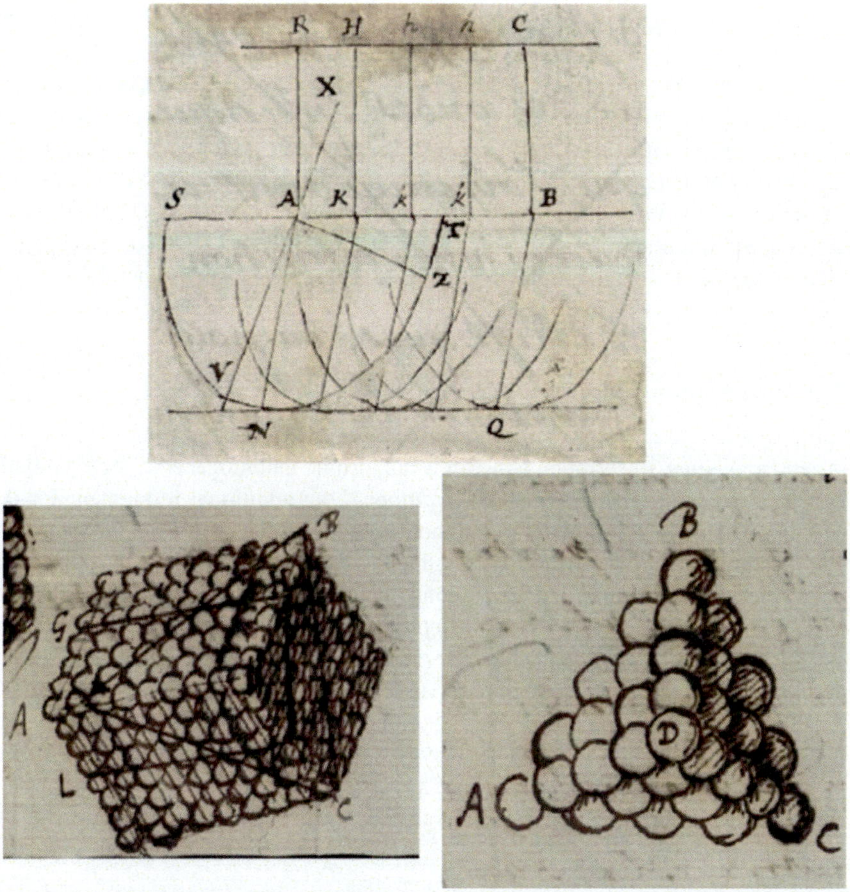

Fig. 4.14 Huygens' principle and anisotropy. Top: construction of the "extraordinary" wave. Bottom left/right: explanation of anisotropy in terms of crystal lattice (from manuscript of *Traité de la Lumière*) [with permission from Leiden University Library, HUG 31, ff. 255r.-312v.]

However, both wavefronts are parallel to the horizontal; hence the normal vector to each wavefront points straight down. In the case of the o wave, the ray and the normal to the front both point in the same direction (vertically down in Fig. 4.13). In the case of the e wave, the ray and the normal point in different directions: the ray points at angle θ and the normal points straight down.

When something is viewed through a transparent anisotropic medium like calcite, two images are seen. One is due to o waves (which are traveling in the same direction as the wavefront) and the other is due to the e waves (which are traveling in a different direction). It is a difficult concept to grasp because it occurs rarely in everyday experience. If ocean waves worked like this (they don't), it would be possible for two surfboarders starting at point A in Fig. 4.13 to approach the shore on different azimuths. The one riding the o wave would come in along line AB. The other riding the *e* wave would approach along line *AC*. Figure 4.14 shows diagrams from Huygens' *Traité de la Lumière* related to the construction of the extraordinary wave and Huygens' explanation of anisotropy in terms of crystal lattice.

The elegant simplicity of Huygens explanation gave (and still gives) it broad intuitive appeal; but he was unable to put it into a final satisfactory mathematical form and this left his explanation open to criticism. His chief critic was none other than one of the greatest mathematicians of the time, Newton. Mathematical development in the nineteenth century (Fresnel, Green, Helmholtz, Kirchhoff) would prove Huygens right. More on this can be found in Chapter 7 "Huygens-Fresnel principle".

Inhomogeneous Media

Apart from being anisotropic, a medium can also be inhomogeneous, or even both. Inhomogeneity poses a *raffinement additionnel*, an additional refinement for the wave propagation. If the medium is inhomogeneous the wave velocity depends on the location and the wavefront will gradually change its shape as it propagates. If this is taken into account, Huygens' construction still applies. Secondary wavelets can still be constructed at any point on a primary wavefront, but because their rate of expansion varies from place to place, the envelope wave may no longer be equidistant to the primary wavefront (see Fig. 4.15 and compare it to Fig. 4.3).

Let us give some examples, starting with atmospheric optics, which deals with light propagation in the atmosphere. Huygens used the principle to explain light ray bending in the atmosphere (Fig. 4.16). The speed of light increases as the air density decreases and therefore increases with altitude. Hence light rays bend towards the

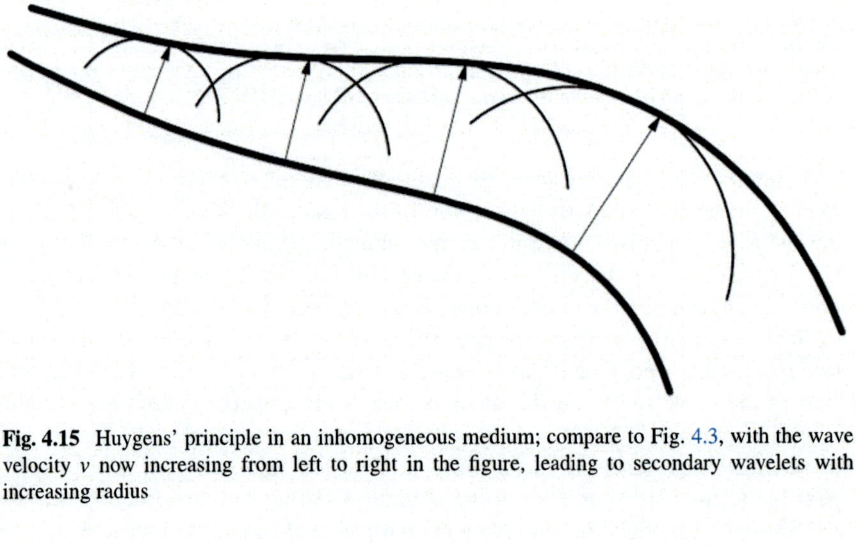

Fig. 4.15 Huygens' principle in an inhomogeneous medium; compare to Fig. 4.3, with the wave velocity v now increasing from left to right in the figure, leading to secondary wavelets with increasing radius

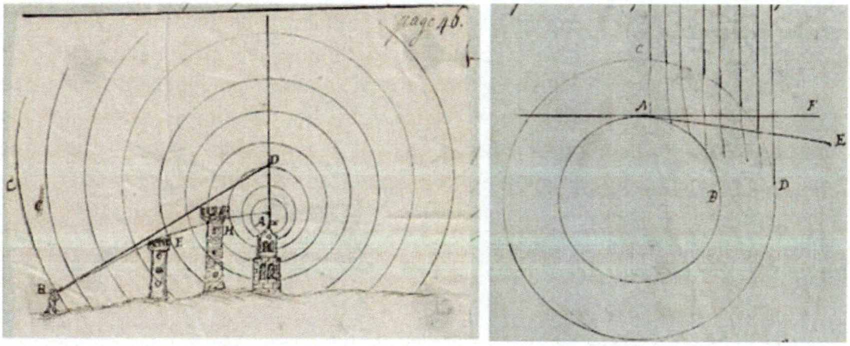

Fig. 4.16 Huygens' principle and light ray bending in the atmosphere [with permission from Leiden University Library, HUG 31, ff. 255r.-312v.]

Earth's surface. As a result, sunrise occurs later and sunset occurs earlier than we see it.

The opposite happens in atmospheric acoustics, which deals with sound propagation in the atmosphere. The speed of sound is proportional to air temperature and therefore decreases with altitude. So sound rays are deflected upwards, and much more during the day than at night (when even an inversion can occur and sound rays bend towards the surface of the Earth). Therefore sound propagates much further along the surface during night. This effect is known as surface diffraction. More on diffraction can be found in Chap. 11 "What Huygens Could Have Written on Diffraction".

Sound propagation in the ocean, studied in ocean acoustics, can be extremely complicated. It is characterized by two types of inhomogeneities: regular and random, and both can have strong effects on the sound field. Sound velocity in the ocean depends on the temperature and the salinity of the water, hence on depth. The "sound fixing and ranging" (SOFAR) or deep sound channel (DSC) occurs at a depth of about 1 km. Here the sound velocity has its minimum, below which it increases again with depth. As a result acoustic waves may be trapped in a so-called wave guide, and may travel very long distances (Brekhovskikh and Lysanov, 2003).

Elastic seismic wave propagation in the Earth, studied in seismology, can be even more complicated by the existence of discontinuities, which can reflect, refract and diffract waves. Longitudinally and transversally polarized seismic shear waves occur and can be converted into each other at discontinuities. Elastic Earth models are often both inhomogeneous and anisotropic. The Earth has a low-velocity zone in its upper mantle at 80 to 220 km depth (Aki and Richards, 1980). In Chap. 10 "Huygens and Curvature" we give an example of a seismic velocity depth profile in which both the rays and wavefronts are circles. More on seismic wave propagation and Huygens' principle can be found in Chap. 12 "Huygens and Geophysics".

All these effects can be explained by Huygens' principle, or the kinematically equivalent Fermat's principle: rays tend to minimize their travel time and therefore deviate their paths so as to travel through regions with higher velocity. Inhomogeneity can also be described in terms of Snell's refraction law if the medium is assumed to be piecewise constant with the pieces infinitely thin (see Chaps. 10 and 12).

Finally, according to Einstein's general theory of relativity, light bends in a curved space due to the presence of massive bodies. A medium is not even needed to cause light ray bending (likewise, no ether is needed). Huygens' and Fermat's principles equally apply here. Huygens' principle does not need a medium, it only needs a wave equation to solve. More on the general theory of relativity can be found in Chap. 9 "Huygens and Centrifugal Force".

Significant Limitations

Despite its obvious and considerable success in solving most wave phenomena at the time, Huygens' principle as formulated originally still had some significant limitations. These had first of all to do with its heuristic and qualitative rather than quantitative character. There was no mathematical foundation yet. Most importantly, there was no notion yet of *interference* (constructive and destructive). As a result, Huygens was not able to explain why the secondary waves would interfere constructively to the new wavefront ($C - E$ in Fig. 4.1 at the beginning of this chapter). Even more, there was no explanation why the secondary waves would interfere destructively right behind the wavefront ($d - d - d - d - d - d - d$ in Fig. 4.1). Recall that we quoted Huygens there, claiming that the waves such as *bb*, *dd*, etc., will be "infinitely feeble" only as compared with the wave DCF. However, without invoking interference this was still wishful thinking.

Moreover, there was no reason why a wave would propagate only in a forward direction, and not in a backward direction at the same time. Consider Fig. 4.17 in the crudest form of Huygens' principle, a wavefront W will at a later instant not only generate a forward propagated wavefront W_1 but also a backward propagated wavefront W_0. Huygens artificially avoided the backward propagation by assuming only forward propagating hemispheres (see Fig. 4.3 of this chapter).

Likewise, *polarization* was not accounted for; Huygens assumed that light propagates as longitudinal waves, just as acoustical sound (note that this very assumption was already called into question by Huygens' experiments on Icelandic crystal).

It was Fresnel and Kirchhoff who solved these problems in the nineteenth century, thereby completing Huygens' principle. More on this in Chap. 7 "Huygens-Fresnel principle".

Another fundamental limitation to Huygens' principle, not foreseen in the seventeenth century, is that Huygens' principle is valid only in spaces of odd dimensions. It is valid in three dimensions, but most importantly, it is invalid in two dimensions, and is for instance not capable of explaining the propagation of surface waves on

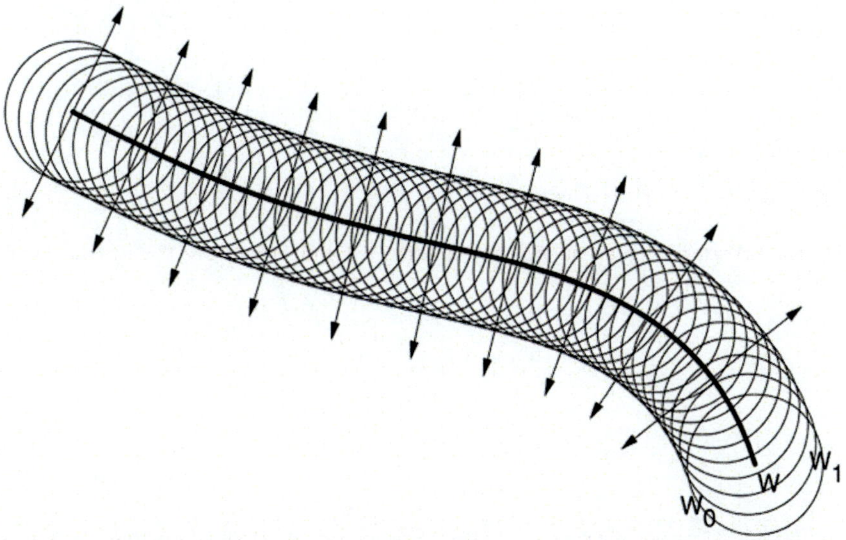

Fig. 4.17 Huygens' principle in its original formulation: forward and backward propagation

water. In one and three dimensions, a transient wave (with a finite duration) propagates and leaves no trace behind it. The passage of the wave creates a temporary disturbance but afterwards there is perfect calm just as there was before the arrival of the wave. In contrast, in two dimensions the passage of a transient wave leaves a dampening trail behind it. A stone thrown in a quiet pond will cause a strong primary wave followed by a trail of waves with decreasing amplitude (see Fig. 4.2). It was the French mathematician Hadamard (1923) who pointed this out and gave a solution for two-dimensional wave propagation in terms of three-dimensional cylindrical waves, so that Huygens' principle can still be used (called the 'method of descent', see Baker and Copson, 1939).

Chapter 5
Huygens and the Telescope

If you can look into the seeds of time, And say which grain will grow, and which will not, Speak then to me.
— *William Shakespeare (1556–1616), Macbeth, Act I, Scene 3.*

The Telescope

Christiaan Huygens was the greatest instrument maker of his time. His most famous invention is the pendulum clock. He was also an accomplished and innovative telescope maker. He invented the first telescope with a built-in micrometer, which revolutionized positional astronomy. He also invented a new type of compound ocular eyepiece—called "Huygens' eyepiece" for use together with a "tubeless telescope," which he used to investigate Titan, the largest moon of Saturn. In his widely read *Cosmotheoros* written in 1695, but published in 1698, he argues that "in the Copernican world system, the earth holds no privileged position among the other planets. It would therefore be unreasonable to suppose that life should be restricted to earth alone". He continues "We may mount from this dull earth," he writes, "and viewing it from on high, consider whether Nature has laid out all her cost and finery upon this small speck of dirt."

About 1250 the magnifying glass was invented in Europe. About 1300 the first form of eyeglasses was produced in Italy. They were shaped like two small magnifying glasses and set into a bone, metal, or leather mounting that could be balanced on the bridge of one's nose. About 1600 Dutch lens crafters invented the refracting telescope and the microscope. On 2 October 1608 in Middelburg (capital of Zeeland, one of the Seven Provinces of the Dutch Republic), Hans Lipperhey (c 1570–1619, also known as Lippershey or Lippersheim) applied for a patent for the first telescope. It was described as an instrument "for seeing things far away as if they were nearby." A few weeks later another Dutch instrument maker, Jacob Metius (1580–1628) also applied for such a patent. Lipperhey's original design had only 3× magnification. Soon after, telescopes were made in the Netherlands in considerable numbers and were used all over Europe.

T. J. Moser and E. A. Robinson, *Walking with Christiaan Huygens*, History of Physics,
https://doi.org/10.1007/978-3-031-46158-3_5

With refracting telescopes, glass lenses are used to refract light rays at the objective to a single focus point. The problem is that glass refracts the different wavelengths of light differently, resulting in different focus points for different colors. This is known as chromatic aberration. It gives fuzzy edges to objects. Another problem was the actual quality of the glass lenses. Defects in the glass were common. Niccolo Zucchi, Marin Mersenne, Bonaventura Cavalieri, and Laurent Cassegrain took up the idea of reflecting telescopes. Niccolo Zucchi (1586–1670) built a reflecting telescope as early as 1616. It used mirrors instead of lenses. Reflecting telescopes could achieve image magnification through optimal focusing by their mirrors and they did not have the problems of chromatic aberration and defects in lenses. Different shapes were suggested for the mirrors (such as spherical or paraboloid). Unfortunately, reflecting telescopes had other problems that were even more serious than those of refracting telescopes. The tolerances for mirrors are much more critical than for lenses. A good reflecting telescope required a parabolic or hyperbolic metallic mirror, the shapes of which were beyond the capability of the craftsmen of the day. Although the precision grinding and polishing of glass for lenses had been practiced in Europe since about 1300, the shaping of metal mirrors to precision was something new. In 1668, Isaac Newton invented a reflecting telescope using a concave primary mirror and a flat diagonal secondary mirror. Newton is often considered the inventor of the reflecting telescope because his design was supposedly functional. However, it could not compete with refracting telescopes of the time. It would be many decades after Newton's passing before the technology and materials to construct reflecting telescopes that could compete with refracting telescopes became available.

In June 1609, Galileo Galilei was told of the "Dutch perspective glass," described as a spyglass that made distant objects appear nearer. The telescope was a great advantage to the military and to the navigation of ships. Galileo spent his time improving the telescope, and finally made one with 23× magnification. By November 1609, Galileo discovered the four largest moons of Jupiter, hills and valleys on the Moon, and the phases of Venus. He also observed sunspots by using a projection method instead of direct observation.

At that time the lenses had spherical surfaces because they were easier to manufacture. A problem with a spherical lens is spherical aberration. This happens when the light rays passing through the lens fail to come together at a single focal point. See Fig. 5.1.

For an ideal lens, all the light rays should come together at a single focal point. See Fig. 5.2

As it happens, for small angles, light striking near the center of a spherical lens will come very nearly to a perfect focus. If an astronomer so designs his telescope that only the central section of the objective lens is used, spherical aberration is very nearly eliminated. Galileo had discovered this property by trial-and-error. Soon other astronomers learned this property and produced better telescopes even though the spherical surfaces on the lenses were retained. The developments were remarkably fast in the tiny span between 1608 and 1611.

Lens systems can be composed of convex and concave lenses. A convex lens converges light rays, a concave lens diverges them. Plano-convex and plano-concave

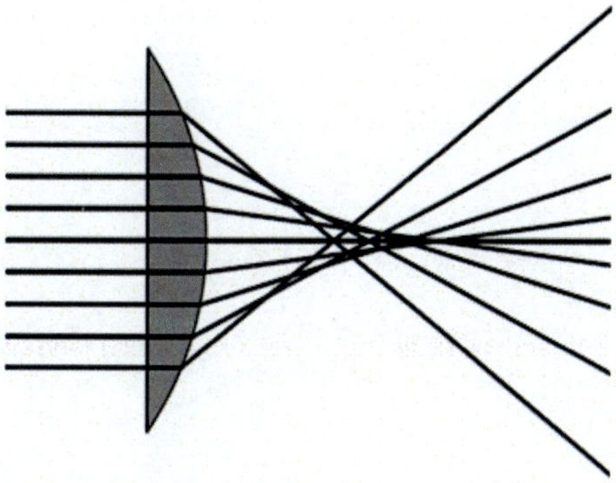

Fig. 5.1 Rays out of focus

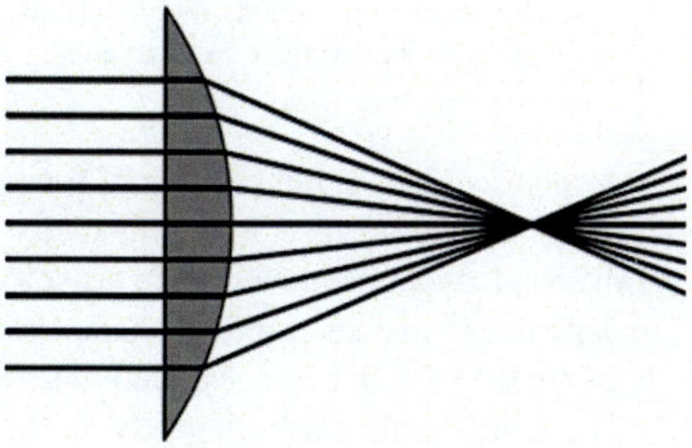

Fig. 5.2 Rays in focus

lenses have a planar surface on one side (the lens shown in Fig. 5.2 is plano-convex). The Keplerian telescope, invented by Johannes Kepler in 1611, is an improvement on Galileo's design. See Fig. 5.3 A and B. It uses a convex lens as the eyepiece instead of Galileo's concave one. The advantage of this arrangement is that the rays of light emerging from the eyepiece are converging.

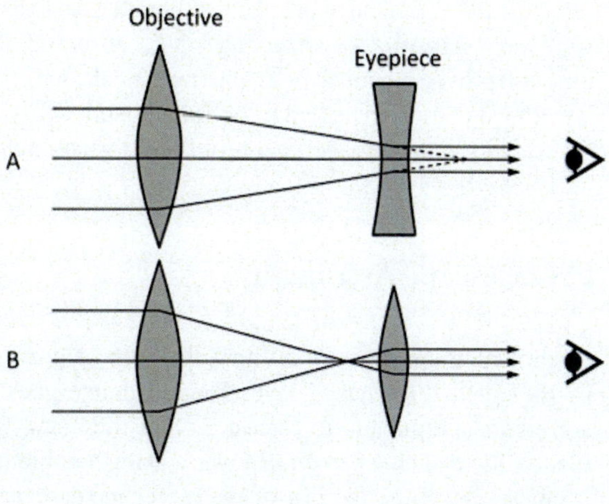

Fig. 5.3 **A** Galilean telescope. **B** Keplerian telescope (also known as a refraction telescope)

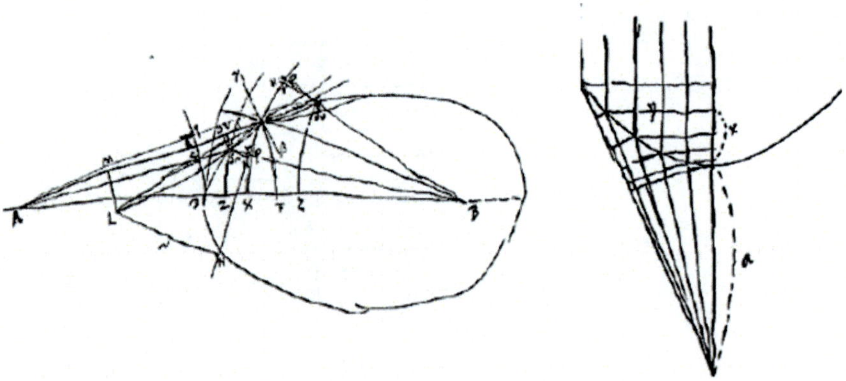

Fig. 5.4 Huygens' constructions to demonstrate that rays for elliptic (concave) and hyperbolic (convex) lenses both converge on a focal point

A spherical lens is not capable of focusing light rays in one point, and hence the situation shown in Fig. 5.2 is not exactly achievable with a spherical convex lens. Christiaan Huygens showed that in order for refracting rays to focus at one focal point the lens must have an elliptical shape (in the case of a concave lens) or hyperbolic shape (in the case of a convex lens). See his original diagrams in Fig. 5.4.

The focusing of a lens depends on the refractive index of the lens (often glass versus air) and its shape (the eccentricity of the ellipse or hyperbola). For a given refractive index a certain shape or eccentricity is needed to focus the rays (Maesumi, 1992). The limiting case of no refraction corresponds to eccentricity equal one, which is a parabola. Therefore, a parabolic lens is ineffective. In contrast, a parabolic *mirror* is capable of focusing *reflecting rays* into one point.

Descartes had high hopes for the hyperbolic lens. Huygens went to great lengths to implement the hyperbolic lens, on the one hand out of interest in building a telescope without spherical aberration, and on the other hand because of his general reverence for Descartes' optical work. But although Christiaan and his brother Constantijn were very skillful at constructing lenses, grinding a hyperbolic-shaped lens at the time still proved too much of a challenge and the lenses were unsatisfactory. Therefore a practical compromise had to be found using spherical lenses. We will come back to this later in this chapter and turn first to what could be seen *without* a telescope.

Pre-telescope Astronomy

If something can be seen without the help of a telescope, it is said that it is seen by the unaided (or naked) eye. The advantage goes to those with perfect eyesight and good instruments to measure what they saw. Tycho Brahe (1546–1601, see Fig. 5.5) was such a person. He was a Danish nobleman, astronomer, and writer. He is considered to be the last of the major naked-eye astronomers, working without telescopes for his observations. In fact, his observations are considered to be the most accurate possible before the invention of the telescope. Tycho Brahe utilized mainly the instruments used at that time for the navigation of ships; namely the compass and a sextant or quadrant. A sextant (from Latin 'the sixth part') is a doubly reflecting instrument which determines the angle between the viewing directions to distant objects, especially the elevation angle of the Sun and heavenly bodies for the purpose of celestial navigation. Similarly, a quadrant is an instrument used for taking angular measurements of altitude in astronomy and navigation, typically consisting of a graduated quarter circle and a sighting mechanism. A micrometer, also known as a micrometer screw gauge, is a device incorporating a calibrated screw widely used for accurate measurements. Micrometers are used in telescopes or microscopes to measure the apparent diameter of celestial bodies or microscopic objects. The micrometer used with a telescope was invented about 1638 by William Gascoigne (1612–1644), an English astronomer

There is yet another person who also accomplished observational feats without a telescope. He is Johannes Hevelius (1611–1687, see Fig. 5.5), a member of a noble family and mayor of Danzig (nowadays in Poland Gdańsk). In 1630 Hevelius went to study jurisprudence at the University of Leiden in the Dutch Republic. On the voyage to Leiden he made observations of a solar eclipse. He frequently published his observations in the *Philosophical Transactions of the Royal Society*. At the university he also studied mathematics and its applications to mechanics and optics. Hevelius had exceptionally keen eyesight. He was able to see stars of the seventh magnitude. In 1641 he built an observatory on the roofs of his three connected houses in Danzig, equipping it with fine instruments of his own making. Following

Fig. 5.5 Portraits of Tycho Brahe (left) and Johannes Hevelius (right)

the lead of Tycho Brahe, Hevelius constructed very large measuring instruments, and managed to improve the accuracy of measured naked-eye stellar positions down to one minute of arc on a routine basis. It was claimed that Hevelius exceeded even Tycho in accuracy. Although Hevelius built and used telescopes, he preferred to measure celestial positions without the aid of lenses. In 1679 the English astronomer Edmond Halley visited Hevelius in Danzig and compared the use of a sextant having telescopic sights with Hevelius' sextant with open sights. Hevelius showed that he could determine stellar positions about as accurately without a telescope as Halley could with one.

The resolving power of a telescope is the angular distance between two stars that are just barely visible through the telescope as separate images. The primary purpose of an astronomical telescope is light gathering, not magnification. If two points on, say, the Moon, are within the resolving power, you will not be able to tell them apart no matter how much you try to magnify. Stars are so far away they are well within the resolving power of any telescope. No matter how much you try to magnify a star it will look like a point of light. Actually, since no telescope is perfect, a star will show as a small "blur", the better the telescope, the smaller the blur. Increasing the magnification will just enlarge the blur.

Hevelius studied the Moon for four years. In 1647 he published a book, *Selenographia sive Lunae descriptio (Selenography, or the description of the Moon)*, in which he reproduced his sketches of the Moon's surface (see Fig. 5.6). Galileo, in his earliest telescopic studies of the Moon, had sketched some of its features as seen by his imperfect telescope. However his sketches did not match the naked-eye sketches of Hevelius. Hevelius was the first to draw a map of the Moon good enough to be recognized as such by modern astronomers. Hevelius' sketches displayed mountain ranges and so-called "seas" on the Moon and gave to them familiar names such as Alps and Apennines, and even Pacific Ocean ("Mare Serenitatis"). Because the telescope did not magnify the images of the stars but only brightened them, he did not use the telescope for stars. Instead, Hevelius observed the positions of stars visible to the unaided eye. He made a catalog of the precise position of 1564 stars. This catalog was the last important work done in astronomy without a telescope.

Huygens would repeat for Mars what Hevelius did for the Moon, using his self-constructed telescope (see Fig. 5.7).

Huygens' Contribution to the Telescope

It is often thought that Christiaan Huygens took Galileo's primitive telescope and simply improved it. Actually Huygens came into a well-developed field that had reached an impasse. Huygens made great practical and theoretical contributions to the development of the telescope. More specifically, he devised the aerial (tubeless) telescope. He invented what is known as the Huygens ocular. And he was first to reveal the potential of William Gascoigne's micrometer.

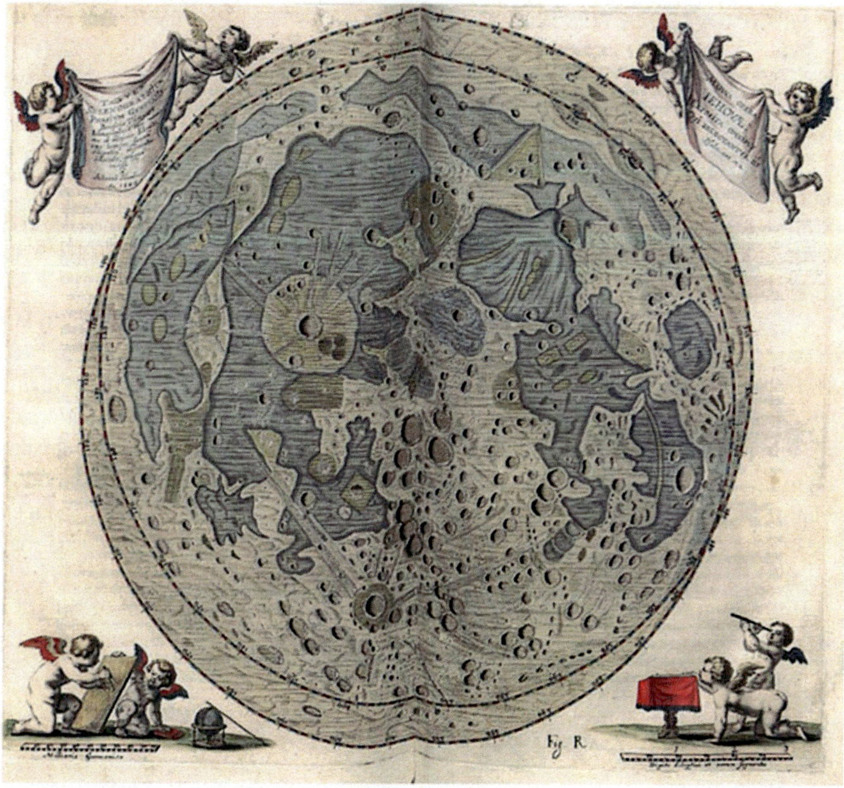

Fig. 5.6 Map of the Moon engraved by astronomer Johannes Hevelius in his book "Selenographia" (1647)

Constantijn Huygens Sr was particularly interested in optical innovations. In 1653 Christiaan Huygens and his brother Constantijn Huygens Jr experimented with a telescope made by a certain 'Master Paulus' of Arnhem (a town in Gelderland, one of the Seven Provinces of the Dutch Republic). Because they found this instrument disappointing, they decided to make one themselves. Instrument maker Caspar Calthof of Dordrecht provided the Huygens brothers with grinding plates and other necessary equipment. The Huygens brothers began grinding objective lenses in 1654. By the spring of 1655, the brothers ground the objective lens which they used to construct a telescope of length 12 feet (3.7 m). On 13 June 1655 Christiaan wrote to the English mathematician John Wallis: "I recently built a telescope of 12 feet in length, and I believe you would be hard put to find a better one, as I am sure that no one before has seen the wonder that I observed with it recently." In his letter he used his well-known anagram (already quoted in Chap. 1)

"ADMOVERE OCVLIS DISTANTIA SIDERA NOSTRIS, VVVVVVVC CCRRHN-BQX" ,

which would read "*Saturno lvna sva circunducitur diebus sexdecim horis quatuor*" ("*Saturn is orbited by its moon in sixteen days and four hours.*").

John Wallis, not to be outwitted, answered Huygens on 1 July 1655 with an anagram pertaining discoveries that he had made in England. Only in 1659 he disclosed to Huygens that he was joking. Huygens admitted to being confused by Wallis' cryptic expression "*quasi lunando vehitur*" ("*as if riding on the moon*"), but never doubted the priority of his discovery.

Huygens had observed Saturn and so discovered a new moon, which would later be named Titan. He published his discoveries in 1659 in a book entitled *Systema Saturnium.* This publication also contains his first description of the workings of a *micrometer* in an astronomical telescope. In this telescope, the objective lens creates a true image which is viewed through the ocular lens. By fitting a distance gauge onto the true image, the apparent position on the celestial sphere of one heavenly body in relation to another could be determined with great accuracy (of course the distance to the body requires further considerations, which were not feasible in Huygens' days much beyond the known Solar System).

Huygens announced his discovery of the true nature of the rings of Saturn in System Saturnium in 1659. At about the same time Huygens made with this same telescope (though he may also have used his new 23-foot telescope) another most

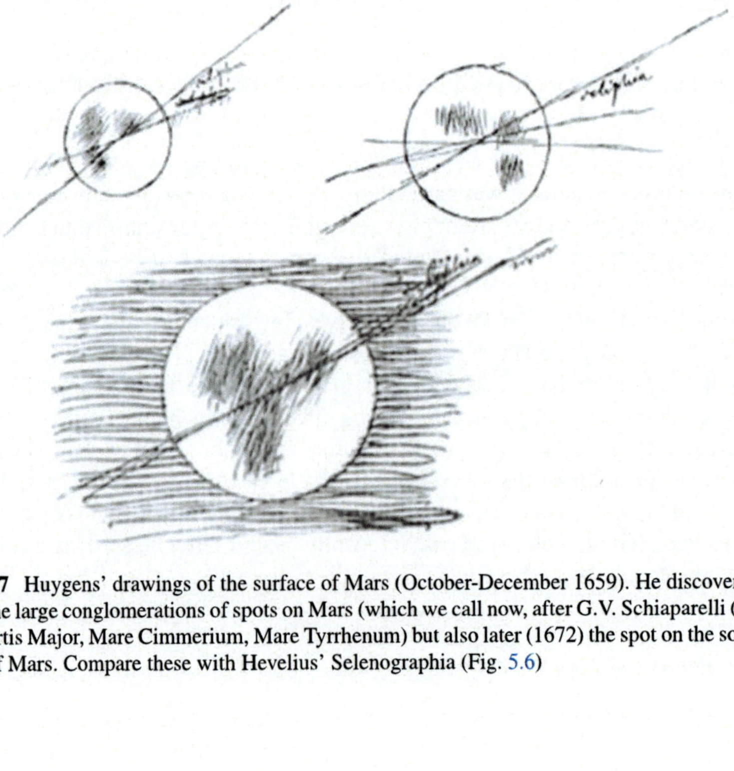

Fig. 5.7 Huygens' drawings of the surface of Mars (October-December 1659). He discovered not only the large conglomerations of spots on Mars (which we call now, after G.V. Schiaparelli (1880), the Syrtis Major, Mare Cimmerium, Mare Tyrrhenum) but also later (1672) the spot on the southern pole of Mars. Compare these with Hevelius' Selenographia (Fig. 5.6)

impressive discovery: he discovered the true nature of the puzzling appearance of the mysterious two ‘attachments’ (ears) of the planet Saturn. Huygens solved the puzzle by examining old telescopic observations made by other astronomers during the first half of the seventeenth century. These astronomers had telescopes with insufficient resolving power so they never truly understood what they were seeing. The appendages regularly disappeared for a short time, and then reappeared for several years. Huygens grasped what he was seeing through his telescope. He saw a three-dimensional view of a flat ring, encircling Saturn, which revealed why the appendages regularly disappeared and reappeared, as the motion of Earth and Saturn changed the angle of viewing.

The Longest Telescope

The occurrence of image distortion was a particular problem with astronomical telescopes. Objective lenses with only a slight curvature gave the best images. However, such objective lenses inevitably required long focal lengths, resulting in almost unmanageably long telescopes. Only very experienced users were able to handle such telescopes. The success of Huygens inspired Hevelius to build long telescopes of his own and by 1673 he had constructed one that was 150 feet (46 m) long.

Hevelius was a person of extremes. He could see the smallest speck (tiny spot) in the sky without a telescope and yet he built the longest “tubed” telescope before the advent of the tubeless aerial telescope. In order to minimize the aberrations, astronomers used lenses that curved only very gently, so that light struck them at only a small angle. Such gently curving lenses bend light only slightly and it takes a long distance for those slightly bent rays of light to come to a focus. Since the eyepiece in a Keplerian telescope must be put at the other side of the focus, the distance between objective and eyepiece became longer as curvature became more gentle. As a result, the Keplerian telescope became longer. They were very long, skinny, flimsy instruments. While Galileo’s telescopes were no more than 4 feet (1.2 m) long, Hevelius made Keplerian telescopes that were up to 12 feet (3.6 m) long and which could magnify up to fifty times. Not satisfied, Hevelius created more powerful telescopes that were 60 and 70 feet long. IIis greatest achievement was a 150 foot (46 m) telescope. It was too long to be encased in an expensive and heavy iron tube. Instead, Hevelius used a wooden trough suspended from a 90-foot pole. Workmen operated it from the ground by using ropes and pulleys. The ropes had to be constantly adjusted because of stretching and shrinking of the hemp ropes. Unfortunately, the telescope would shake in a breeze and the wooden planks tended to warp. The unsteadiness also made it difficult to line up the lenses for observations. Due to these difficulties, this huge telescope was not practical. The telescope required constant adjustment and it was hardly ever used (Fig. 5.8).

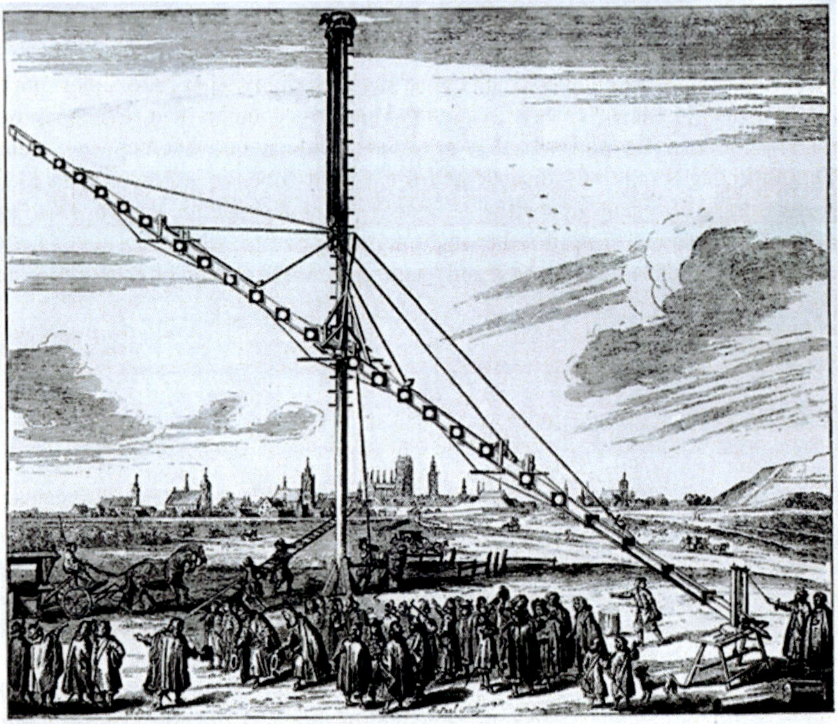

Fig. 5.8 The impractical 150-foot telescope devised by Hevelius

The Aerial Telescope

Long telescopes suffered from problems of wind sensitivity and from difficulties of handling. As for Huygens, he eventually avoided the difficulty introduced by the weight of something even as light as wood in his longest telescopes by dispensing altogether with tubes. In this way, Christiaan Huygens invented the so-called 'aerial' (tubeless) telescope. He published a description in the book *Astroscopia compendiaria* in 1684 (see Fig. 5.9 and Fig. 3.4 in Chap. 3).

Huygens limited this instrument to a large objective lens and an eyepiece. The objective lens was hoisted up a mast in a holder and joined to an eyepiece or ocular lens by a rope. A small oil lamp was used to make it easier to aim. The objective lens was mounted in a short metal tube and attached to a high pole where it could be manipulated from the ground. The eyepiece was in another small tube resting on a wooden support. Between the two was a length of cord which, when taut, aligned the lenses properly. The aerial telescope had its own difficulties. Adjusting the two lenses so that one could look through both was a delicate matter. In addition, stray light found its way into the lenses entirely too easily, tending to wash out the image. However, the long telescope was used in astronomy for many years. In 1722,

CHRISTIANI HUGENII
CONST. F
ASTROSCOPIA
COMPENDIARIA,
TUBI OPTICI
MOLIMINE LIBERATA.

HAGÆ-COMITUM,
Apud ARNOLDUM LEERS, Bibliopolam.
cIↄ. Iↄ. c. LXXXIV.

Fig. 5.9 Huygens' Astroscopia Compendiaria and sketch of the tubeless telescope

the English astronomer James Bradley (1693–1762) used a telescope 212 feet (65 m) long. The Frenchman Adrien Auzout (1622–1691) believed that with a 1000 foot (305 m) telescope it would be possible to see animals roving on the Moon. Long telescopes had reached their limits. Because of technological advances in lens making, shorter telescopes were again used.

Huygens Eyepiece

An eyepiece, or ocular lens, is a lens that is attached to telescopes and microscopes. It is the lens that is closest to the eye when a person looks through the device. The objective lens collects light and brings it to focus creating an image. An eyepiece is placed near the focal point of the objective lens to magnify this image. In 1662 Christiaan Huygens greatly improved the eyepiece for telescopes. His invention is now commonly known as the Huygens eyepiece. It consists of two positive lenses with different focal lengths, separated from each other by a certain distance. A Huygens eyepiece gives an improved and wider field of view and it fully removes lateral chromatic aberration.

The Huygens eyepiece consists of two plano-convex lenses with the planar sides towards the eye separated by an air gap. The lenses are called the eye lens and the field

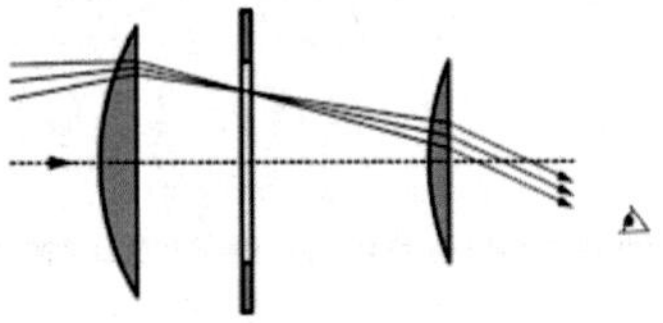

Fig. 5.10 Huygens telescope (top, drawing by Huygens) and Huygens eyepiece

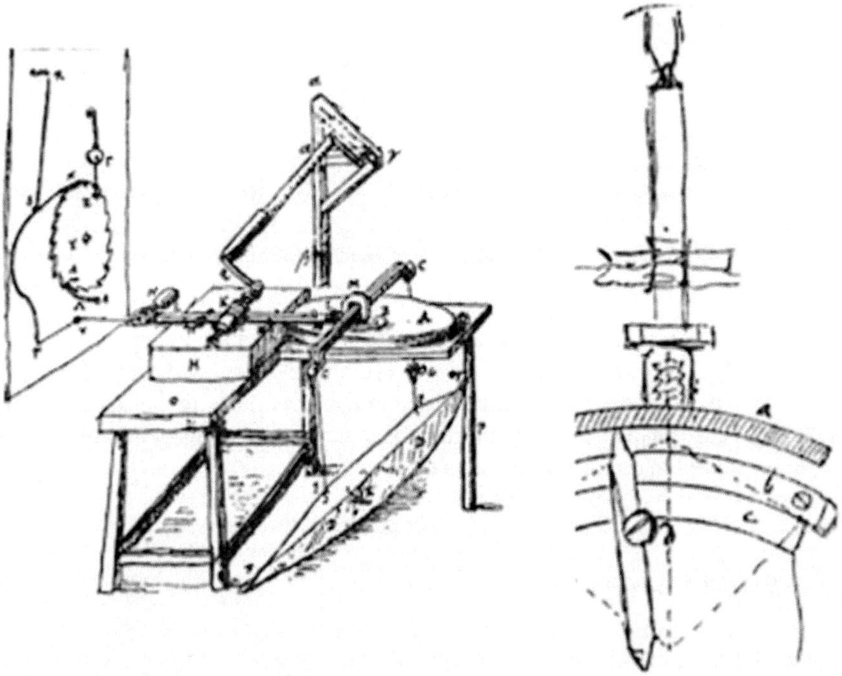

Fig. 5.11 Huygens' lens polishing devices

lens. The focal plane is located between the two lenses. It was the first compound (multi-lens) eyepiece.

Huygens noted that a sensible combination of lenses can have great advantages. Paraphrasing him (Proposition III *Dioptrica,* 1653):

> Although the lenses should not be multiplied unnecessarily, because a lot of light is lost due to the thickness of the glass and by repeated reflections, experience has shown, however, that it is advantageous to do so here. Because if we add to the large lens (the objective) two oculars having between them a certain relation and a determined distance, not only is the field of the telescope admirably enlarged, so that with a single glance we can see much more than when the instrument is built with only one ocular lens, but also the images appear less

> distorted. Finally, any impurity due to the irregularities of the eyepieces is subtracted from the gaze so completely that, although there are two lenses, it cannot be seen at all while in the case of a single lens it greatly affects the sharpness of the images. This may not be the best combination of lenses among all those that are possible, which would also be difficult and perhaps impossible, but at least the one that experience has shown us to be useful.

In 1666 Huygens assumed that he had found the optimal theoretical combination of a convex lens together with a concave eyepiece. However, after he had asked his brother Constantijn to grind the lenses, the result was still unsatisfactory. It is speculated that this is in part due to the fact that Huygens might have suffered from a mild case of myopia, or near-sightedness, and that he compensated this condition by building telescopes that over-magnified by a factor of 3.5 (Pietrow, 2023). In 1669, by trial-and-error, he finally discovered that two air spaced plano-convex lenses can be used to make an eyepiece with zero transverse chromatic aberration. Huygens eyepieces work well with the very long focal length telescopes. Many variants have been proposed in later times. In Huygens' days they were used with single-element refracting telescopes that had long focal lengths and did not use achromatic lenses. These telescopes included aerial telescopes with very long focal lengths. Huygens eyepiece is known as the negative eyepiece because the real inverted image formed by the objective lies behind the field lens and this image acts as a virtual object for the eye lens. See Figs. 5.10 and 5.11.

Huygens probe, Which Landed on Titan

In 1655, using his telescope, Huygens discovered Titan, the largest moon of Saturn. The Cassini-Huygens, U.S.-European space mission to Saturn, was launched on October 15, 1997. The mission consisted of the Cassini orbiter, which was the first space probe to orbit Saturn, and the European Space Agency's Huygens probe, which landed on Titan. The Huygens probe touched down on a frigid floodplain surrounded by icy (methane ice) cobblestones after a gradual descent that took more than two hours. The Huygens probe achieved the first landing on an outer solar system moon by doing this.

A planet outside the Solar System is called an exoplanet or extrasolar planet. As of late 2021, about 4980 exoplanets have been discovered, and the number of discoveries continues to rise. The discovery of exoplanets has intensified interest in the search for extraterrestrial life. The region surrounding a star known as the habitable zone is where temperatures are just right for liquid water to exist on the planets that orbit that star. It is possible for life to exist on planets that orbit in a star's habitable zone. For example in our solar system, Pluto is so far away from the Sun that all surface water would freeze. Mercury, on the other hand, is so near the Sun that any surface water would evaporate. The ideal distance for water to remain a liquid is the radius at which the Earth orbits the Sun. Rocky exoplanets located in their stars' habitable zones are better candidates for detection of liquid water on their surfaces. Liquid water is necessary for life as we know it.

Saturn's moon Titan is not in a habitable zone. Its surface temperature is about minus 179 °C, too cold for any liquid water. However, methane and ethane are liquids on Titan's surface. Radar images of Titan's northern polar region show a variety of irregular black shapes; these are thought to be liquid methane-ethane lakes. They reveal an astounding world with a complete liquid cycle, much like the hydrologic cycle on Earth, but with methane and ethane instead of water. The rivers and lakes of methane-ethane act like their counterparts on Earth. The lakes evaporate to form clouds, the clouds rain back down onto the surface, flowing through rivers and collecting in lakes. It is the only planet or moon in our solar system other than Earth where a liquid cycle like this takes place. Hydrocarbons and water are the essential ingredients for our life on Earth. Methane and ethane are the simplest hydrocarbon molecules. Hydrocarbons are adaptable. It is possible that hydrocarbons on Titan turned inanimate matter to some form of life, but not life as we know it.

It is currently unknown and the subject of scientific analysis and research whether there is life on Titan. Titan is much colder than Earth, but it is the only celestial body in our Solar System known to have liquids on its surface in the form of rivers, lakes, and seas, with Earth being the only other one. Carbon compounds abound in its thick atmosphere, which is chemically active. There are bodies of liquid methane and ethane on the surface, and it is likely that a layer of liquid water exists beneath the ice shell. Some scientists theorize that these liquid mixtures may have provided pre-biotic chemistry for living cells that are different from those on Earth.

As stated it is significant that Huygens discovered the one other body in the solar system that could support life.

In his book *The celestial worlds discover'd: or, conjectures concerning the inhabitants, plants and productions of the worlds in the planets*, in short *Cosmotheoros*, written at the end of his life, Huygens prophesied on the possibility of life on other planets. Based on his firm conviction that the Earth and our Solar system are not unique in the universe, he projected the terrestrial circumstances onto other planets. Exoplanets will be similar to the planets of our Sun, they are rigid and have a mass. They will have life, animals and plants with the same diversity as on Earth. There will be no shortage of water. There will be animals that use reason. They will have senses, a face, vision and hearing, smell and taste. There will be animals similar to humans. They will have a society. They will practice science, art, music and singing. They will have invented logarithms and the (Huygens') 31-tone system.

Who knows how far Huygens was it right? (Fig. 5.12).

In his last letter to Huygens, on 1 July 1695 (unfortunately too late to reach Huygens still alive), Leibniz wrote:

> I would very much like to see your philosophical treatise, which is said to look at particular considerations on the constitution of other planets or worlds. You could hardly undertake a subject more beautiful and more worthy of you. Mr Mariotte told me that you should one day be one of the inhabitants of Saturn, because it owes it to you to become better known to us. And if it loves fame, it must be sensitive to it. I would not disapprove of this change of domicile provided you do it very late. *Serus in coelum redeas diuque Laetus intersis populo petenti* (Horatius, Carminum, Lib.I, 2., which translates to 'May you return to heaven late and be happy among the people for a long time requested').

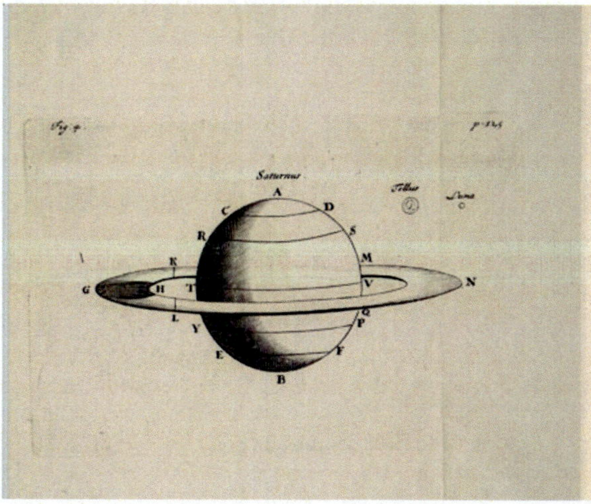

the Planetary Worlds. 125

Suppose then, agreeable to what has been said, the Globe of *Saturn*, whose Poles are A, B. G N is the Diameter of the Ring, as you view it sideways, representing a narrow Oval. Those that live about the Poles within the Arches C A D, E B F, each of which are 54 Degrees, (if the Cold will suffer any Body to live there) never have a Sight of the Ring. From all other parts it is continually to be seen for fourteen Years and nine Months, which is just half their Year. The other Half it is hid from their View. Those then that dwell between the Polar Circle C D, and the Equator T V, all that time that the Sun enlightens the Part opposite to them; have every Night the Sight of a Piece of it H G L, much in the Shape of a shining Bow, which comes from the Horizon, but is darken'd in the Middle by the Shadow of *Saturn* G H, which reaches most commonly to the outermost Rim of it. But after Midnight that Shadow by little and little begins to move towards the right Hand to those in the Northern, but the Left

Book 2. Fig. 4.

The Appearances of the Ring in Saturn.

I 2 to

Fig. 5.12 Basic astronomy for Saturnians (see also Chap. 2). Page from Cosmotheoros, English edition (1689), where Huygens speculates on what inhabitants of Saturn might see of their own planet's ring system

Chapter 6
Huygens and the Clock

...the power of this line [the cycloid] to measure time.
— *Christiaan Huygens, Horologium Oscillatorium*

Navigation and Timekeeping

Christiaan Huygens made major improvements to the two most important navigational tools of the seventeenth century, the telescope and the clock. As described in previous chapters, Huygens, together with his brother Constantijn, greatly improved the telescope, becoming the first to make significant astronomical observations. As we have seen, using his new telescope, Huygens discovered Saturn's moon Titan, and by carefully observing Saturn's periodic changes in shape, he was able to demonstrate that Saturn must be surrounded by a ring.

Astronomy and nautical navigation both require accurate timekeeping. In navigation, the determination of both latitude and longitude is of crucial importance for positioning at sea and mapping the globe. Dead reckoning is a technique whereby the current position of the ship is calculated by keeping track of the changes in speed and direction, as well as elapsed time. This involves an integration of the speed over time, which is prone to cumulative errors along the journey. It is said that Christoffer Columbus navigated on dead reckoning while sailing to America.

Celestial navigation uses the celestial bodies, such as stars, planets and the Sun to calculate the ship's position. Celestial navigation is less prone to errors than dead reckoning, as the measurements can be done independently from past measurements. Still, to determine longitude an accurate clock is needed.

Latitude measures how far a point is north or south of the equator. Since the Earth is approximately round, the distance from the equator is measured in angular degrees. The equator is at 0°. Latitude is measured in degrees north when in the northern hemisphere and degrees south when in the southern hemisphere. The northernmost point (the North Pole) is at 90° north. The southernmost point (the South Pole) is at 90° south.

T. J. Moser and E. A. Robinson, *Walking with Christiaan Huygens*, History of Physics,
https://doi.org/10.1007/978-3-031-46158-3_6

Longitude measures how far a point is east or west of the "prime meridian", a great circle arc passing through both poles, and historically set to pass through Greenwich Observatory, near London. Again, since the Earth is round, the distance from the prime meridian is measured in angular degrees with the prime meridian being at 0° longitude. As you move east or west, the longitude measures up to 180° from the prime meridian. Latitude and longitude are both terms originating in antiquity, when latitude and longitude referred to the breadth and length of the world known at the time. Since the known world essentially encompassed the Mediterranean, latitude came to be used for the north–south coordinate and longitude for the east–west coordinate.

Latitude is relatively simple to measure, namely by the highest altitude of the Sun during the day or (on the northern hemisphere) by the altitude of the Polaris star at night, taking into account seasonal fluctuations. In the seventeenth century, the central problem of navigation was the determination of longitude. The correct longitude is required to establish the east–west location of a ship at sea. Because the Earth rotates around its axis every day, the determination of longitude depends upon the accurate measurement of time. The longitudinal distance between two points can be calculated if one knows the time difference between the two locations. This requires an accurate clock and the improvement of the clock's accuracy was at the center of scientific attention. The existing mechanical clocks, which were accurate to only about 15 min a day, were unsuitable for navigation. For example, it took Dutch navigator Jan van Riebeeck (1619–1677) 104 days to travel 12,000 km from Holland (departure 24 December 1651) to Cape Town (arrival 6 April 1652). Sailing 120 km per day, he could have been 1.25 km off per day, or 130 km over the entire trip.

The word *isochronous* (from the Greek roots for "equal" and "time") refers to something performed in equal time increments. It has the characteristics of a uniform rate of operation or periodicity. Galileo thought that a pendulum is isochronous, since its period, the time it needs for a full swing, is independent of how far it swings. The isochronous property would make the pendulum suitable for the design of a clock. However, in 1636 Descartes found evidence that a pendulum is in fact not isochronous.

Huygens published two books on the pendulum clock with similar titles. These were *Horologium* (1658) and *Horologium Oscillatorium* (1673). In *Horologium* Huygens examined isochronism and its implications for the development of the pendulum clock. He described a pendulum controlled clock for their Lordships the Governors of Holland with a view to establishing his priority of invention. He showed that Galileo's isochronism is valid only for very small amplitudes of the swing angles. For not-so-small amplitudes, the period of a pendulum of a given length increases as the amplitude of the swing angle increases. The problem is that a simple pendulum (consisting of a bob suspended by a wire of fixed length attached at a fixed point) describes a circle. The circle makes it mathematically incompatible with isochronism.

In addition, Huygens showed that a pendulum swinging through the arc of the circle appears isochronous only when the length of the arc is quite short relative to the length of the pendulum. This property gives a clock with a long pendulum an advantage over a clock with a short pendulum. However, the pendulums of the early clocks were kept short and light to minimize the amount of energy needed to keep them in motion. As a result, the early pendulum clocks had very wide pendulum swings, which decreased their accuracy. This was due to the type of escapement (the mechanism for regulating the clock's speed) that was used. At the time of the early pendulum clocks, the verge escapement was used, which required a large amplitude. The anchor escapement, invented later, needed only a small amplitude, which was more economical (see below in this chapter).

Horologium (1658) was a precursor to Huygens' classic work on clocks, *Horologium Oscillatorium* (1673), in which he presented the complete mathematical pendulum theory and demonstrated how isochronism can be achieved. His derivations are a beauty of geometrical skills and we will come back to this in detail in this chapter.

The pendulum clock designed by Huygens revolutionized time measurement. A driving weight or mainspring provides the force for the pendulum, and an escapement counts its swings. A mechanical clock's escapement functions as a feedback regulator to manage speed. Typewriters also employ an escapement.

As he put it himself in *Horologium* (1658), on Christmas Day 1656 Huygens devised a new method for measuring time. He actually never built a clock himself. He collaborated with Salomon Hendricksz Coster, a clockmaker born in Haarlem around 1620 and based in The Hague, until Coster's sudden death in December 1659. The pendulum patent was given to Coster in June 1657 and the pendulum tests in Scheveningen (December 1657–January 1658, already mentioned in Chap. 1) were conducted under Coster's supervision. The world's first pendulum clock is in Museum Boerhaave (Leiden, Netherlands) and was built by Salomon Coster. In the second quarter of 1658, the clockwork mechanism in Utrecht's Dom tower was modified with a pendulum, and Coster was commissioned with this modification. The role of Coster and the Huygens-Coster collaboration was of enormous importance: Huygens was the scientist and Coster the master watchmaker, who could translate Huygens' thoughts into a working timepiece. They complemented each other well.

Until the invention of the pendulum clock, spring-driven clocks had a fusee, an invention of Leonardo da Vinci. The fusee is a conical cylinder with a spiral groove on which a traction device is wound that transmits the driving force from the clockwork to the going train. It was necessary to avoid irregular movement, which occurs when spring tension decreases. With the invention of the pendulum clock and its inherent isochronism, a fusee was no longer necessary.

The pendulum clock was accurate to 10 s a day, as opposed to the previous accuracy of 15 min per day. This was precise enough to measure the orbital periods of Jupiter's satellites over the course of a calendar year. For the following 300 years, pendulum clocks would remain the most precise clocks available.

Huygens went to great efforts to invent a pendulum-like mechanism that would be isochronous. If successful, the pendulum could swing through a large angle. In turning to mathematics, Huygens originated the branch of mathematics known as differential geometry. Huygens had to solve the problem that the arc of a pendulum is circular and the pendulum is not isochronous. The pendulum had to be modified so that its arc had a shape that would make the pendulum isochronous. What could be the shape of that arc?

The answer is the cycloid. The modified pendulum is isochronous if its arc is cycloidal. The cycloid is intimately related to the circle. In fact, the cycloid is the locus of a point on the rim of a circle rolling along a straight line. In *Horologium* (1658) and correspondence with his contemporaries Huygens already hinted at the cycloid; in *Horologium Oscillatorium* (1673), he gave the complete mathematical description of the improved pendulum clock, called a cycloidal clock. The cycloidal clock is isochronous because its pendulum is forced to swing in a cycloidal arc.

Let a pendulum (that is, a bob suspended by a wire) swing against metal cheeks in the shape of a cycloid. As the pendulum swings to and fro, the wire partially wraps itself against the metal cheeks. It can be shown that the path of such a pendulum bob is also a cycloid, and hence its period is independent of its amplitude. Therefore we can say that this cycloidal pendulum is isochronous or, equivalently, tautochronous. We will come back to this topic in detail later on in this chapter.

Unfortunately, the movement of the pendulum against the metal cheeks caused an excessive amount of friction. Robert Hooke (1635–1703), an English physicist well known because of "Hooke's Law" (which relates stress and strain), and the English watchmaker William Clement (1638–1704) developed the anchor escapement for a pendulum clock. Subsequently, Clement used the anchor escapement in his invention of the grandfather clock in 1680. The anchor escapement, which consisted of an escape wheel in the shape of a ship's anchor, required a smaller angle of swing than the angle required by the escapements of the early pendulum clocks. As a result, loss of movement due to friction was minimized and pendulum clocks became so accurate that the cycloidal clock went out of regular use. Even although his cycloidal clock did not stand the test of time, another Huygens invention certainly did. In 1675, Huygens designed a clockwork that used a balance spring instead of a pendulum. This invention was an important one and opened the way for accurate and stable longitude determination. However, ironically Huygens was somehow still too much preoccupied with a sea clock equipped with a pendulum (or perhaps too fascinated by its mathematics?) to exploit the huge potential of the spiral clock invention. Nevertheless, until the invention of the quartz crystal oscillator in the twentieth century, balance springs found widespread use in nearly every mechanical watch. These small timepieces, commonly worn in a gentleman's waistcoat pocket on a chain, or, on a strap on the wrist, since their appearance in the early nineteenth century, relied on balance springs for their timekeeping.

The cycloidal clock is historically important because it is the first successful design of an intricate apparatus based on the use of higher mathematics. In this case, the key concept was another discovery of Huygens, the evolute (the locus of the centers of curvature along a curve). While inventors from antiquity to renaissance, such as Heron of Alexandria and Leonardo da Vinci, used mechanical principles in order to design their inventions, they did not employ much mathematics beyond Euclidean geometry. The introduction of the use of higher mathematics to accomplish mechanical design gives Huygens a strong claim as one of the fathers of modern technology.

Galileo

From time immemorial the sky was used to measure time on Earth. The rotation of the Earth and the movements of the Moon and Sun against the stars made it possible to mark off days, months, and years. A year is the time taken by planet Earth to make one revolution around the Sun. The solar day is the Earth's rotation period relative to the Sun (from solar noon to next solar noon). For periods shorter than a day such simple methods were not available. In the seventeenth century Christiaan Huygens changed the course of history. According to Greek philosopher Plutarch (46–120 AD), no beginnings of things, however small, are to be neglected because continuance makes them great. One such small beginning was Christiaan Huygens looking into Galileo's earlier research into pendulums. The result was that Huygens invented the pendulum clock in 1656. The continuance was that the pendulum clock became the world's most accurate timepiece and remained so for about 300 years. Our modern world rests upon our ability to measure time. By Huygens' invention of the pendulum clock, people could accurately measure the passage of time. In particular, Huygens changed astronomy into a quantitative science. The times at which the Sun, Moon, planets, and stars rose and set at could be accurately recorded. As a result, much was learned about the movements of heavenly bodies.

From ancient times onward, people told time by the changing length and position of the shadow cast by a rod stuck in the ground as the Sun moved from east to west. Shadow clocks or sundials (*solarium* in Latin) indicate the time by casting a shadow or light onto a surface known as a dial face. The place where shadow or light falls on the dial face tells the time. Sundials were first used in Egypt as long ago as 3500 BC. Constructed with sufficient delicacy, they could be made to allow for the changing height of the Sun with the progress of the seasons. However, sundials were useless at night and on cloudy days. Even on sunny days, sundials could at best only give the time with accuracy of a sizable part of an hour.

Other early clocks used devices like as a burning candle, sand sifting through an orifice (hourglass), or water dripping through a small hole (clepsydra or water clock). These methods could work indoors and at night as well as by day but, even at best, they were no more accurate than sundials; moreover, they needed to be tended continuously.

The mechanical clock, a major improvement over earlier clocks, was in use by the fourteenth century. The clock was driven by a falling weight attached to gear trains. The weight was slowly dragged downward by the pull of gravity. The weight turned a hand around a circular dial divided into twelve equal parts. The hand made two complete turns per day. Because the mechanism was affected by friction, it was imprecise by as much as two hours per day.

For his experiments with falling bodies, Galileo relied on water clocks or the beats of his pulse to measure the passage of time. These methods barely met Galileo's needs, not being very reliable. With his newly improved telescope Galileo observed the four largest moons of Jupiter in 1610–1611. At about the same time as Galileo did, the German astronomer Simon Marius also observed these moons and named them Io, Europa, Ganymede, and Callisto. Galileo thought that it might be possible to tell time by means of the regularity of the movements of these four moons. Galileo believed that he could devise a pattern that would be sufficient to split time accurately into small units. Unfortunately, the pattern was complicated. The technology of those days could not accurately determine the times when the moons are eclipsed, data he needed as the basis for measuring time on Earth.

What was needed was a mechanism that oscillates with a short-period. Actually, Galileo had discovered such a mechanism. In 1581, when studying medicine at the University of Pisa, Galileo watched a swinging chandelier. In effect, the chandelier was a pendulum. He realized that it swung back and forth in approximately the same period, whether through a wide arc or a narrow arc. Galileo did not stop there. In about 1602 he investigated the pendulum. A simple pendulum is a weight suspended from a pivot so that it can swing freely. When displaced from its resting position, a pendulum will oscillate about the equilibrium position, swinging back and forth. The time for one complete cycle (that is, a left swing and a right swing) is the period. The period depends on the length of the pendulum and to a lesser degree on the amplitude (that is, the width of the pendulum's swing). *Isochronism* implies having a uniform period of vibration, independent of the amplitude. The key property that makes pendulums useful timekeepers is that they are approximately isochronous, which means that the period of swing of a pendulum of given length is approximately the same for different sized swings.

The back-and-forth movement of a pendulum as shown in Fig. 6.1 is a type of motion that is very frequently observed in nature. It is a vibratory motion. Other examples of vibratory motion are the up-and-down motion of a bobbing spring, the oscillations of the prongs of a tuning fork, or the clicks of a "metronome". A metronome is a tool that emits a click or other audible sound at regular intervals that can be set by the user, usually in beats per minute. Musicians use the device to practice playing to a regular beat. It is parodied in Beethoven's 8th symphony. Metronomes typically include synchronized visual motion (that is, a swinging pendulum or blinking lights). In fact, vibratory motion is typical of all waves, whether seismic waves, water waves, visible light, radio waves, x-rays, or gamma-rays. The simplest form of vibratory motion is sinusoidal motion, also known as simple harmonic motion. Sinusoidal motion can often provide fairly good approximations of more complicated wave motion. Two important questions about sinusoidal motion are: What is the period (or equivalently, the frequency) of vibration? And, secondly, what is the displacement of a vibrating body (measured from the equilibrium position) at a certain time?

In the forewords to *Horologium* and *Horologium Oscillatorium*, Huygens paid tribute to Galileo as a great genius, who had invented the pendulum and discovered isochronism. It is occasionally said that Galileo invented the pendulum clock (in 1637) and that his son Vincenzo began its construction but that they did not complete it before their deaths in 1642 and 1649, respectively. Whether this is historically accurate is doubtful, as no documentation is known. There is a mechanism (no clock yet) in a drawing by Galileo in which a pendulum keeps moving by means of wheels and manual force. See Fig. 6.2 (the drawing was sent to Ismael Boulliau, who forwarded it to Huygens in a letter of 9 January 1660). That was an ingenious invention by

Fig. 6.1 An ideal pendulum

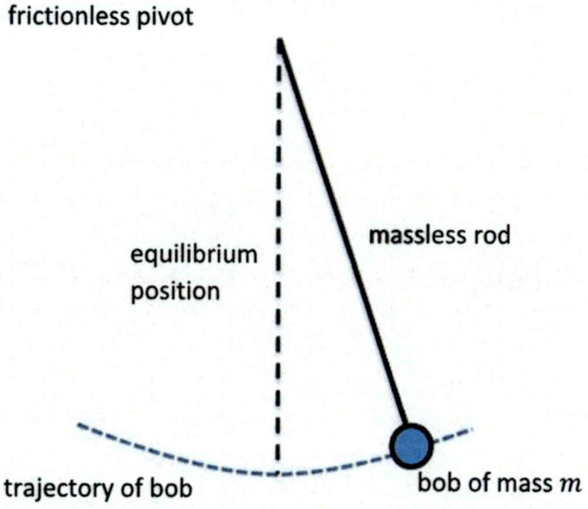

Galileo because until then a pendulum attached to a thread had to be kept moving somehow in order to be able to carry out physics experiments, which was quite troublesome. However, Galileo's mechanism is to be qualified as a counting mechanism rather than a clock. It was Huygens who first came up with the idea of connecting the pendulum to a clockwork. Salomon Coster delivered the first pendulum clock in September 1657, which was sent to Grand Duke Ferdinand de Medici in Florence as early as 25 September 1657 (see Hordijk and Memel, 2021).

In summary, Galileo discovered the crucial property that makes pendulums useful as timekeepers. This property is called *isochronism.* Isochronism means that the period of the pendulum is approximately independent of the amplitude or width of the swing. He also found that the period is independent of the mass of the bob, and proportional to the square root of the length of the pendulum. He first employed free-swinging pendulums in simple timing applications. It was Huygens and Coster who took over from there. Together they constructed the first pendulum clock, in which the cycloid was one of the key design components. But before we turn to the mathematics of the cycloid, let us first mention another attempt to solve the amplitude problem in the realm of instrument making, the so-called OP-construction.

OP Construction

An early solution to the amplitude problem, pursued by Huygens, was the so-called OP-construction. This was a mechanical construction, which consisted in an additional pair of perpendicular wheels aimed at minimizing the amplitude of the swing of the pendulum (see Fig. 6.3, right).

The term OP construction, or OP gearing, originates from a display in *Horologium* (1658) where the wheels are labeled with letters O and P. Huygens wrote that he

Fig. 6.2 Galileo's pendulum, but no clock yet

chose a ratio for the gears of 1:2 or 1:3, as this would experimentally result in the most equalized motion. The OP construction was used in the Dom Tower in Utrecht, among other places (see Hordijk and Memel, 2021, and mentioned earlier in this chapter).

Because of the mechanically reduced swing, the pendulum would be approximately isochronous. Unfortunately, the OP construction did not work well because it caused too much friction and therefore irregular motion. After Huygens invented the cycloid, he abandoned the OP construction. It fell into oblivion and became merely a footnote in the history of timekeeping. There is a curious (romanticized?) nineteenth century painting by the Dutch painter Lambertus Lingeman (1829–1894)

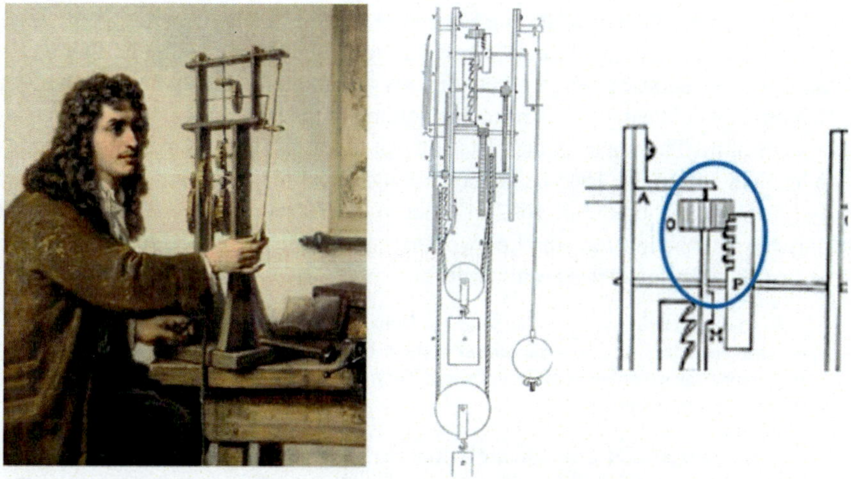

Fig. 6.3 Huygens experimenting with the movement of a clock using the OP construction (left, nineteenth century painting by L. Lingeman) and a drawing by Huygens from *Horologium* showing the OP-construction (with the ellipse on a zoom of it) [with permission from the Collectie Stichting Museum en Archief voor Tijdmeetkunde, de Zaanse Tijd, Zaandam]

showing Huygens experimenting with movement of a clock using OP construction (see Fig. 6.3, left).

Blaise Pascal

Blaise Pascal was born in Clermont, France, on June 19, 1623. Pascal became interested in geometry at age 12, and taught himself quickly about the properties of geometrical figures. By age 14, Pascal was meeting weekly with prominent French mathematicians; the group that would form the foundation for the French Academy. Blaise Pascal's first contribution to mathematics was composed at the age of 16. It was what is now called Pascal's theorem. It states that pairs of opposite sides of a hexagon inscribed in any conic section meet in three collinear points. The theorem was widely praised. Mathematics indisputably became Pascal's great passion.

The normal method of calculation in ancient Greece and Rome was by moving pebbles (*calculi* in Latin) on a table and counting the results. Later, an instrument (namely, the abacus) was developed by sliding beads along rods and counting the results. The abacus with useful modifications became widely used in Europe up to the nineteenth century. In 1642, when he was only 19 years old, Pascal conceived an entirely new idea; a method that would help his father in the tedious work of calculating many numbers for the collection of taxes. Instead of people counting pebbles or beads, Pascal wanted a machine do the counting for them. Pascal, still a teenager, was the first person to visualize a digital computer. Pascal set out to build such a machine. Others were quick to point out the faults in such a contraption, but Pascal struggled on. He built a calculating machine (named the Pascaline) that could add and subtract two numbers directly and multiply and divide by repetition. The device would carry digits from one column to the next. Pascal worked on many versions of the Pascaline. In fact, Pascal went through 50 prototypes before settling on his final design. By 1645 he had a working model of the machine. It was the size of a glove box. It took until 1652 in order to have a standard model in production. Blaise Pascal had invented the first working mechanical digital calculator; in effect, the first digital computer. Pascal wrote:

> We owe a great debt to those who point out faults. For they mortify us. They teach us that we have been despised. They do not prevent our being so in the future; for we have many other faults for which we may be despised. They prepare for us the exercise of correction and freedom from fault.

Pascal realized that a mechanical machine expends much energy in operation. With this in mind, he designed his machine to use the force of gravity as much as possible. The "sautoir" is the centerpiece of the carry mechanism on the Pascaline. When it is time to propagate a carry, the sautoir, under the sole influence of gravity, is thrown toward the next wheel without making any contact between the wheels. During its free fall the sautoir behaves like an acrobat jumping from one trapeze to the next, without the trapezes touching each other ("sautoir" comes from the French verb *sauter*, which means to jump). All the wheels (including gears and sautoir) would have the same size and weight independently of the capacity of the machine. Pascal used gravity to arm the sautoirs. Pascal noted that a machine with 10,000 wheels would work as well as a machine with two wheels because each wheel is independent of the others.

Pascal exhibited unwavering dedication to his goal of discovering a method to operate devices without dependence on external sources of energy. Would it be possible to invent a machine that uses gravity alone? In other words, the machine would continue to operate without drawing energy from an external source. It would be a perpetual motion machine (of the first kind); namely, a mechanism that could resist friction and spin indefinitely without an energy source. In Pascal's days, a perpetual motion machine was well within the bounds of theoretical possibility. The law of conservation of energy, although numerous physicists (Empedocles, Simon Stevin, Galileo, Huygens, Newton, Leibniz, to name a few) had developed the idea, was not yet commonplace in the seventeenth century.

For his perpetual motion machine, Pascal thought of a disc and a rotational movement; namely, the cycloid. However, every attempt of Pascal to devise a perpetual motion machine failed. Incredibly, in 1665, one of his failed attempts was a nearly frictionless spinning wheel. Thomas Edison said, "Just because something does not do what you planned it to do does not mean that it is useless." Indeed it turned out that the nearly frictionless wheel was not useless. Little is known of the path that this wheel took in becoming the basis for the first official roulette games recorded in 1796, but it is unanimously agreed that the roulette wheel is the invention of Pascal. A roulette wheel has 37 slots, numbered 0, 1, 2, ..., 36. Slot 0 is colored green, 18 of the other slots are red, and the remaining 18 slots are black. The croupier spins the roulette wheel in one direction. He then spins a ball in the opposite direction along a grooved track running the circumference of the wheel. The wheel is carefully balanced so that the ball is equally likely to land in any slot when the wheel slows. The winning number and color are announced, and bets are paid out appropriately.

The Pascaline was extraordinarily expensive. This was a barrier to sales. In 1654 production ceased. Meanwhile, Pascal continued to work in analytical geometry as well as in physics. He created the arithmetical triangle, the famous Pascal's triangle of binomial coefficients, which arises in probability theory, combinatorics and algebra. Together with Fermat, Pascal. created the calculus of probabilities, which deals with the mathematical modeling and analysis of uncertainty and randomness. He also began a series of experiments on atmospheric pressure. Pascal demonstrated how the weight of the Earth's atmosphere balanced the mercury in the barometer (invented earlier by Evangelista Torricelli). Pascal's experiments with the barometer proved the now familiar facts that atmospheric pressure (as shown by the height of the mercury in the barometer) decreases as altitude increases, and also changes as the weather changes. To honor his contributions, the name pascal was adopted for the metric unit for pressure, newton per square meter (N/m^2), by the 14th General Conference on Weights and Measures in 1971.

By 1654, Pascal had moved from science to the study of religion and philosophy. His book "Pensées" became a classic. It was Pascal's most influential work. In the book, Pascal discusses philosophical paradoxes. Included are topics such as infinity and nothing, faith and reason, soul and matter, death and life, meaning and vanity. Pascal's "Pensées" is among the most profound and beautifully written works of genius.

In 1658 Pascal found himself one day suffering from insomnia and a painful toothache. In desperation he began to look at the cycloid. Quickly the pain abated. Pascal took this as a divine sign that he was permitted once again to engage in mathematical pursuits. He proceeded to study the cycloid, an endeavor he delved into with unwavering intensity. Pascal published a book on the "History of the Cycloid", using the pseudonym Amos Dettonville. He rediscovered most of what had already been learned about the cycloid, and discovered several results that were new. Pascal proposed a contest to solve various problems involving the cycloid.

Pascal died in Paris in 1662 at the young age of 39 and was buried in the cemetery of Saint-Etienne-du-Mont. Pascal wrote: "All our dignity consists, then, in thought. By it we must elevate ourselves, and not by space and time which we cannot fill. Let us endeavor, then, to think well; this is the principle of morality."

Huygens was among the scientists who accepted Pascal's challenge.

Cycloid

Let us describe 3 types of curves: roulette, its subtype trochoid and *its* subtype cycloid.

1. A *roulette* is a curve generated by a curve rolling on another curve. More specifically, a roulette is a curve traced by a point (called the *generator* or *pole*) attached to one curve as it rolls, without slipping, along a second fixed curve.
2. If the rolling curve is a circle and the fixed curve is a line, then the roulette is called a *trochoid*.
3. If, in case (2), the point lies on the circle then the trochoid is called a *cycloid*. In other words, a cycloid is the path traced by a point on the boundary of a circular disk that rolls without slipping along a straight line. It forms a sequence of arches resting on the line, as seen in Fig. 6.4.

Cycloids have been a subject of interest to leading mathematicians for some five centuries. They seem to have gone unnoticed in antiquity, but were discovered by Nicolas Cusanus in 1450. Galileo studied them in detail and coined the name cycloid. When he was old and blind, Galileo suggested that the cycloid would give a graceful form for the arch of the bridge that was projected over the Arno river at Pisa. The Italian mathematician Vincenzo Viviani (1622–1703), a disciple of Galileo, worked on the geometry of the cycloid. Marin Mersenne (1588–1648), a French polymath, publicized the cycloid among his group of correspondents, including the young Gilles Personne de Roberval (1602–1675). By the 1630s Roberval had determined many of the major properties of the cycloid, such as the interesting fact that the area under a complete cycloidal arch is exactly three times the area of the rolling circle (this was already experimentally found by Galileo). Roberval's solution was not published until 1693, whereas in *Horologium Oscillatorium* (1673) Huygens gave a simple derivation. In doing so, Huygens avoided touching the problem of the quadrature of the circle, the impossibility of which was proved only two centuries later. Because of its beauty and ensuing priority disputes on who discovered what with the cycloid, it was called the "Helen of Geometry" in reference to Helen of Troy, who was the cause of the Trojan War. Huygens' work on the cycloid nicely illustrates his way of deriving results in a geometrical manner ('*more geometrico*'). We follow examples from the *Complete Works* XIV p 347–351 represented by J. Aarts, *Christiaan Huygens, Het*

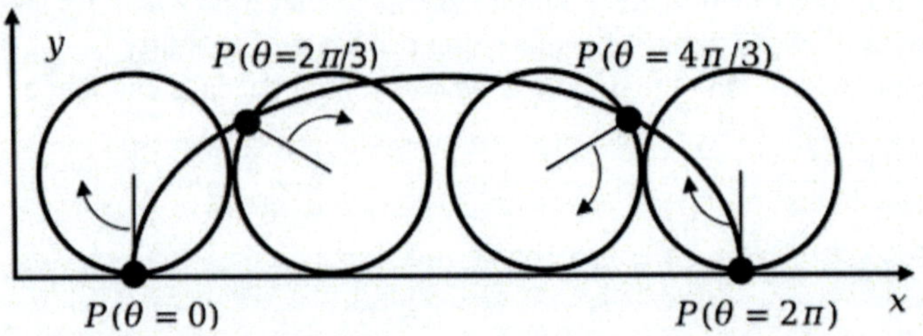

Fig. 6.4 Definition of cycloid: the trajectory of a point P on a circle rolling along a straight line

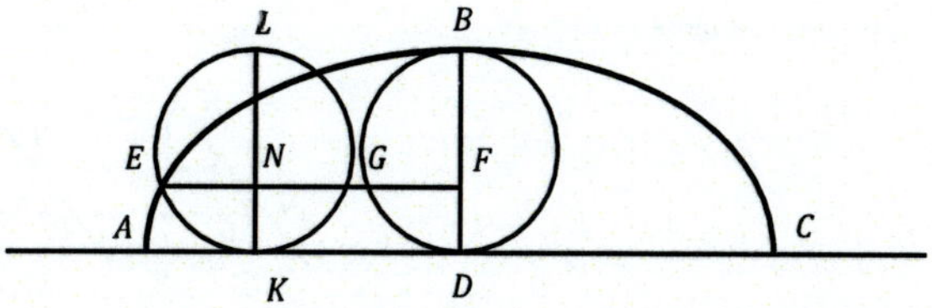

Fig. 6.5 Auxiliary property: line segment EG and circular arc BG have equal lengths

Slingeruurwerk (2015) which has a very detailed collection of Huygens' derivations related to the pendulum clock.

The coordinates x, y of a point on the cycloid are defined in terms of the parameter θ by:

$$x = \theta - \sin\theta, \quad y = 1 - \cos\theta \,. \tag{6.1}$$

As shown in Fig. 6.4, the cycloid is defined as the curve traced by a point P which is fixed on a circle rolling over a straight line.

Figure 6.5 demonstrates the following useful property of the cycloid: if E is a point on the cycloid and G a point on the generating circle such that EG is parallel to AD, then the straight length EG is equal to the circular arc BG (in Aarts' notation (2015): $\overline{EG} = \widehat{BG}$). This follows by construction: as soon as the point B has reached the location E, the circle touches the base AD at K. A being another point on the cycloid, the straight length of AK is equal to the circular arc EK ($\overline{AK} = \widehat{EK}$; this follows from the definition of the cycloid). Since the circle KL can be obtained by translation of the circle BD over the distance DK, the circular arcs EK and GD are equal ($\widehat{EK} = \widehat{GD}$), and the straight-line segments EK and GD are parallel. Therefore $DK = GE$, and the straight length AK is equal to the circular arc DG ($\overline{AK} = \widehat{DG}$). Therefore the completions of these two to the half circle are also equal: the straight length KD = circular arc GB ($\overline{KD} = \widehat{GB}$). Therefore the straight length EG = circular arc BG ($\overline{EG} = \widehat{BG}$, which was to be proved, *quod erat demonstrandum, Q.E.D.*).

Figure 6.6 illustrates how Huygens used the above property to construct the tangent to the cycloid. If B is a point on the cycloid and a line BE is drawn parallel to the base to the first intersection E with the generating circle AD, then Huygens claims that the line through B which is parallel to EA touches the cycloid. His argument is as follows: first take a point H above B, draw a line from H to the circle which is parallel to the base, which cuts the cycloid in L, the line AE in M and the circle in K. The straight length KL equals the circular arc $KA\left(\overline{KL} = \widehat{KA}\right)$, because of the above proven auxiliary property, and the straight length KM is shorter than the straight length KE, which is in turn shorter than the circular arc $KE\left(\overline{KM} < \overline{KE} < \widehat{KE}\right)$ (the first inequality is nontrivial and left to the reader). Therefore, the straight length ML is shorter than the circular arc AE which is itself equal to EB and HM, again because of the above proven property ($\overline{ML} < \widehat{AE} = \overline{EB} = \overline{HM}$). Therefore, the point H lies *outside* the cycloid. Second and similarly, take a point N

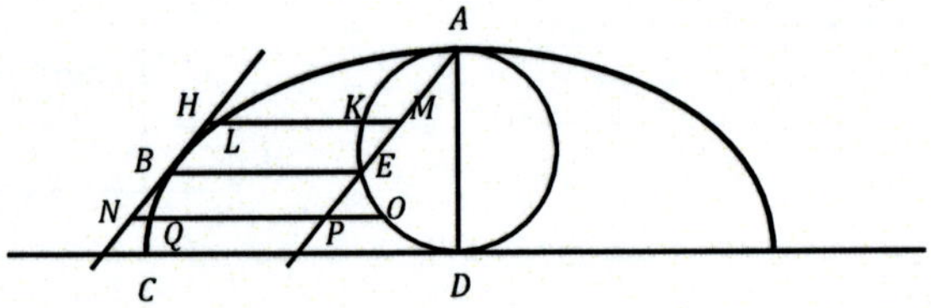

Fig. 6.6 Tangent to cycloid

below B and draw a line from N to the circle which is parallel to the base, cuts the cycloid in Q, the line AE in P and the circle in O. Now the straight length OQ equals the circular arc OA ($\overline{OQ} = \widehat{OA}$) and because the straight length OP is larger than the circular arc OE ($\overline{OP} > \widehat{OE}$), we have that the straight length $NO = NP + PO$ is larger than the straight length BE + circular arc OE together. Because the latter is equal to the circular arc OA and the straight length OQ $\left(\overline{NO} = \overline{NP} + \overline{PO} > \overline{BE} + \widehat{OE} = \widehat{OA} = \overline{OQ}\right)$, it follows that the point N lies *outside* the cycloid.

With this ingenious juggling Huygens shows that all points on the line are outside the cycloid, except point B itself. Therefore HBN is the tangent to the cycloid at B. The interesting aspect of this derivation is that it is purely based on geometry and does not involve any limiting process such as we are used to, when we take the derivative of a function. Such a process would result in

$$\frac{\mathrm{d}y}{\mathrm{d}x} = \frac{\mathrm{d}y}{\mathrm{d}\theta} \Big/ \frac{\mathrm{d}x}{\mathrm{d}\theta} = \frac{\sin\theta}{1-\cos\theta} = \sqrt{\frac{2}{y} - 1}\,. \tag{6.2}$$

The slope of the cycloid is vertical for $y = 0$ ($\theta = 0$ or $\theta = 2\pi$) and horizontal for $y = 2$ ($\theta = \pi$, at the top). We also note that Huygens only presented the textual proof, not the above formulas between brackets. The modern reader may appreciate that the formulas are easier to follow than Huygens' text.

Figure 6.7, another example, is concerned with the area of the cycloid, more precisely, of one arch of the cycloid. The area under one arch of the cycloid consists of the generating circle BGD and the two cheeks $BEKNADMGB$ on the left and a mirrored one on the right. From the above proved property we know that the straight length EG equals the circular arc GB, and straight length NM = circular arc MB. Let H be the midpoint between B and D, and choose points F and L on BD such that $HF = HL$. Then the circular arcs GB and MD are equal ($\widehat{GB} = \widehat{MD}$) and the straight lengths EG and NM add up to AD ($\overline{EG} + \overline{NM} = \overline{AD}$), which equals half the perimeter of the circle. This process can be repeated for all points F and L. It follows that the area of the left cheek is equal to that of the circle, and therefore the area under the cycloid is three times the area of the circle.

In modern notation we have

$$\text{area} = \int_0^{2\pi} y\,\mathrm{d}x = \int_0^{2\pi} (1-\cos\theta)^2 \mathrm{d}\theta = 3\pi\,. \tag{6.3}$$

Fig. 6.7 Area of cycloid

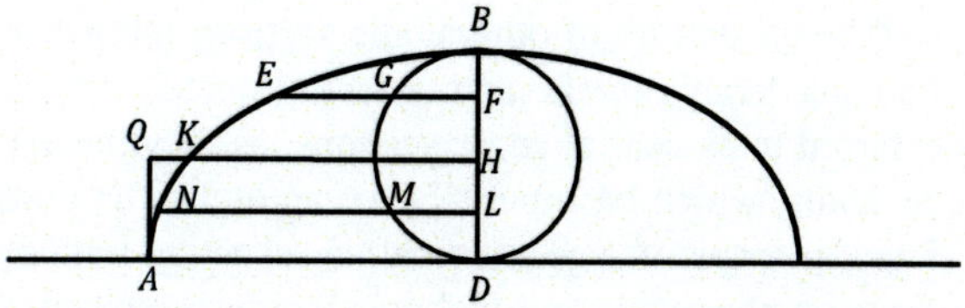

Huygens and π

In his evaluation of the area of the cycloid, Huygens carried out an integration, again without a limiting process like taking the antiderivative of a function (or forming the integral using a Riemann procedure, which consists of taking the limit of blocky approximations to the function using smaller and smaller subintervals). Note that in relating the area of the cycloid to the area of the generating circle, Huygens did not actually provide a formula for the latter nor did he touch the problem of the quadrature of the circle here, whose impossibility would be proved by the German mathematician Ferdinand von Lindemann (1852–1939) in 1882. Yet, Huygens still had much to say in the ongoing debate on the quadrature of the circle.

Quadrature problems, or the problem of finding the area of a geometrical object, have a very long history, going back to antiquity. Archimedes found the quadrature of the parabola (see Fig. 6.8, left). He proved that the area of a parabolic segment is 4/3 times the area of an inscribed triangle. In modern notation it is relatively simple to compute the area under a parabola. For instance when the parabola is given by an expression $y = x^2$, the integral I between 0 and 1 is given by (see Fig. 6.8, right):

$$I = \int_0^1 x^2 \mathrm{d}x = \frac{1}{3}x^3\Big|_0^1 = \frac{1}{3}\,. \tag{6.4}$$

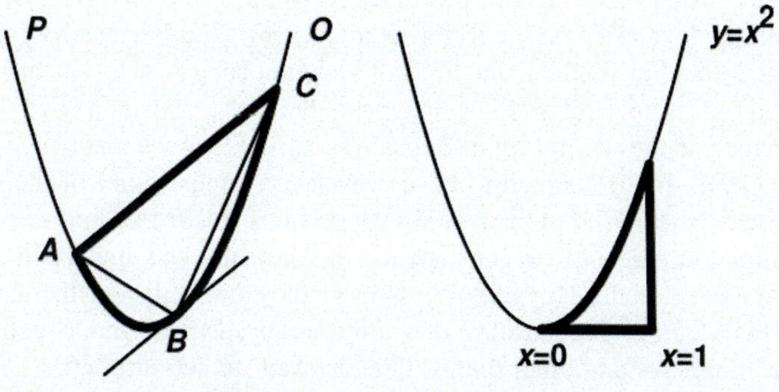

Fig. 6.8 Left: Archimedes' quadrature of parabola (*PABCO*): the area of the black segment *ABC* is equal to 4/3 of the area of the inscribed triangle *ABC*, provided that the tangent to the parabola at *B* is parallel to *AC*. Right: quadrature of parabola by integration

The quadrature of other conic sections (ellipse, circle and hyperbola) proved to be considerably harder and in fact ultimately impossible. These conic sections are referred to as central conic sections, as they have a geometrical center (unlike the parabola, which has none). Since antiquity the problem of quadrature was formulated in terms of a geometrical construction with a compass and straightedge in a finite number of steps. Algebraically this is equivalent to using only four basic arithmetic operations (addition, subtraction, multiplication and division) and square root extraction. The resulting expression for the quadrature would be only in terms of simple and known functions, and therefore easy to compute.

Finding the quadrature of the circle, or alternatively proving its impossibility, has been a challenge for many centuries and would secure the solver of eternal fame. One of those who claimed to have solved the quadrature of the circle was the Flemish Jesuit and mathematician Grégoire de Saint-Vincent (1584–1667). Jesuits were scientifically highly regarded (Stein, 1941). One of them, Joannes Ciermans (1602–1648), who belonged to the mathematical school of De Saint-Vincent in Antwerp, was involved in a vivid correspondence with René Descartes, whom he considered as a second Columbus. De Saint-Vincent's main book was *Opus geometricum quadraturae circuli et sectionum coni decem libris comprehensum*, published in Antwerp, in 1647. Apart from his work on the circle, as noted in Chap. 3, De Saint-Vincent solved Zeno's paradox by showing that the infinite sum $1 + 1/2 + 1/4 + \ldots$ converges to 2. Leibniz considered him to equal to Fermat and Descartes.

Still, several contemporary scientists were skeptical of his claims of solving the quadrature of the circle, among whom Marin Mersenne and Christiaan Huygens. Huygens and de Saint-Vincent corresponded intensively on the subject. Their letters show the high regard they felt for each other. They also show Huygens trying to convince de Saint-Vincent of the flaws in his arguments. De Saint-Vincent answered Huygens very kindly, praising his ingenuity, all the while not admitting that he was wrong. Huygens, it must be said, became impatient at times. In 1651 Huygens published his *Theoremata de Quadratura Hyperboles, Ellipsis, et Circuli, ex dato portionum gravitatis centro. Quibus subjuncta est Exetasis Cyclometriae Cl. Viri Gregorii a S. Vincentio,* or in short form *Exetasis,* the examination of the quadrature of the circle ("exetasis" is Greek for examination). The Exetasis is considered a courteous refutation of De Saint-Vincent's claim. They met in Ghent on 13 July 1652. The Exetasis would establish Huygens' fame as a mathematician.

Conversely to the attempts of de Saint-Vincent, the Scottish mathematician James Gregory (1638–1675) claimed to have proven the impossibility of the quadrature of the circle. In 1667 he published his conclusions in *Vera Circuli et Hyperbolae Quadratura.* Here again, Huygens spoiled the pleasure, and argued that Gregory's proof was flawed. In this Huygens was supported by the English mathematician John Wallis (1616–1703). The ensuing debate between Huygens and Gregory became heated at times. Among other things, Gregory had studied diffraction of light by a bird feather. There is more on diffraction in Chap. 11.

During his life, Huygens was often asked about alleged propositions. Concerning an unknown claimant of the quadrature of the circle, he wrote in a letter to W. Wichers on June 15, 1694 (quoted by Nobel laureate Pieter Zeeman in 1929 in a speech on

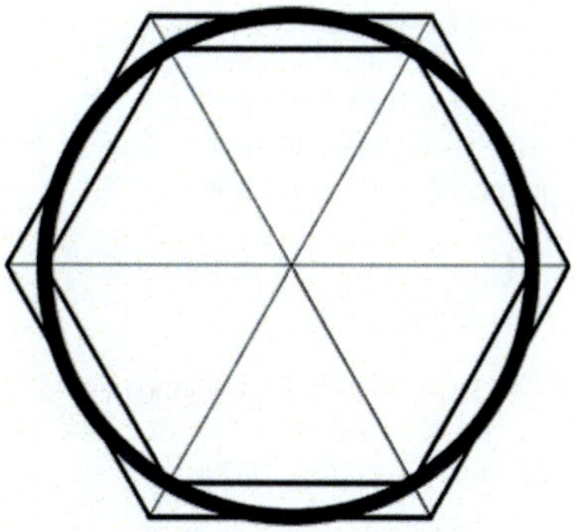

Fig. 6.9 Circle with inscribed and circumscribed regular polygons

the occasion of Huygens' 300th birthday): "His knowledge of geometry cannot be much, because he finally concludes that the ratio of the circumference of the circle to its diameter is 16:5, which even if it were said by an Angel from Heaven, it certainly would not be accepted by me, so sure I know the opposite by examinations by others and myself. So that it is not worth seeking what error he may have committed, which would otherwise be easy to find."

The quadrature of the circle is closely related to its *rectification.* Rectification involves the problem of expressing its circumference in terms of simple functions. Archimedes had already devised a method to do this approximately by circumscribing and inscribing the circle using a regular n-sided polygon (see Fig. 6.9). The circumference of the circle ($2\pi r$ where r is its radius), and hence the number π can then be included by the circumference of the inscribed and circumscribed polygons, which can both be easily computed in elementary terms. By increasing the number of corners n, Archimedes derived a recurrent system of inequalities (each inequality following from the previous one) that approximated π with increasing accuracy. Using 96-cornered polygons he obtained the interval inclusion: $3\frac{10}{71} < \pi < 3\frac{1}{7}$, which has two correct digits and prevailed for many centuries.

Willebrord Snellius (1580–1626) obtained a more accurate approximation, which he put forth in his book *Cyclometricus* (1621). Snellius observed that the recurrence works twice as fast for the inscribed polygons as it does for the circumscribed polygons; he obtained 7 correct digits of π using Archimedes' recurrence for $n = 96$, and 34 digits for $n = 2^{30}$. Ludolph van Ceulen (1540–1610) obtained 35 digits of π, which were engraved in his tombstone in Leiden. Sometimes π is referred to as the Ludolphian number. Adriaan Metius (1571–1635) found the approximation $355/113$, which is easy to remember, has 6 correct digits and is sometimes called Metius' number; the same approximation was first discovered by the Chinese mathematician and astronomer Zu Chongzhi (429–500). It is the best approximation using less than 4 digits in the enumerator. The symbol π was introduced in 1706, before that time it would be described in full as the ratio between a circle's circumference and diameter.

It was Huygens who provided a rigorous proof of Snellius' conjecture in 1654 in his book *De circuli magnitudine inventa Accedunt Eiusdem Problematum*

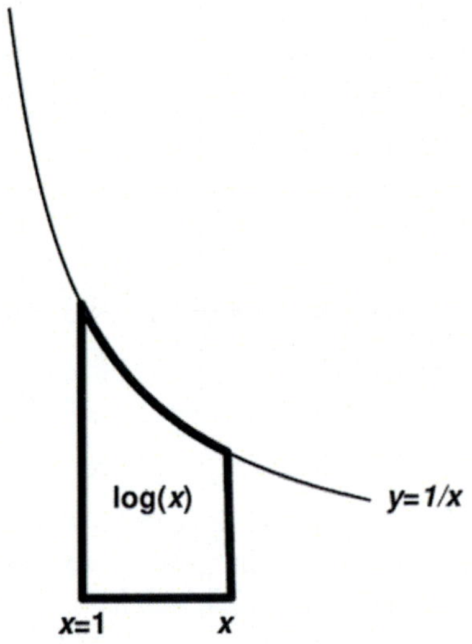

Fig. 6.10 Quadrature of the hyperbola: the area under the hyperbola is equal to $\log(x)$, which is a transcendental number for all x except for $x = 1$

Quorundam Illustrium Constructiones (Discoveries on the magnitude of the circle, after which follow constructions of certain famous problems). From a sixty-cornered polygon ($n = 60$) and using the inscribed polygon only, he obtained the interval inclusion $3.1415926533 < \pi < 3.1415926538$ whereas what he referred to as "the old recurrence" (using both inscribed and circumscribed polygons) would merely give $3.140 < \pi < 3.145$. With *De circuli magnitudine inventa*, Huygens continued to build on his reputation as a mathematician initiated with the Exetasis. Huygens may be characterized as the last mathematician who obtained approximations to π using geometrical methods, and he did so with great skill. The correspondence on π and the quadrature of the circle by Huygens and his contemporaries is very lively and makes for entertaining reading, especially for puzzle enthusiasts. With the advent of calculus methods, promoted by Leibniz, Taylor and others, much faster algorithms for the approximation of π have been developed. Since then, the race for more digits of π has continued incessantly. Nowadays π is known with trillions of digits, and still counting.

The impossibility of the quadrature of the circle, or the impossibility of expressing π by a finite number of simple arithmetic operations, was proved in the nineteenth century. As a first step, the Swiss mathematican and physicist Johann Heinrich Lambert (1728–1777) proved in 1761 that π is not a rational number: it cannot be expressed as the ratio of two integers. Finally, Lindemann proved in 1882 that π is a transcendental number, a term coined by Leibniz meaning that it cannot be a solution of an algebraic equation with integer coefficients. Therefore the quadrature of a circle with compass and straightedge is impossible. In addition, Lindemann showed that $\log(x) = \int_1^x dx/x$ is transcendental except for $\log(1) = 0$, which implies the impossibility of the quadrature of the hyperbola (see Fig. 6.10). These results solved some of the most pressing questions of Huygens' time.

Pendulum as Harmonic Oscillator

Let us return to the pendulum clock. As we have seen, in the seventeenth century, the central problem of navigation was determining longitude. Longitude can, in effect, be measured by time because if the difference in local time at two points is known, the longitudinal distance between them can be computed, as a fraction of the Earth's circumference. However, in the first part of the seventeenth century, this was not a practical option because the existing mechanical clocks were not sufficiently accurate. Galileo had discovered that a pendulum could be used as a frequency-determining device. The implication of this discovery was that the pendulum could be used for a clock and, although many consider him the father of this scientific breakthrough, he never built such a clock.

The pendulum clock was invented by Christiaan Huygens in December 1656, when he was 28 years old. The escapement on the pendulum clock counts the swings and a driving force (weight or mainspring) provides the push. In effect, the escapement is a feedback regulator that controls the speed of this type of a mechanical clock. The pendulum clock was much more accurate than any other contemporary clock. It is not an overstatement to contend that it was one of the great technological breakthroughs in history. The pendulum clock became the world's standard timekeeper. By the twentieth century, it had achieved an accuracy of about one second per year. It was superseded as a time standard by the quartz clock in the 1930s. In conclusion, from 1656 on, there existed a reliable method of measuring minutes and seconds of time. Clocks had two hands, the hour hand and the minute hand, and eventually a third hand for seconds. The invention by Huygens of the first accurate clock can be considered the beginning of the modern world, which is based on science and technology, because it permitted much more sophisticated experiments and detailed measurements.

Huygens was able to design such a clock because of his investigations of the mathematics of the circle, and this would lead him to additional discoveries that are of major importance in modern science. As we have seen above, Galileo believed that a pendulum is isochronous; in other words, that the period of a pendulum does not depend on the amplitude of its swing. Huygens was able to prove that a pendulum swinging through the arc of the circle is not isochronous.

Huygens first had to solve the so-called *tautochrone* problem in order to make a perfect clock. Tautochronism, derived from the Greek roots for "the same" and "time," refers to the concept of maintaining consistent or equal time intervals, in this case for falling objects. More specifically, Huygens had to determine the curve along which a mass would descend to the bottom in an equal amount of time under the influence of gravity, regardless of its starting point. It turns out that a curve is isochronous if it is tautochronous. The cycloid satisfies both properties, and, curiously, also the brachistochronous property, (which we discuss later in Chap. 10). In short,

$$\begin{aligned} \text{isochronous curve} &= \text{tautochronous curve} = \text{brachistochronous curve} \\ &= \text{cycloid.} \end{aligned}$$

As stated already earlier in this chapter, Huygens used traditional geometrical methods because calculus had not yet been invented. Geometrical derivations could be lengthy and complicated, but Huygens excelled at them. He showed that the required curve was a cycloid and not the circular arc of a pendulum's swing. Therefore, conventional pendulums are not isochronous.

Isochronism pertains to processes that require timing coordination to be successful. A sequence of events is called isochronous if the events occur regularly, or at equal time intervals. Isochronism is important in clocks. A property of simple harmonic motion is that the period T does not depend on the amplitude. Thus simple harmonic motion is *isochronous*. In other words it takes the same time to make the same sinusoidal oscillation regardless of the amplitude of the oscillation. The theory of Fourier series and integrals rests upon this property. A mechanical clock is considered *isochronous* when it keeps running at the same rate, regardless of changes in its drive force. This means that it maintains a constant periodic time (T) as its mainspring unwinds. As we have just seen, a conventional pendulum is not isochronous. However, if the maximum angle θ_0 of swing is small, then conventional pendulum is nearly isochronous.

Figure 6.11 juxtaposes the swing of the cycloidal and the conventional pendulum. Huygens realized that isochronism being limited to small swings comes from the fact that the conventional pendulum swings in the arc of a circle. As a result, he designed a clock (the Huygens cycloidal clock) whose pendulum swings in the arc of a rolling circle. A rolling circle (that is, a cycloid) is the curve that is generated by a point on the circumference of a circle as it rolls along a straight line, as discussed in the section *Cycloid.* The cycloidal clock of Huygens gives the correct time, no matter what the size of the largest angle attained by its pendulum. In other words, the cycloidal pendulum is a simple harmonic oscillator, unlike the usual pendulum. In *Horologium Oscillatorium* (1673), Huygens gives a complete mathematical description of his cycloidal clock with a pendulum forced to swing in an arc of a cycloid. Huygens did this by suspending the pendulum (made up of a bob on a wire string) at the cusp of the cycloid. Such a cycloidal pendulum is isochronous, regardless of amplitude. The cycloidal clock was extremely accurate, but unfortunately the movement caused an excessive amount of friction.

Harmonic Oscillator

The pendulum was the first harmonic oscillator used by man. The pendulum is not a simple harmonic oscillator. However, if the swing is small, then the pendulum is approximately a simple harmonic oscillator. A simple harmonic oscillator is a system that, when displaced from its equilibrium position, experiences a restoring force F proportional to the displacement x; that is,

$$F = -\alpha x \,, \tag{6.5}$$

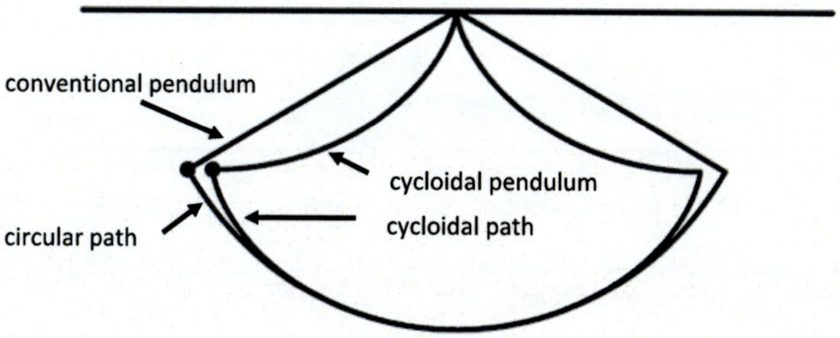

Fig. 6.11 The conventional straight pendulum swings in a circular path, whereas the cycloidal curved pendulum swings in a cycloidal path. Note that for small swings, the circular path is essentially the same as the cycloidal path

where α is a positive constant. Such a system undergoes simple harmonic motion. The phrase "simple harmonic motion" is a traditional term used to denote a sinusoidal oscillation about the equilibrium point with constant amplitude and constant frequency (which does not depend on the amplitude). Let t denote time. Simple harmonic motion may be represented by a cosine function with amplitude A, angular frequency ω and initial phase ϕ, as given by

$$f(t) = A\cos(\omega t + \phi)\,. \tag{6.6}$$

A function $f(t)$ is called periodic if there is a nonzero constant T for which

$$f(t + T) = f(t)\,. \tag{6.7}$$

The constant T is called the period. The period of the cosine function Eq. (6.6) is $T = 2\pi/\omega$. It is easily verified that the sum, difference, product, or quotient of two periodic functions with the same period T is again a periodic function of period T. If a periodic function is plotted on a closed interval with length equal to the period, then the entire graph of the function can be obtained by periodic repetition of the portion of the graph corresponding to this interval.

The use of sines and cosines allows us to define frequency. We can think about a wheel rotating at a rate of 10 times a second. Each rotation through 360° (which is 2π radians) represents one cycle, so we say that the cyclic frequency is 10 cycles per second. The word hertz (Hz) stands for cycles per second, so alternatively we say that the vector has a frequency of 10 Hz. The frequency is customarily denoted by the symbol f.

"Period" is a general word that is used to denote an interval of time, but the word "period" also has special meaning in frequency analysis. By the *period* of a repeating process, we mean the time-interval during which the process exactly repeats itself. In our example of the wheel, the motion exactly repeats itself 10 times a second, so we say that it has a period $T = 10$ seconds. The period T and the frequency f are

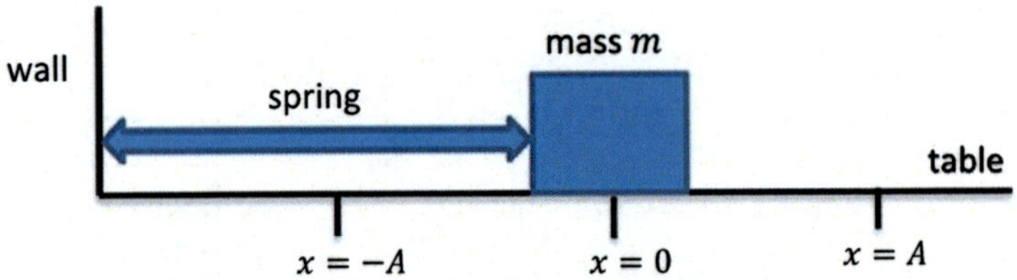

Fig. 6.12 A block attached to a spring is placed on a frictionless table. The equilibrium position, where the spring is neither extended nor compressed, is marked as $x = 0$

reciprocals of each other; that is, $T = 1/f$. The angular frequency ω is measured in radians per second. Angular frequency is related to the cyclic frequency f by the fundamental relation $\omega = 2\pi f$. The 2π factor in this equation comes from the fact that there are 2π radians (or 360°) in each complete rotation.

Let us illustrate, as is done in Fig. 6.12, simple harmonic motion by means of a mechanical problem. A mass attached to a spring is placed on a frictionless table. The other end of the spring is attached to the wall. When the spring is stretched or compressed, it exerts a force on the mass, as in Eq. (6.5). The equilibrium position of the mass, when the spring is neither stretched nor compressed, is marked as $x = 0$.

(a) The mass is displaced to a position $x = A$ and released from rest.
(b) The mass accelerates as it moves in the negative x-direction, reaching a maximum negative velocity at $x = 0$.
(c) The mass continues to move in the negative x-direction, slowing until it comes to a stop at $x = -A$.
(d) The mass now begins to accelerate in the positive x-direction, reaching a positive maximum velocity at $x = 0$.
(e) The mass then continues to move in the positive direction until it stops at $x = A$.

The point mass continues to move back and forth along a straight line under the action of a restoring force provided by the spring. The restoring force may be written as in Eq. (6.5).

The spring provides the restoring force F, which is directed towards the origin. The distance x is positive if the point mass lies to the right of the origin, and x is negative if the point mass lies to the left of the origin. That is, the usual positive direction is assigned to the line. Therefore, Newton's second law, which says that mass times acceleration equals force, becomes:

$$m\frac{d^2x}{dt^2} = -\alpha x \,. \tag{6.8}$$

This equation shows that the point mass, which moves back and forth along the straight line, has an acceleration which is proportional to the displacement x and is at all times directed toward the equilibrium position $x = 0$. This mechanical scheme represents a simple harmonic oscillator. Equation (6.8) can be written as the ordinary differential equation:

$$\frac{\mathrm{d}^2x}{\mathrm{d}t^2} + \omega^2 x = 0\,, \tag{6.9}$$

where we have defined $\omega = \sqrt{\alpha/m}$. It is easily verified that the solution of this differential equation is the function:

$$x = A\cos(\omega t + \phi)\,, \tag{6.10}$$

where A and ϕ are unknown constants. These unknown constants can be found from a knowledge of the position and velocity of the point mass m at the initial time $t = 0$. Thus, under the action of the restoring force F, the point mass m undergoes sinusoidal motion, or simple harmonic motion. The amplitude A is the maximum deviation of the point mass m from the origin. The period is $T = 2\pi/\omega$ seconds/radian. The quantity ϕ is the initial phase and characterizes the initial position of the point, since for $t = 0$ we have:

$$x_0 = A\cos\phi\,. \tag{6.11}$$

This completes the description of the harmonic oscillator.

Approximation of the Pendulum

See Fig. 6.13. Consider the motion of a simple pendulum, ideally just a point mass m at the end of a rigid and massless support of length L, with no friction or air resistance. Figure 6.13 visualizes this situation, leading to the question: can it be determined whether the simple pendulum does indeed execute simple harmonic motion? Huygens answered this question. He determined that the simple pendulum does not execute simple harmonic motion. However, he also found that for small enough displacement, the simple pendulum does approximately execute simple harmonic motion.

The net force (F in the figure) is drawn as the resultant of the tension S in the weightless string or rod, and of the weight W of the bob. Note that F must be perpendicular to the string, and thus to S. (If this point offers difficulty, try to think what would have to happen if F were not perpendicular to S. Any component parallel to the string will extend it, move it, or break it.) Therefore, F equals $W\sin\theta$. The weight W equals mg, where g is the gravitational acceleration.

On the other hand, since the bob is moving, it will obey Newton's second law which says that $F = ma$, where a is the acceleration. The acceleration is related to the second derivative of the arclength s along its trajectory with respect to time t. The arclength is itself equal to $L\theta$, where the angle θ (which depends on t) is measured in radians and L is the length of the string (which is constant). Putting this together, we obtain:

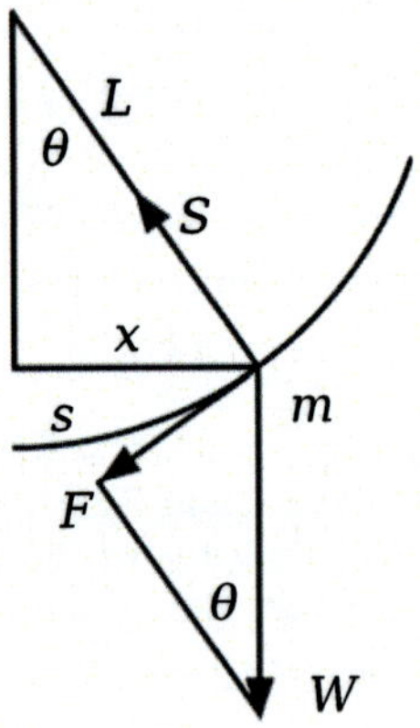

Fig. 6.13 Net force on the bob pendulum

$$F = -mg\sin\theta = mL\frac{d^2\theta}{dt^2}\,, \tag{6.12}$$

or

$$\frac{d^2\theta}{dt^2} + (g/L)\sin\theta = 0\,. \tag{6.13}$$

This differential equation is non-linear and very hard to solve, because of the $\sin\theta$-term. Modern mathematics offers solutions to this problem (in terms of elliptic integrals), but these do not make the pendulum a harmonic oscillator, nor solve the problem that it is not isochronous.

Huygens made the approximation $\sin\theta = \theta$ for small angles θ, so that Eq. (6.13) becomes linear and easy to solve. For such an approximation, we assume that the path of the bob is (nearly enough) a straight line and that the net force on the bob is proportional to the displacement, and always directed to the center (equilibrium) position. If the motion of the bob is restricted to suitably small angles, the straight line condition is approximately met (and x corresponds to arc length s).

The drawing in Fig. 6.13 also shows that F is always directed to the equilibrium position, which means that F and x have opposite sign. As we know, a force which has algebraic sign opposite to that of x is called a restoring force. Thus, with the replacement of arclength s by horizontal displacement x, we have $F = -Wx/L$ and

$$F = -Wx/L = m\frac{d^2x}{dt^2}\,, \tag{6.14}$$

or

$$m\frac{d^2x}{dt^2} + (W/L)x = 0\,. \tag{6.15}$$

As we have seen in the previous section, (6.15) is a linear differential equation which is straightforward to solve. Hence, the bob will execute simple harmonic motion for small angles of swing. As we know $F = -\alpha x$ and $\omega = \sqrt{\alpha/m}$ for simple harmonic motion. Thus for the present problem:

$$\alpha = \frac{W}{L} \text{ and } \omega = \sqrt{\frac{W}{mL}} . \tag{6.16}$$

Therefore the approximate period of a pendulum is

$$T = \frac{2\pi}{\omega} = 2\pi\sqrt{\frac{mL}{W}} . \tag{6.17}$$

Note again that $W = mg$, where g is the acceleration due to gravitation. It follows that the approximate period of a pendulum is given by

$$T = 2\pi\sqrt{\frac{mL}{mg}} = 2\pi\sqrt{\frac{L}{g}} . \tag{6.18}$$

This formula, which was first derived by Huygens, holds as long as the swing is not too large, say less than six degrees of arc. The formula proves that three statements are true (again for small amplitude):

1. The period does not depend on amplitude,
2. The period does not depend on mass,
3. The period is proportional to the square root of the length of the pendulum.

These results are somewhat unexpected, for they imply that except for air resistance, a heavy and a light pendulum bob will both swing side by side in step if suspended from equally long strings. Also, the approximate period is the same for a medium-sized swing as for a very small one. This property is referred to as the *approximate* isochronism of the pendulum, a phenomenon which Galileo is reputed to have discovered experimentally when a young student. Indeed, the above statements can be very easily checked experimentally (for instance at home).

Looking at the exact Eq. (6.13) for the circular pendulum and observing that $\sin\theta \leq \theta$ for $\theta \geq 0°$, we can also deduce that the period T will be longer for larger amplitudes: replacing $\sin\theta$ by θ has a similar effect as reducing the ratio g/L. As a result, by virtue of Huygens' formula, the period T will be somewhat longer than predicted based on the assumption that the pendulum is isochronous.

Let us now look at the effect of the gravitational acceleration. Let g_1 and g_2 represent the values of the gravitational constant g in Paris and Cayenne, French Guiana, respectively. Similarly, let T_1 and T_2 denote the corresponding periods, respectively, for the same clock. Thus,

$$4\pi^2 L = g_1 T_1^2 = g_2 T_2^2 . \tag{6.19}$$

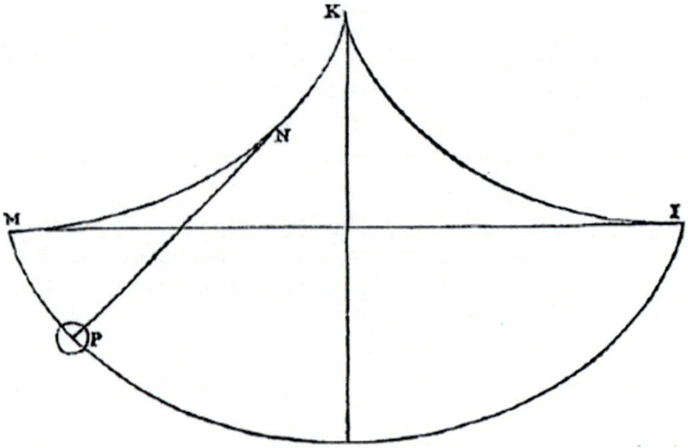

Fig. 6.14 Diagram from *Horologium Oscillatorium*

When he wrote the *Principia*, Newton knew that in 1672 the French astronomer Jean Richer (1630–1696) had taken a pendulum clock from Paris to Cayenne to help in astronomical observations. There the clock lost 2 and a half minutes each day, so he could calculate T_2 from T_1. From g_1 for Paris, Newton thus found g_2 for Cayenne. It was this observation which drew general attention to the distinction between the concepts of mass m and weight, which would be mg_1 in Paris but would be mg_2 in Cayenne. Until then mass and weight were not distinguished as separate entities. Since then mass is interpreted as a measure of the amount of matter in an object (measured in kilograms—kg), while weight is a measure of the force that gravity exerts on the object (measured in newtons—N).

Incidentally, in Huygens' Eq. (6.18) the two expressions for mass m are not conceptually identical: m in the numerator refers to inertial mass; whereas m in the denominator refers to gravitational mass. The inertial mass plays a role in the equation of motion of the pendulum; the gravitational mass plays a role in the counterforce exerted by the as the string or rod holding the bob. Only if these two have the same numerical value, can we cancel them. Conversely, if Eq. (6.18) is found to hold experimentally, as it does, then we can deduce that these two values are really identical. Newton made pendulum bobs of a great variety of substances. He discovered that this equation holds, demonstrating the equivalence of the two types of masses. Einstein's general theory of relativity, asserting the equivalence of gravitational and inertial mass, was already looming in the background.

Incidentally again, Huygens' formula is famous in the history of science for being one of the first recorded physics formulas, where for the first time physical quantities were represented by mathematical symbols and as such related to each other by a formula. (Exactly which of Huygens' physics formulas is truly the first one is still an open question; the same claim could go to his formulas describing collision and the centrifugal force).

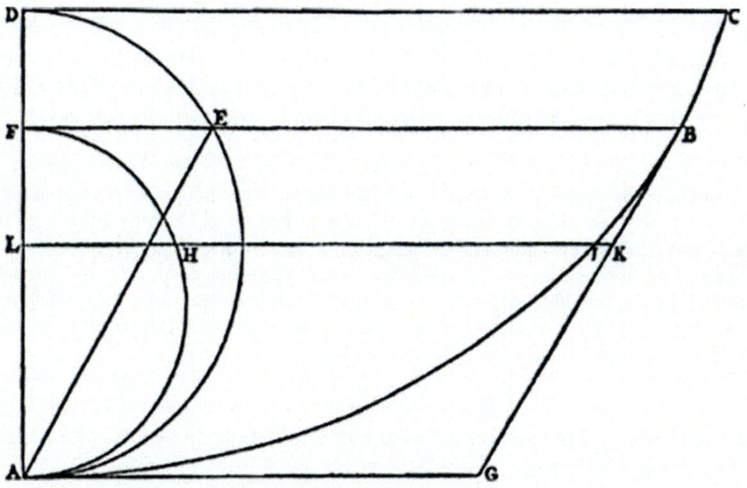

Fig. 6.15 Figure 51 from *Horologium Oscillatorium*. Objects released from any point on the cycloid *CBIA* will all reach the bottom point in the same time

The pendulum clock became the world's standard timekeeper, and after subsequent refinements, finally did achieve an accuracy of about one second per year. It was superseded as a time standard by the quartz clock in the 1930s. In conclusion it can be said that from 1656 on, there existed a reliable method of measuring minutes and seconds of time. Galileo did not construct a pendulum clock. The first to do so was Huygens in 1656. His greatest difficulty lay in the fact that the pendulum does not have a period that is completely independent of the length of the swing. As shown above, long swings take a *little* longer than short swings. His first solution was to arrange to have the pendulum beat through a very small swing, since the smaller the swing, the less the deviation from equal periods as the length of the swing varies slightly. Huygens showed the trouble to be that a pendulum bob marks out a segment of a circle as it swings. If, somehow, the bob could be made to follow a cycloid then its period would not vary at all with the length of swing.

Cycloidal Clock

As we have seen, Descartes first discovered that a simple pendulum (that is, a pendulum whose bob describes a circular arc) is not isochronous. Huygens, through his invention of the theory of evolutes (on which more in Chapter 10), showed that a cycloidal pendulum (that is, a pendulum whose bob describes a cycloidal arc) is isochronous. Huygens faced the problem of manufacturing a pendulum whose bob describes a cycloidal arc. By 1659, Huygens solved the problem by not allowing the pendulum to swing freely (see Fig. 6.14). Instead, the top of the cord moved against a curved stop (made up of cycloidal arcs KM and KI), so that less and less of the cord was truly free to swing as the bob moved outward. The stop curved in such a way on either side of the cord that the bob swung in a cycloidal arc MPI. Here are the words of Christiaan Huygens in *Horologium Oscillatorium (Part One)*:

> Truly, in order that the wonderful nature of the line and the effect may be understood better, the whole semicycloids KM and KI, here seen to be expressed by another diagram [see Fig. 6.14], between which the pendulum KNP is suspended and moving, [of length] twice the diameter of the generating circle, and the oscillations of any amplitude, as far as the largest of all through the arc MPI will be made in the same times; and thus, so that the center P of a sphere hung on, is always moving to and fro on the line MPI which is part of a whole cycloid. I do not know of any other line with this conspicuous quality, except for this given line, as truly it describes its own evolute.

The central theorems of *Horologium Oscillatorium* to prove the validity of the cycloidal clock are Proposition XXV of Part II and Proposition VI of Part III.

Proposition XXV reads:

> In a cycloid with a vertical axis and whose vertex is at the bottom, the times of descent in which a mobile, starting from rest at any point of the curve, reaches the lowest point, are equal to each other, and have at the time of the vertical fall along the entire axis of the cycloid a ratio equal to that of the half-circumference of a circle to its diameter.

This proposition states the tautochrone property of the cycloid. See Fig. 6.15. Huygens draws an upside-down cycloid $CBIA$, and proves that an object released at *any* point (C, B, I or any other point on the cycloid) falling down along the cycloid will reach its bottom point A in the *same* time. Moreover, the ratio of this falling time to the falling time along the vertical axis DA is equal to the ratio of the diameters and half-circumference of the circles drawn in the figure (which is π). In formula:

$$T(CA)/T(DA) = DEA/DA = T(BA)/T(DA) = FHA/FA = \pi\ . \tag{6.20}$$

Fig. 6.16 Pair of cycloids at the Hofwijck estate: objects released simultaneously from different heights will all reach the end point at the same time [with permission from the Huygensmuseum Hofwijck Voorburg]

The proof of this remarkable property takes several pages of dense derivations for which we refer to *Horologium Oscillatorium* or the book by Aarts (2015). In Huygens' Hofwijck estate near The Hague, now a museum, two cycloids are on display where the property can be tested. To the surprise of some visitors, objects released from different starting points will *always* reach the bottom after the same time (Fig. 6.16).

The crux is of course that an object released at a higher point acquires more speed while falling than one released from a lower point. Falling time depends on the falling speed integrated over the curve. Of all conceivable curves, *only* for the cycloid the falling times turn out to be equal. Note, at the risk of repeating, that this is *not* the case for a circular falling trajectory.

The tautochrone property of the cycloid has been studied by later mathematicians such as Lagrange, Euler, forming the topic of infinitesimal calculus. The Norwegian mathematician Niels Henrik Abel (1802–1829) proved that the tautochrone curve must be a cycloid using what now is called Abel's integral equation.

Proposition VI reads:

> By the evolution, from the vertex, of a half-cycloid, another half-cycloid is described, equal and similar to the first, whose base coincides with the straight line which touches the cycloid developed at its vertex.

See Fig. 6.17 and turn it upside down. With this proposition, Huygens shows that a cycloidal path AFN can be achieved for the bob of a pendulum fixed in the point C if its wire $EKBC$ is wound around a cheek $ABCN$ which is *itself* a cycloid with the same dimensions. The winding refers to the properties of the evolute; for more on evolutes see Chap. 10.

For this set up, Huygens' pendulum formula

$$T = 2\pi\sqrt{\frac{L}{g}} \tag{6.21}$$

is *exact*, for *any* amplitude. The cycloidal pendulum is an exact harmonic oscillator, without any approximations. In this case, the pendulum length L is interpreted as the line $CBKE$ wound around the cycloidal cheek $ABCN$. The pendulum length L is also equal to the radius of curvature of the cycloid at the point F and equal to the radius of the circular pendulum path if there were no cheeks. L is four times the radius of the rolling circle defining both the cycloids of the cycloidal pendulum path and the cycloidal cheeks. The circular pendulum path is the osculating (best fitting) circle of the cycloidal path at its top (for details on the osculating circle see Chap. 10). If we turn Fig. 6.17 upside down and combine Propositions XXV and VI, we are back at Fig. 6.14 and all has been proven.

Huygens considered his result important enough to encrypt it in the form of the following anagram, before submitting it to Henry Oldenburg, secretary of the Royal Society in London on 4 September 1669 (as mentioned in Chap. 1):

a b c d e f g h i l m n o p q r s t u x y

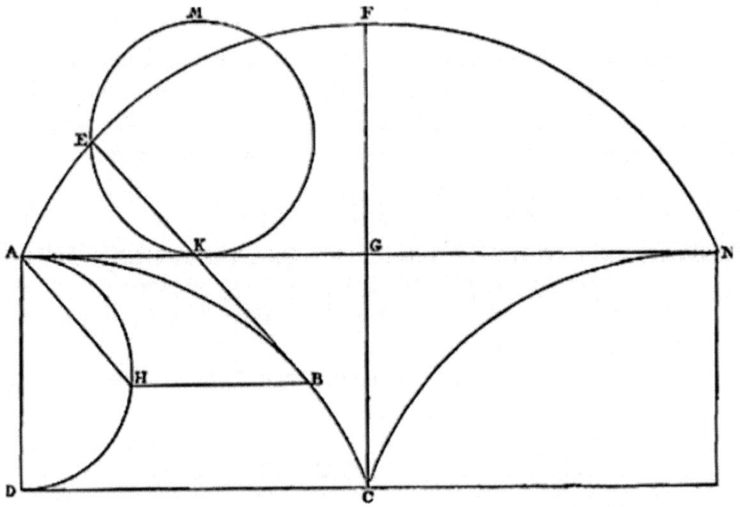

Fig. 6.17 Figure 58 from *Horologium Oscillatorium.* Turn it upside down so that gravity points from C to F. The bob of the pendulum $CBKE$ describes a cycloid (and is therefore isochronous according to Proposition XXV) if it is wound around a cheek $ABCN$ which is *itself* a cycloid

2 1 4 3 7 0 0 0 4 2 3 2 3 3 1 0 8 5 5 0 1

4 0 5 4 10 1 0 0 7 1 6 2 0 1 0 5 5 3 6 1 0

"Tempus descensus a quolibet puncto semicycloidis est ad tempus descensus per axem ejusdem ut semicircumferentia circuli ad diametrum" (*"The time of descent from any point of the semicycloid is the time of descent through the same axis as the semicircumference of the circle to the diameter"*). In other words, the formula is approximate for the circular pendulum, but becomes exact for the cycloidal pendulum. The rest is history.

As an aside, another takeaway from this analysis is that it would be better to design children's swings with cycloidal cheeks than swinging freely.

Epilogue: After the Cycloidal Clock

In mechanical watches and clocks, an escapement is a mechanical linkage that sends impulses to the timekeeping element and periodically releases the gear train to allow the clock's hands to advance. The escapement is the very reason that the clock makes the familiar ticking sound. Due to their verge escapements, early clocks had pendulum swings as wide as 80°. Huygens demonstrated that large swings caused the pendulum to be inaccurate, causing its period, and thus the rate of the clock, to vary due to unavoidable variations in the driving force provided by the movement. Only pendulums with modest swings of a few degrees are roughly isochronous. This shortcoming was solved, when Robert Hooke and William Clement developed the anchor escapement, reducing the pendulum swing to about 5°. The anchor escapement evolved into the common escapement used in pendulum clocks. In addition to improving accuracy, the anchor's narrow pendulum swing made it possible for the clock's case to accommodate longer, slower pendulums that consumed less power and exerted less strain on the movement. Today, all mechanical clocks are equipped with an anchor escapement: the grandfather-clock, table- and wall clocks.

Huygens and his clockmaker Salomon Coster extensively worked on the utilization of a long pendulum that beat in a one-second rhythm, providing more regularity than the short-pendulum clocks initially employed by Huygens. The entire mechanism was enclosed in a case to prevent movement that could interfere with an object not as long as a pendulum. Suppose a pendulum were replaced by a highly elastic material that was bent into a loose spiral. This spiral (a "hair-spring") would take up very little room and could be made to pulse in and out with great regularity. Just as a pendulum could be powered by a falling weight, a hairspring could be powered by a slowly opening larger "mainspring". The mainspring-powered clock was called a "watch". Both clocks and watches were periodically wound. In the clock, the fallen weight is lifted again by the mechanical turning of a key; in the watch, the loosened mainspring is tightened again by turning a knob. However, the main and obvious advantage of a timepiece with a balance spring is its ability to function regardless of its position. Gravitation has minimal influence on it, unlike a clock with a pendulum.

The new timepieces were immediately used by astronomical observatories as methods of measuring time. As early as in the first half of 1657, Huygens informed the astronomer Samuel Kechelius (1611–1668) of the Leiden Observatory of his invention. Two other astronomers to take advantage of the instruments were the Frenchman Jean Picard (1620–1682), who headed the Paris Observatory, and Giovanni Cassini (1625–1712), whom he brought to Paris. In 1676, the observatory at Greenwich, England, was using two clocks with pendulums 4 m (13 feet) long, with two-second beats. At the other extreme, the Scottish astronomer James Gregory (1638–1675) was using a pendulum with a third-of-a-second beat as early as 1673.

The clocks showed that the "regular" motions of the heavenly bodies were not necessarily very regular. The pendulums and balance springs kept time more accurately. It was shown, for instance, that the interval between successive periods in which the Sun passed over the meridian to indicate noon was not invariable. It changed slightly in the course of the year. This was not because Earth's rate-of-rotation changed. It was because certain complications were introduced by the fact that Earth's orbit about the Sun was elliptical rather than circular and that the Earth's axis was tipped 23.5° from the perpendicular to the plane of its revolution.

Huygens' Planetarium

Huygens' fascination for automatic machines did not stop at the pendulum clock or the horologium. His next dream was a machine that could predict the position of celestial bodies as a function of time, the planet machine. Indeed he designed and built one and wanted to sell it to French minister Colbert. Unfortunately, Colbert died in 1683 before they could conclude the deal.

Huygens' description of the planetarium published posthumously in *Descriptio Automati Planetarii* in *Opuscula Postuma* in 1703 combines technical ingenuity as well as the latest astronomical discoveries and the latest mathematical inventions. He used the term "machine planétaire" in French, "automaton planetarium" in Latin. In English, this is referred to as an "orrery". In modern usage, the term "planetarium" typically refers to a complete theater or dome that is used for astronomical presentations, but historically it has been used more specifically to refer to any tool or device used for representing the motions of celestial bodies—we will stay with the term "planetarium".

Huygens notes that it was only in his time that the knowledge of planetary bodies became reasonably complete. Thus, his Planetarium contains the Sun together with the five known planets and their moons: Mercury, Venus, Earth (with its Moon), Mars, Jupiter and Saturn (with their four and five moons known at the time, respectively). The movements of these five planets and the Earth's Moon are represented by a small number of interlocking wheels parallel to the surface of a plane table. The moons of Jupiter and Saturn are fixed to their respective wheels. The whole machine has an octogonal shape with a diameter of two feet and a depth of six inches (see Fig. 6.18 and Fig. 6.19).

The front of the planetarium shows the orbits of the five planets and their moons in actual proportion. For better visibility their diameters are not shown in proportion, the small circular diagram at the bottom illustrates their actual proportions. The planetary movements are generated by a clockwork driven by a spring and a system of gears with teeth in relation to their orbital times. The spring has to be wound once every week. By putting the planetary orbits slightly eccentric to the Sun, the elliptical motion would be represented, in accordance with Kepler's laws.

The planetarium could run automatically to show the current location of planets, or by hand to run forward in the future by up to 200 years or backward in the past by up to 100 years—compare this to modern age digital planetariums available as computer applications.

The number of teeth in the wheels needed to accurately approximate the planetary orbital times presented Huygens with an interesting mathematical problem. He explains this in detail for Saturn. The annual movement of Saturn is $12^{\circ}13'\,34''18'''$, the annual movement of the Earth is $359^{\circ}45'\,40''\,31'''$. Reduced to one-sixtieth of a second, the orbital periods of the Earth and Saturn around the Sun have a ratio of 2640858:77708431. It would be technically infeasible to provide the wheels with so many teeth, so an approximation had to be found. Huygens resorted to the mathematical theory of *continued fractions*, which was still in its infancy in his days. He

Fig. 6.18 Huygens' planetarium, left frontside, right backside (pictures in *Beschrijving van het Planetarium*, translation by J.A. Volgraff and D.A.H. van Eck in Planetarium-boek Eise Eisinga, 1928) [with permission from Rijksmuseum Boerhaave Leiden]

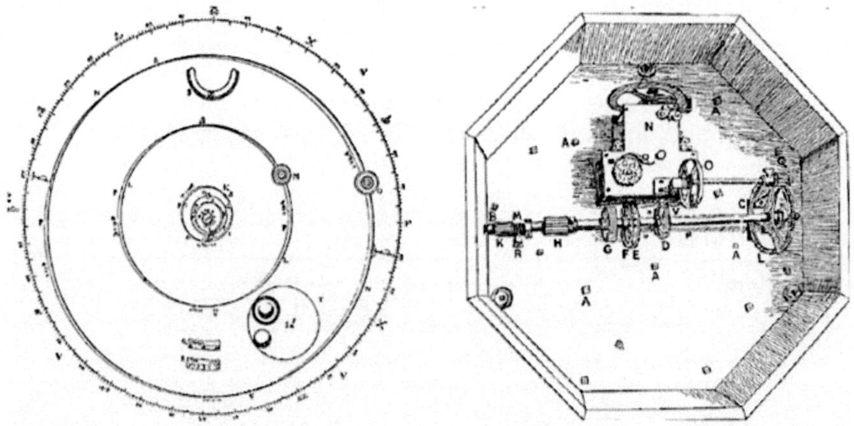

Fig. 6.19 Huygens' diagrams of the planetarium: frontside (left) and interior (right)

proceeded to repeatedly divide the two numbers in the following way:

$$
\begin{aligned}
77708431/2640858 &= 29 + 1123549/2640858 \\
2640858/1123549 &= 2 + 393760/1123549 \\
1123549/393760 &= 2 + 336029/393760 \\
393760/336029 &= 1 + 57731/336029 \\
336029/57731 &= 5 + 47374/57731 \\
57731/47374 &= 1 + 10357/47374 \\
10357/5946 &= 1 + 4411/5946 \\
5946/4411 &= 1 + 1535/4411 \\
4411/1535 &= 2 + 1341/1535 \\
1535/1341 &= 1 + 194/1341 \\
1341/194 &= 6 + 177/194 \\
194/177 &= 1 + 17/177 \\
177/17 &= 10 + 7/17 \\
17/7 &= 2 + 3/7 \\
7/3 &= 2 + 1/3 \\
3/1 &= 3 + 0/1
\end{aligned}
\tag{6.22}
$$

This sequence of divisions forms a recursion, with a repetition on $a/b = c + d/b$ and $a = b$, $b = d$, until $d = 0$. Therefore, $\frac{77708431}{2640858}$ can be represented by the expression:

$$
\frac{77708431}{2640858} = 29 + \cfrac{1}{2 + \cfrac{1}{2+\cfrac{1}{1+\cfrac{1}{5+\cfrac{1}{1+\cfrac{1}{4+\cfrac{1}{1+\cfrac{1}{1+\cfrac{1}{2+\cfrac{1}{6+\cfrac{1}{1+\cfrac{1}{10+\cfrac{1}{2+\cfrac{1}{2+\frac{1}{3}}}}}}}}}}}}}}, \tag{6.23}
$$

which is a finite *continued fraction.* Approximations to $\frac{77708431}{2640858}$ are found by truncating the fraction after a number of steps. These approximations are called *convergents.* The convergents after truncation at the subsequent steps are:

$$
\begin{gathered}
29/1\\
59/2\\
147/5\\
206/7\\
1177/40\\
1383/47\\
6709/228\\
8092/275\\
14801/503\\
37694/1281\\
52495/1784\\
352664/11985\\
405159/13769\\
4404254/149675\\
9213667/313119\\
22831588/775913\\
77708431/2640858
\end{gathered}
\tag{6.24}
$$

Huygens' achievement was to prove that each of these convergents gives the best approximations to the final ratio: no smaller numbers would give better approximations. For Saturn he selected the approximation 206/7. This would lead to an error of only $0°1'\,34''$ after 20 years of operating the planetarium. For the other planets he found approximations with similar accuracy in the same way.

Huygens noted that continued fractions can be used to represent irrational numbers as well (they are infinite in this case): for π he gave

$$
\pi = 3 + \cfrac{1}{7 + \cfrac{1}{15 + \cfrac{1}{1 + \cfrac{1}{292 + \cfrac{1}{1 + ..}}}}}\,. \tag{6.25}
$$

The convergent 355:113 is the already mentioned approximation to π found by Adrianus Metius. The theory of continued fractions was further developed in the eighteenth century by Leonhard Euler (1707–1783) and Joseph-Louis Lagrange (1736–1813). The Frisian amateur mathematician Eise Eisenga (1744–1828) built the oldest still functioning planetarium, driven by a pendulum clock, in his home town of Franeker.

Chapter 7
Huygens-Fresnel Principle

To every thing there is a season, and a time to every purpose under the heaven. A time to break down, and a time to build up.
—Bible, *Ecclesiastes 3:1–3*

In the following sections we continue the development of Huygens' principle from Huygens' first formulation started in Chap. 4 into a full modern description.

As we have seen in Chap. 4, Huygens' principle asserts that every point on a propagating wavefront acts as a secondary source of spherical wavelets, or elementary waves, or particular waves as Huygens called them. These wavelets then combine to form a new wavefront that represents the wave's progression over time. We also established that Huygens' principle in its original formulation is essentially incomplete, as it does not account for interference and therefore suffers from a number of artifacts.

For this chapter we concentrate on regular wave propagation. In Chap. 11 we deal with the phenomenon of diffraction around obstacles. This separated treatment is somewhat uncommon in usual textbooks, but as we will point out in Chap. 11, Huygens rather mysteriously did not recognize diffraction as a significant wave phenomenon. Moreover, Huygens' principle, as generalized into the Huygens-Fresnel principle and completed by Kirchhoff leads to a wave description which is free of approximations and associated artifacts. By contrast, diffraction theory requires a set of intricate approximations to make it work.

For further reading into the physics and mathematics of wave propagation and Huygens' principle, we recommend two very accessible text books: Baker and Copson 1939. *The Mathematical Theory of Huygens' Principle*, Oxford at the Clarendon Press and Born and Wolf 1999. *Principles of optics*, Cambridge University Press. In this chapter we adopted several of the derivations from these two text books.

T. J. Moser and E. A. Robinson, *Walking with Christiaan Huygens*, History of Physics,
https://doi.org/10.1007/978-3-031-46158-3_7

Huygens' Principle After Huygens

The time directly following Newton and Huygens, spanning roughly the eighteenth century, is characterized by slow progress in the development of wave theory. Newton's corpuscular theory for light, which described light as composed of particles, enjoyed unbeatable and widespread popularity. This popularity hindered the progress of wave theory. Wave theory fell into obscurity during this time, especially in the Anglo-Saxon world. Huygens' theory faded away altogether.

The mathematical tools required to describe wave and light propagation were still under development. In 1746, Jean-Baptiste le Rond d'Alembert (1717–1783) derived the one-dimensional wave equation by utilizing Hooke's theory of elasticity. He also found a solution to this equation (more details in Chap. 10). Later, Leonhard Euler (1707–1783, see Fig. 7.1), a proponent of the wave theory for light, generalized d'Alembert's wave equation to three dimensions. In three dimensions the wave equation reads:

$$\nabla^2 u - \frac{1}{c^2}\frac{\partial^2 u}{\partial t^2} = 0,\ \nabla^2 = \frac{\partial^2}{\partial x^2} + \frac{\partial^2}{\partial y^2} + \frac{\partial^2}{\partial z^2}\,, \tag{7.1}$$

where u is the wave function, c the medium velocity, x, y, z are spatial coordinates and t the propagation time. The wave equation constitutes a linear partial differential equation of second order. The presence of the second time-derivative alongside the second derivative in spatial coordinates allows the equation to describe an oscillating and propagating wavefield. The wave equation is fundamental equation in physics and plays a crucial role in describing various wave phenomena, including sound waves, electromagnetic waves, and seismic waves. It provides a mathematical framework to study the behavior of waves and their interactions with different media. Solutions to (7.1) (obtained by Poisson, Green, Helmholtz and others, see later in this chapter) are available subject to specified initial conditions when the medium velocity c is considered homogeneous.

Fig. 7.1 Jean-Baptiste le Rond d'Alembert, Leonhard Euler

Interference and the Huygens-Fresnel Principle

During 1800–1804 Thomas Young (1773–1829, see Fig. 7.2) presented a series of lectures on vision, sound, light and colors at the Royal Society of London. In these lectures, he described two major contributions to wave theory: *interference* and *diffraction*. Here we concentrate on interference, in Chap. 11 we deal with diffraction. His double-slit experiment (Fig. 7.3) showed that if a light beam (a) is split into two beams (b) and (c), this results in a regular pattern of "fringes" (alternating zones of light and dark) on the observation screen (S3). Young attributed this to the wave character of light propagation and was the first to measure the wave length of a light beam. He was also the first to refer to this pattern as "interference", in the Syllabus for his Royal Institution lectures, published early in 1802. Young demonstrated the same interference effect on water waves (Fig. 7.4), where the waves add up constructively along a specific set of lines (constructive interference) and cancel each other along another set of lines (destructive interference). In the caption to Fig. 7.4 we give some quotes from Young himself.

While Young's ideas were still largely qualitative, his main achievement was in establishing the wave theory of light, thereby challenging Newton's old view that light consists of particles. His work *A Course of Lectures on Natural Philosophy and the Mechanical Arts* (1807) provides an interesting insight into the state of physics at that time, with generous references to Huygens, not only in the theory of light but also in various other areas of Huygens' interest. Young maintained a very active and collegial correspondence with his French contemporary Augustin Jean Fresnel (1788–1827, see Fig. 7.2), who took his ideas further and expanded them into a mathematical treatment of wave propagation. At this point Huygens' principle was still dormant and ignored by physicists. Euler, Young and Fresnel in his early career had ignored Huygens' principle. Young and Fresnel still used ray paths to support their arguments. Fresnel was the first, in 1818, to dust off Huygens' principle and

Fig. 7.2 Thomas Young, Augustin Jean Fresnel, George Green

Fig. 7.3 Young's double-slit experiment. Wave beams passing through two slits b and c form an interference pattern at the screen S3 (shown by the black and white recording below screen S3)

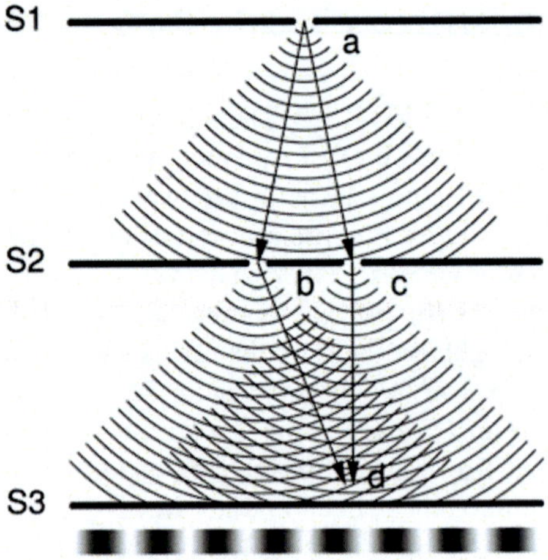

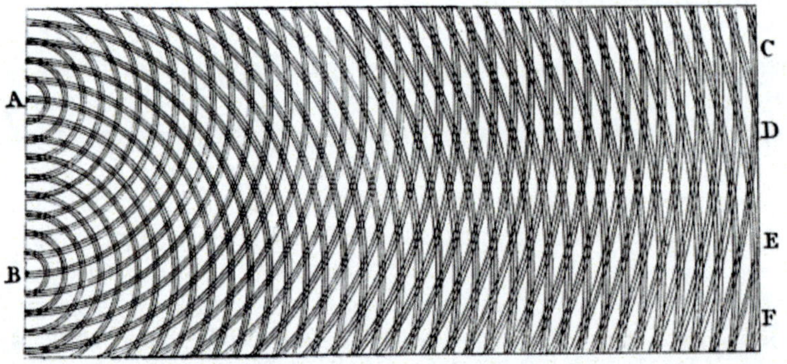

Fig. 7.4 Young's lectures, showing the pattern created by interference of two waves. Quoting Young: 'Two equal series of waves, diverging from the centres A and B, and crossing each other in such a manner, that in the lines towards C, D, E, and F, they counteract each other's effects, and the water remains nearly smooth, while in the intermediate spaces it is agitated.'

cast it in the form of an integral. See the diagram of Fig. 7.5 and details in Born and Wolf (1999).

Considerable progress had been made in complex number theory since the seventeenth century. Euler had established the relation $e^{ix} = \cos(x) + i\sin(x)$, where i is the imaginary number $i = \sqrt{-1}$. As a result, wave functions could now be conveniently expressed in terms of exponential functions e^{ix}. These wave functions represent elementary solutions to the wave Eq. (7.1), and a cleverly designed combination of them allows for the superposition of wave functions, which accounts for interference. Integral calculus had also matured since it was introduced by Leibniz in the seventeenth century. Integration of wave functions originating at a primary wavefront now allowed to mathematically express Huygens' principle. The idea of

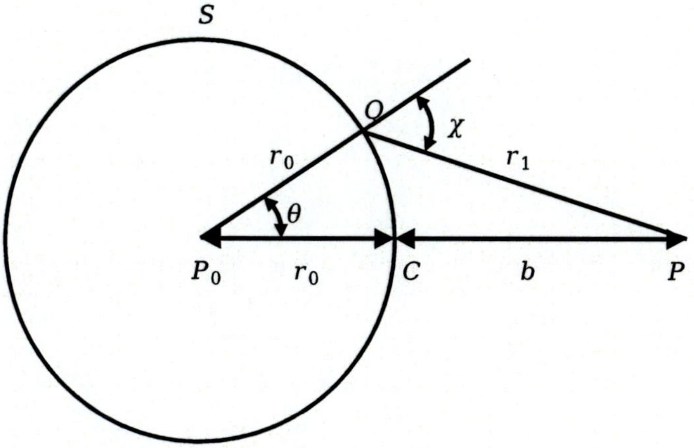

Fig. 7.5 Derivation of Huygens-Fresnel principle

elementary wave solutions would be later formalized to Green's functions (see later in this chapter).

A point source at P_0 generates a spherical wave with front S at a certain instant (see Fig. 7.5, where the two-dimensional circle represents the three-dimensional sphere). According to Huygens' principle a new wavefront is formed by the envelope of secondary waves at S (see Chap. 4). Fresnel replaced the geometrical envelope construction by an algebraic integration of wavelets. To measure their contribution to the wave recorded at an observation point P, Fresnel set up the following integral:

$$U(P) = \frac{e^{ikr_0}}{r_0} \int_S \frac{e^{ikr_1}}{r_1} K(\chi) dS \, . \tag{7.2}$$

Here $U(P)$ is the resulting wave observed at P, e^{ikr_0}/r_0 the primary wave from P_0 measured at S, and e^{ikr_1}/r_1 are the secondary waves generated at S. The point Q acts as the engine behind the integration, as it moves all over the sphere S. The primary and secondary waves are both expressed by generic exponential functions: e^{ikr}. Here, r is the propagation distance and k is the wave number, which is related to the wave length λ by $k = 2\pi/\lambda$.

The medium is still considered homogeneous without obstacles. In such an obstacle-free space the integral (7.2) would have to be equal to the undisturbed direct wave $e^{ik(r_0+b)}/r_0 b$ traveling along the straight line from P_0 via C to P. Fresnel showed how this weighted integral that accounts for interference can accomplish this, thereby achieving the rectilinear propagation from P_0 to P. Considering that Newton's main criticism of the wave theory was that it could not easily explain rectilinear propagation and that this was a major challenge for Huygens, this was one of Fresnel's major achievements.

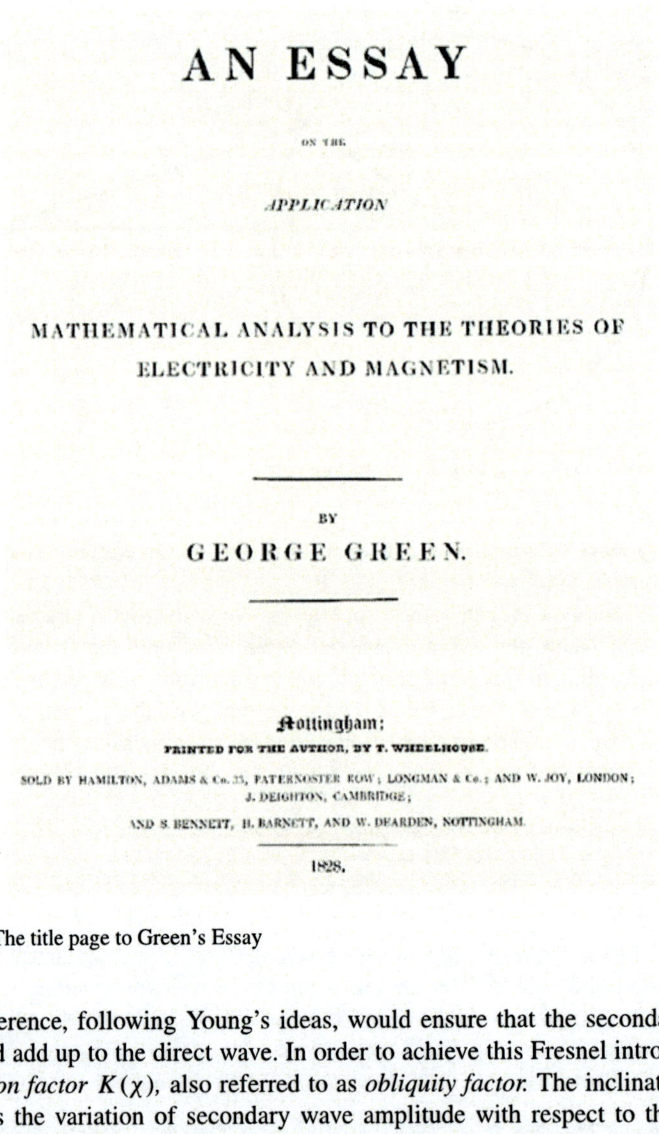

AN ESSAY

ON THE

APPLICATION

OF

MATHEMATICAL ANALYSIS TO THE THEORIES OF

ELECTRICITY AND MAGNETISM.

BY

GEORGE GREEN.

Nottingham:

PRINTED FOR THE AUTHOR, BY T. WHEELHOUSE.

SOLD BY HAMILTON, ADAMS & Co. 33, PATERNOSTER ROW; LONGMAN & Co.; AND W. JOY, LONDON;
J. DEIGHTON, CAMBRIDGE;
AND S. BENNETT, H. BARNETT, AND W. DEARDEN, NOTTINGHAM.

1828.

Fig. 7.6 The title page to Green's Essay

Interference, following Young's ideas, would ensure that the secondary waves all would add up to the direct wave. In order to achieve this Fresnel introduced the *inclination factor* $K(\chi)$, also referred to as *obliquity factor.* The inclination factor describes the variation of secondary wave amplitude with respect to the diffraction angle χ (between the normal P_0Q and the direction QP); it is assumed to be maximum for the forward direction $\chi = 0$ and rapidly decreasing for increasing χ. By introducing the inclination factor, Fresnel achieved two goals: 1. he was able to explain rectilinear wave propagation (along P_0CP) by Huygens' principle and 2. he was able to exclude the backward propagating wave which as we saw was part of Huygens' principle in its original form (Fig. 4.17 in Chap. 4).

Fresnel was thus able to correctly predict the propagation of a spherical wave in free space, using the combination of Huygens' principle, Young's interference and the mathematical expression (7.2), which since then is usually referred to as the *Huygens-Fresnel principle.* However, the assumption of the inclination factor was still purely heuristic and not based on physical principles. Fresnel correctly concluded that K has a maximum value at $\chi = 0$, but incorrectly assumed that K is equal to zero for $\chi = \pi/2$. Throughout the nineteenth century the correct inclination factor for Huygens' principle would keep haunting physicists. In 1849, the Irish physicist George Gabriel Stokes (1819–1903) obtained the correct form of the inclination factor $K(\chi) = (1 + \cos\chi)/2$. Stokes' derivation still has a heuristic character matching the observations $K(0) = \max$ and $K(\pi) = 0$. Finally, Kirchhoff would give a full analytical derivation for the correct expression (see later in this chapter).

Apart from reintroducing Huygens' principle for wave propagation in general, Fresnel came up with two new insights. He was the first to realize that light behaves like a transverse wave, with its displacement being perpendicular to its direction of propagation, as opposed to its previously assumed longitudinal wave nature. In addition, he recognized that polarization, a property of transverse waves, played a crucial role in understanding Huygens' observations on the Icelandic crystal (as discussed in Chap. 4).

The word "wavelet" (French: *ondelette*), apparently first used in 1813 in a poem by Shelley, is the diminutive of the word wave; that is, a wavelet was a little wave or a ripple. Over time, however, the word wavelet was used as a synonym for the secondary wave in the statement of Huygens' principle. Thomas Young appears to have been the first to use the word wavelength ("length of an undulation") and Fresnel the first to use its symbol familiar to us, the Greek letter λ, in his mathematical exposition. When referring to wavelets, Huygens used the term *particular waves*, Fresnel used the term *elementary waves*, while Kirchhoff did not speak of elementary waves at all. As we will see in the next section, Huygens' wavelets are closely connected to Green's functions.

Green's Function

George Green (1793–1841, see Fig. 7.2) was born in Nottinghamshire, England, and lived there for most of his life. His father was a baker. Children were often expected to work for their living. Green began working daily in his father's bakery at the age of about five. In 1801 at age eight George Green went to Robert Goodacre's school. Green excelled at mathematics, but in midsummer 1802 at age nine, he left the school to work full time at his father's bakery. In 1807 his father built a brick wind corn-mill at Sneinton, just outside Nottingham. The mill stood 16 m high. It was technologically impressive for its time, but required nearly twenty-four-hour maintenance. The mill was to become George Green's burden for the next twenty years. However, Green found time to work on mathematics entirely on his own through the years that he worked at his father's mill. His place of study was on the top floor of the mill.

Despite his difficult circumstances, in 1828 Green published on his own account one of the most important mathematical works of all time. Its title was "An Essay on the Application of Mathematical Analysis to the Theories of Electricity and Magnetism." Green coined the term "potential" to denote the results obtained by adding the masses of all the particles of a system, each divided by its distance from a given point. The *Essay* begins with introductory observations emphasizing the central role of the potential function. The general properties of the potential function are subsequently developed and applied to electricity and magnetism. In the *Essay* the formula connecting surface and volume integrals, now known as Green's theorem, was introduced, as was "Green's function", the concept now extensively used in the solution of partial differential equations (see below in this chapter).

The great importance of the *Essay* was not realized by anyone at the time of its publication. Nobody with sufficient mathematical skills to appreciate its importance had seen the work. Green carried on working his mill. In 1832, at the age of nearly forty, Green was admitted as an undergraduate at Cambridge. He won the first-year mathematical prize. He graduated in 1838 as the fourth highest scoring student in his graduating class. The reason was that his self-study of Latin and Greek could not compare to the many years of schooling that the other students had undergone. The Cambridge Philosophical Society, on the basis of Green's *Essay* and three other publications, elected Green a fellow. In the next two years (1838–1840), Green published an additional six papers with applications to hydrodynamics, sound and optics. In 1840 Green became ill, and died a year later at age 47. Green's work was never recognized during his lifetime.

William Thomson (later known as Lord Kelvin) (1824–1907) was a British mathematical physicist who did important work in the mathematical analysis of electricity and in the formulation of the first and second laws of thermodynamics. In 1892, he became the first British scientist to be elevated to the House of Lords. Absolute temperatures today are stated in units of Kelvin in his honor. Only a few weeks before Green's death in 1841, William Thomson was admitted to St Peter's College, Cambridge. He noticed a reference to Green's *Essay*, but was unable to find a copy of the *Essay* until after receiving his degree in January 1845. Thomson was responsible for republishing Green's *Essay*. Green was the first to formulate a mathematical theory for electricity and magnetism. The Scottish mathematician and physicist James Clerk Maxwell (1831–1879), William Thomson and others built their theories on the groundwork laid by his theory. Green and Carl Friedrich Gauss (1777–1855), the German mathematician, geodesist and physicist who had a profound impact on mathematics, science and engineering, both worked on potential theory at the same time. Green's work on the motion of waves anticipates the WKB approximation of quantum mechanics, named after Wentzel, Kramer and Brillouin, and related the geometrical-optics solution of the wave equation. Green's efforts on light-waves and the properties of the ether produced what is now known as the Cauchy-Green tensor. Green's theorem and Green's functions are important tools in quantum mechanics, classical mechanics and in signal processing.

Let us now give a brief description of Green's function in modern terminology. We take the case in which the wavefield can be described by a linear equation. In such a representation the effect of independent inputs is additive. Thus, the effect due to a complicated input can be analyzed by expressing the complicated input as

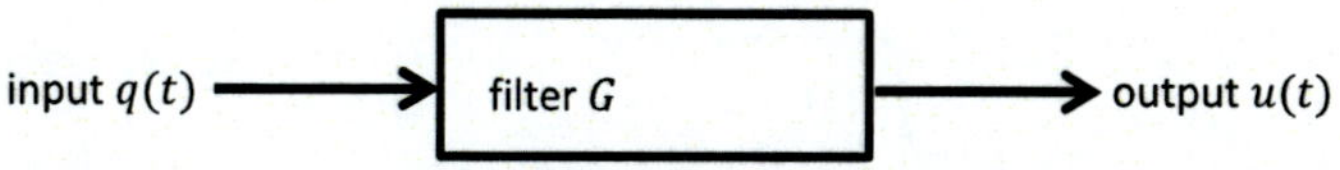

Fig. 7.7 The filter or operator G transforms the input $q(t)$ into the output $u(t)$

a superposition of simple inputs. Once the effect of the simple input is known, then it follows from the property of linear filters that the effect of the complicated input can be found. Consider the filter shown in Fig. 7.7. Let us confine our attention to time filters. The symbol t denotes time.

Let G represent a filter (or operator) that transforms the input signal $q(t)$ into the output signal $u(t)$. In symbols,

$$Gq(t) = u(t) . \tag{7.3}$$

The filter is called linear if it satisfies:

(1) the *superposition property*; namely, the input $q_1(t) + q_2(t)$ gives the output $u_1(t) + u_2(t)$.
(2) the *multiplicative property*; namely, the input $cq(t)$ gives the output $cu(t)$, where c is a constant.

A linear filter G is *time invariant* if it has the property

$$Gq(t - \tau) = u(t - \tau) . \tag{7.4}$$

In other words, if the input $q(t)$ is shifted by τ time units then the output $u(t)$ is also shifted by τ time units.

The Dirac delta function $\delta(t)$ is a function defined to be equal to zero everywhere on the t axis except at $t = 0$. At zero, the delta function is a spike that is infinitely high and infinitesimally thin. However, the total area under the spike is defined to be equal to one:

$$\int_{-\infty}^{\infty} \delta(t) \mathrm{d}t = 1 . \tag{7.5}$$

The delta function, defined by the British physicist Paul Dirac (1902–1984), is very useful in physics. Mathematically, it is difficult to give it a precise meaning, since the above properties are contradictory for normal functions. This issue was finally resolved by the French mathematician Laurent Schwartz (1915–2002) in the framework of distribution theory. The most valuable property of the delta function is its *sifting property*; namely

$$\int_{-\infty}^{\infty} q(\tau)\delta(t-\tau)\mathrm{d}\tau = q(t)\,. \tag{7.6}$$

Let us now introduce the concept of a *Green's function*. The concept applies to linear differential equations (both ordinary and partial). Let the three-dimensional wave equation be represented by the differential operator L. Let the input be q and output be u.

Thus, we have

$$Lu = q\,. \tag{7.7}$$

This equation says that the operator L acting on the output u is equal to the input q. Define G to be the inverse of L; that is,

$$G = L^{-1}\,. \tag{7.8}$$

We have

$$u = L^{-1}q \text{ or } u = Gq\,. \tag{7.9}$$

This equation says that the operator G acting on the input q is equal to the output u.

For simplicity, we will describe Green's function for a time-invariant linear filter G. Let the input be the delta function $\delta(t-\tau)$. See Fig. 7.8. The Green's function $g(t)$ is defined as the resulting output. In symbols,

$$G\delta(t) = g(t)\,. \tag{7.10}$$

Because the filter is time-invariant, we have

$$G\delta(t-\tau) = g(t-\tau)\,. \tag{7.11}$$

Since the filter G is linear, it satisfies the multiplicative property

$$G[q(\tau)\delta(t-\tau)] = q(\tau)g(t-\tau)\,. \tag{7.12}$$

Here $q(\tau)$ is an arbitrary constant, defined to be the value of a time function $q(t)$ at some fixed time τ. The superposition property says that we may integrate over all

Fig. 7.8 The filter or operator G transforms the delta function $\delta(t)$ into Green's function $g(t)$

possible values of τ. As a result we obtain

$$\int_{-\infty}^{\infty} G[q(\tau)\delta(t-\tau)\mathrm{d}\tau] = \int_{-\infty}^{\infty} q(\tau)g(t-\tau)\mathrm{d}\tau \,. \tag{7.13}$$

Because G operates on t, not τ, we can take G outside of the integral. Thus we have

$$G\left[\int_{-\infty}^{\infty} q(\tau)\delta(t-\tau)\mathrm{d}\tau\right] = \int_{-\infty}^{\infty} q(\tau)g(t-\tau)\mathrm{d}\tau \,. \tag{7.14}$$

We can apply the sifting property to the expression within the brackets. The result is

$$G[q(t)] = \int_{-\infty}^{\infty} q(\tau)g(t-\tau)\mathrm{d}\tau \,. \tag{7.15}$$

As we know, the output is

$$Gq(t) = u(t) \,. \tag{7.16}$$

The final result is

$$u(t) = \int_{-\infty}^{\infty} q(\tau)g(t-\tau)\mathrm{d}\tau \,. \tag{7.17}$$

The above Eq. (7.17) is called the *superposition integral.* It states that the output $u(t)$ of a time-invariant linear filter with input $q(t)$ is obtained by (a) finding the Green's function $g(t)$ and then (b) using the superposition integral to determine the output $u(t)$. In electrical engineering, the Green's function is called the impulse response function. In optics the Green's function is referred to as the point spread function of a spatially-varying optical filter. In physics and wave propagation, the Green's function represents the effect of a linear wave at the point r and time t resulting from an impulsive source applied at the point r_0 at time t_0. Although the idea behind the Green's function is the same, the terminology is different for different fields of science. Regardless of the area of application, the fundamental fact is that the Green's function is the response of a system to an impulsive excitation.

Green's functions are used in many areas of physics, specifically in quantum field theory, aerodynamics, aeroacoustics, electrodynamics, statistical field theory, and Born scattering theory. In quantum field theory, Green's functions take the roles of propagators. Green's functions are widely used in electrodynamics and quantum field theory, where the relevant differential operators are often difficult or impossible to solve exactly but can be solved in a perturbative manner using Green's functions. In field theory contexts the Green's function is often called the propagator or two-point correlation function since it is related to the probability of measuring a field at one point given that it is sourced at a different point. Green's functions are powerful tools for obtaining relatively simple and general solutions of basic problems such as scattering and bound-level information. The bound-level treatment gives a clear physical understanding of questions such as superconductivity, the Kondo effect, and, to a lesser degree, disorder-induced localization.

In effect, it was Christiaan Huygens who discovered this impulse response function, at least empirically, as seventeenth-century mathematics was not developed enough to write down its mathematical form. The study of wave propagation in three dimensions can be quite complicated. Huygens' contribution was to discover how to make it beautifully simple. He observed that a point source radiates a spherically symmetric wave-like disturbance, or wavelet. So, we can say that Christiaan Huygens in 1678 anticipated Green's *Essay* of 1828 by 150 years. The Huygens wavelet is a *Green's function* and the Huygens' construction is the *superposition integral.*

For our current purposes, let us consider spatial coordinates only. The Green's function is defined as the impulse response of an inhomogeneous linear differential equation with specified initial and boundary conditions. Consider

$$Lu(x) = f(x)\,, \tag{7.18}$$

where L is the linear differential operator defining the differential equation, x represents the spatial coordinate (or coordinates in more than one dimension) and f the source term ((7.18) is called *inhomogeneous* if f is non-zero). The first step is to solve

$$LG(x, y) = \delta(x - y)\,, \tag{7.19}$$

where $\delta(x - y)$ is the source impulse or delta function (in analogy to 7.5 defined by $\delta(x - y) = 0$ for $x \neq y$ and $\int_{-\infty}^{\infty} \delta(x)dx = 1$) and G is the impulse response or *Green's function.* Once the Green's function has been constructed, the solution of the inhomogeneous (7.18) can be expressed by the simple convolution integral:

$$u(x) = \int G(x, y)f(y)\mathrm{d}y\,, \tag{7.20}$$

where y is the integration variable and the integral is taken over the area where $f \neq 0$ (that is, the *support* of function f).

For the wave equation, the Green's function gives a way to express the wave response of arbitrary source terms by summation, or superposition, of the response of point sources. As a matter of fact, we are back at the Huygens-Fresnel principle. If the operator L represents the wave equation operator, for instance given by the left hand side of (7.1) from the beginning of this chapter:

$$Lu = \nabla^2 u - \frac{1}{c^2}\frac{\partial^2 u}{\partial t^2} \tag{7.21}$$

and point sources with strength f are located at y, then the wavefield recorded at x is given by the above convolution integral (7.20). The Green's function G describes the propagation from the point sources to the recording point. It encapsulates all complexities of wave propagation underway, making the resulting integral looking deceivingly simple.

The next contribution of Green came to be known as Green's (Second) Identity, which relates a volume integral to a surface integral. For any two integrable functions $u(\mathbf{x})$ and $v(\mathbf{x})$ defined in three-dimensional space (bold $\mathbf{x}$ denoting location in three-dimensional space), it gives the following relation between an integral over a volume V and an integral over its surface S (see Fig. 7.10a):

$$\int_V (u\nabla^2 v - v\nabla^2 u)\mathrm{d}V = \int_S (u\frac{\partial v}{\partial n} - v\frac{\partial u}{\partial n})\mathrm{d}S\,, \tag{7.22}$$

where $\nabla^2 = \frac{\partial^2}{\partial x^2} + \frac{\partial^2}{\partial y^2} + \frac{\partial^2}{\partial z^2}$ and $\partial/\partial n = \mathbf{n}\cdot\nabla$ is the normal derivative to the surface S (the dot denotes the vector dot product, $\mathbf{n}$ denotes the normal vector to the surface). Note that Green's Second Identity is reminiscent of integration by parts, and is in fact a higher dimensional version of it:

$$\int_a^b (fg'' - f''g)\mathrm{d}x = (fg' - f'g)\big|_a^b\,. \tag{7.23}$$

The left-hand side of (7) corresponds to the integration over the interval $[a, b]$, the right-hand side over its boundaries. As we will see, both the Green's function and Green's Identity were corner stones for the further development of wave theory.

Further Development—Helmholtz

The next step was made by the German physicist Hermann von Helmholtz (1821–1894, see Fig. 7.9). In 1858, Helmholtz published *On air vibrations in pipes with open ends*. Starting from Green's results, he developed the Helmholtz-Kirchhoff integral for monochromatic waves (waves with only a single frequency). In what follows we briefly summarize the exposition of Baker and Copson (1939). Helmholtz started with the wave (7.1), considered the monochromatic wave $u = v\exp[-ikct]$ (where k is the wave number and c the wave velocity, both taken to be constant) and arrived at what is now called the *Helmholtz equation*:

$$(\nabla^2 + k^2)v = 0\,. \tag{7.24}$$

Fig. 7.9 Hermann von Helmholtz and Gustav Kirchhoff

Here the time t has disappeared. Following Green's recipe (7.20) and taking a delta function $\delta(\mathbf{r}' - \mathbf{r})$ as the source term and right-hand side of (7.24), Helmholtz then derived the Green's function in three dimensions for the monochromatic wave equation:

$$G\left(\mathbf{r}, \mathbf{r}'\right) = G(r) = -\frac{1}{4\pi}\frac{\exp ikr}{r} \,. \tag{7.25}$$

The physical meaning of $G(\mathbf{r}, \mathbf{r}')$ is the wave response at a location $\mathbf{r}$ due to a monochromatic wave generated at a point source at location $\mathbf{r}'$, located at a distance $r = \left\|\mathbf{r} - \mathbf{r}'\right\|$. Note that (7.25) has a singularity for $\mathbf{r} = \mathbf{r}'$ but upon integrating the Green's function around this singularity over a small sphere containing $\mathbf{r}$, the result remains finite.

An interesting property of this Green's function is its *reciprocity* whereby the locations $\mathbf{r}$ and $\mathbf{r}'$ can be interchanged, corresponding to source and measurement location being swapped. If I can see your eyes, you can see mine too; if I can hear you, you can hear me too.

Helmholtz proceeded by noting that if two arbitrary functions v and w both satisfy the monochromatic wave (7.24), then the left-hand side of Green's identity (7.22) vanishes so that it becomes

$$\int_S \left(w\frac{\partial v}{\partial n} - v\frac{\partial w}{\partial n}\right)\mathrm{d}S = 0 \,. \tag{7.26}$$

In particular we can substitute the Green's function for one of them, say $w = \exp ikr/r$, where r is the distance from a fixed point P. We consider the situations where P is inside or outside S separately. If P is located outside S (Fig. 7.10b), we have:

$$\int_S \left(\frac{\exp ikr}{r}\frac{\partial v}{\partial n} - v\frac{\partial}{\partial n}\left(\frac{\exp ikr}{r}\right)\right)\mathrm{d}S = 0 \,. \tag{7.27}$$

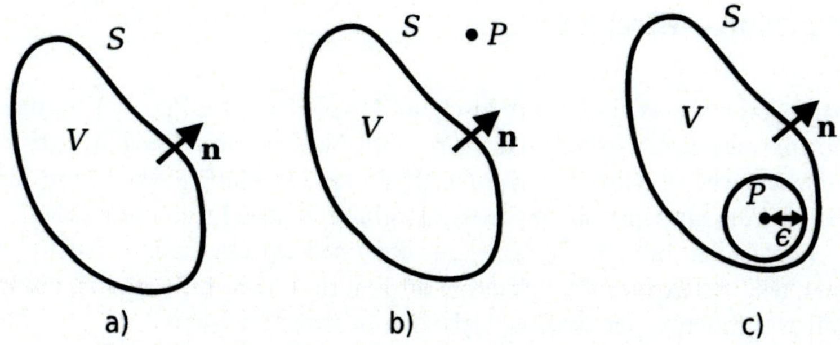

Fig. 7.10 **a** Green's identity (see (7.22)): the integral of a function over the volume V can be replaced by an integral over the surface S. **b**, **c** Helmholtz' theorem, illustrating the terms in (7.27) and (7.28)

On the other hand, if P is inside S, the surface S is complemented by a small sphere of radius ϵ surrounding P (Fig. 7.10c), and we have:

$$\int_S \left(\frac{\exp ikr}{r} \frac{\partial v}{\partial n} - v \frac{\partial}{\partial n} \left(\frac{\exp ikr}{r} \right) \right) \mathrm{d}S + \int_\epsilon \left(\frac{\exp ikr}{r} \frac{\partial v}{\partial n} - v \frac{\partial}{\partial n} \left(\frac{\exp ikr}{r} \right) \right) \mathrm{d}S = 0. \tag{7.28}$$

Take the limit $\epsilon \to 0$, then the second term of (7.28) becomes $-4\pi v(P)$. Therefore

$$v(P) = \frac{1}{4\pi} \int_S \left(\frac{\exp ikr}{r} \frac{\partial v}{\partial n} - v \frac{\partial}{\partial n} \left(\frac{\exp ikr}{r} \right) \right) \mathrm{d}S\,, \tag{7.29}$$

if P is inside S, and $v(P) = 0$ if P is outside S.

This statement is referred to as Helmholtz' theorem. Its implications are first of all that we have an expression of the wavefield at any point inside a closed volume in terms of an integral over its surface. This expression contains the wavefield and its derivative only at the surface S and the Green's function to propagate it to the point P. The expression (7.29) is *exact* for a homogeneous, isotropic medium, there are no assumptions or approximations involved. For more general, inhomogeneous isotropic or anisotropic, media suitable, more general Green's functions need to be used. As Arnold Sommerfeld (1868–1951) formulated it (1949): the Green's function is a mathematical tool, in some sense a probe (German: "Sonde") with which we can investigate the wavefield. By the same token, Sommerfeld could have stated that Huygens' secondary waves can be used as probes to investigate the wavefield.

We could also say that the Green's functions embody all propagation complexities underway. Helmholtz' equation then represents an analytical form of Huygens' principle as originally envisioned by Fresnel. The secondary sources are physical in the sense that their amplitude and phase contribute to the wavefield.

Completion—Kirchhoff

With the German physicist Gustav Kirchhoff (1824–1887, see Fig. 7.9), the mathematization of wave theory, sketched in the previous section and starting in the eighteenth century, achieved its culmination point. Rigor was becoming the key word. To quote Kirchhoff from his paper on the theory of light rays (see Fig. 7.11, Kirchhoff, 1882): "The conclusions, through which one, mostly based on examinations by Huygens and Fresnel, tends to explain the formation of light rays, their reflection and refraction, as well as diffraction phenomena, lack rigor in several respects."

A first step forward to completion of wave theory was the generalization of Helmholtz' theorem for polychromatic waves (waves of arbitrary wave form). To accomplish this, Kirchhoff first derived the Green's function for the three-dimensional wave (7.1):

$$G(r, t) = -\frac{1}{4\pi r}\delta\left(t - \frac{r}{c}\right), \tag{7.30}$$

where c is the propagation velocity and $r = \|\mathbf{r} - \mathbf{r}'\|$. The physical meaning of $G(r, t)$ is the wave response at a location $\mathbf{r}$ due to a delta function generated at a point source at location $\mathbf{r}'$. This is simply a delta function describing the wavefront spreading out in time, and a geometrical spreading factor $1/r$. Because of the argument $t - r/c$, the Green's function is called "retarded". In other words, the solution is non-zero only at times $t = r/c$ exactly, which is an implication of Huygens' Principle (in odd dimensions, that is, see section on Hadamards's syllogism later in this Chapter).

Next, using a Fourier expansion argument, Kirchhoff considered general (polychromatic) solutions to the wave equation

3. Zur Theorie der Lichtstrahlen;
von G. Kirchhoff.

Sitzber. d. k. Acad. d. Wissensch. zu Berlin vom 22. Juni 1882. p. 641.
Wied. Ann. Bd. 18. p. 663. 1883.

Die Schlüsse, durch welche man, hauptsächlich gestützt auf Betrachtungen von **Huygens** und **Fresnel**, die Bildung der Lichtstrahlen, ihre Reflexion und Brechung, sowie die Beugungserscheinungen zu erklären pflegt, entbehren in mehrfacher Beziehung der Strenge. Eine vollkommen befriedigende Theorie dieser Gegenstände aus den Hypothesen der Undulationstheorie zu entwickeln, scheint auch heute noch nicht möglich zu sein; doch lässt sich jenen Schlüssen eine grössere Schärfe geben. Ich erlaube mir, der Academie Auseinander-

Fig. 7.11 Introduction to Kirchhoff's paper (1883)

$$\nabla^2 V = \frac{1}{c^2}\frac{\partial^2 V}{\partial t^2}\,, \tag{7.31}$$

where $V = V(x, y, z, t)$ is a general wave function of coordinates (x, y, z) and time t. V can be represented in the form of a Fourier integral:

$$V(x, y, z, t) = \int_{-\infty}^{+\infty} U_\omega(x, y, z)\exp - \mathrm{i}\omega t \mathrm{d}t\,, \tag{7.32}$$

where $U_\omega(x, y, z)$ satisfies the (monochromatic) Helmholtz (7.24) for frequency ω. Therefore, Helmholtz' theorem can be applied to each Fourier component $U_\omega(x, y, z)$ separately:

$$U_\omega(P) = \frac{1}{4\pi}\int_S \left(\frac{\exp ikr}{r}\frac{\partial U_\omega}{\partial n} - U_\omega\frac{\partial}{\partial n}\left(\frac{\exp ikr}{r}\right)\right)\mathrm{d}S \tag{7.33}$$

Inserting (7.33) in (7.32) results in Kirchhoff's integral formula:

$$V(P, t) = \frac{1}{4\pi}\int_S \left([V]\frac{\partial}{\partial n}\frac{1}{r} - \frac{1}{cr}\frac{\partial r}{\partial n}\left[\frac{\partial V}{\partial t}\right] - \frac{1}{r}\left[\frac{\partial V}{\partial n}\right]\right)\mathrm{d}S\,, \tag{7.34}$$

if P is inside S, and $V(P) = 0$ if P is outside S, as in Helmholtz' solution of he previous section. The square brackets [] are a notation for retarded values, that is, the values of the function taken at the time $t\text{–}r/c$.

Let us make a few notes to Kirchhoff's integral formula (7.34). First of all, the integral formula is exact (as is its monochromatic counterpart the Helmholtz formula (7.29)). There are no assumptions or approximations involved beyond those of the wave equation itself. The formula is valid for a homogeneous medium; for general inhomogeneous anisotropic media the constant-medium Green's function needs to be generalized. The input wavefield at the surface has an arbitrary time dependence. We can therefore model wavelets of any shape and duration, in particular short pulses associated with transient waves. Fresnel and Helmholtz, on the other hand, only worked with monochromatic waves of infinite duration.

Second, a comparison of Kirchhoff's integral formula (7.34) and Fresnel's heuristic integral expression (7.2) shows that the laws defining the contributions from different elements of the surface are considerably more complicated than Fresnel initially assumed. An interpretation of the individual terms in (7.34) involves an intricate interplay of point sources and doublets (see Baker and Copson (1939, Sect. I.5.1) and Born and Wolf (1999, Sect. 8.3.1) for details). In the end, these terms also accounted for the correct inclination factor, which had been chased during the best part of the nineteenth century.

Third, since it is an integral formula, Kirchhoff's formula again embodies the basic ideas of Huygens' principle: the wavefield at an advanced position can be expressed as a superposition (integral) of the wavefield at a previous position. In

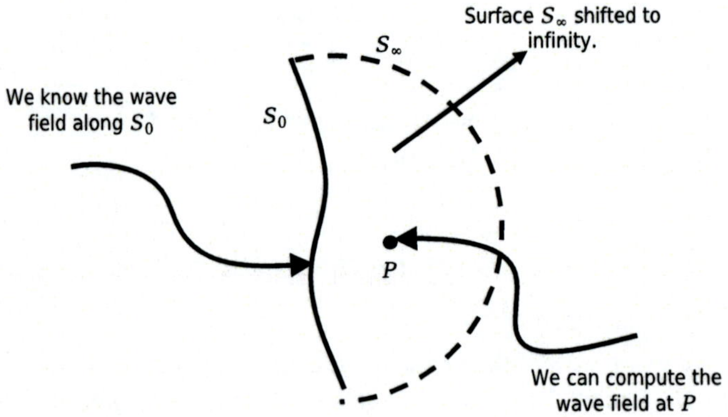

Fig. 7.12 Huygens-Kirchhoff integral

accomplishing this, an ingenious construction of the surface S is needed, consisting of a part where the wavefield is known and another part which is extended to infinity so that the integral over it can be safely assumed to be zero (see Fig. 7.12). The surface S is composed of two parts: $S = S_0 + S_\infty$, where S_0 is the surface along which the initial wavefield is known, and S_∞ is an artificial surface which is shifted to infinity so that the wavefield is zero along it and does not contribute to the Huygens-Kirchhoff integral (Sommerfeld's radiation condition). For any point P inside this construct, the wavefield can therefore be computed from the initial wavefield using the Huygens-Kirchhoff integral.

Modern Anisotropy

In Chap. 4 we have seen how Christiaan Huygens studied the light propagation through Icelandic crystal and was able to explain the double refraction observed there by considering anisotropy. The speed of light depended on its propagation direction, so that the wavefront was no longer spherical but ellipsoidal. Huygens was able to invoke his (Huygens') principle to explain the effect.

Nowadays, anisotropy is studied in many areas of science, such as material science and engineering, medical acoustics, fiber optics and optical communication, non-destructive testing, seismology and seismic exploration. Broadly speaking, there are two main classes of wave propagation involving anisotropy: electromagnetic and elastodynamic waves.

Huygens' studies were concerned with anisotropy of electromagnetic (light) waves in solid materials. In this case Huygens' principle is used to construct solutions to Maxwell's equations, a set of coupled linear partial differential equations that govern electromagnetic wave propagation.

The theory of elastic wave propagation dates mainly from the nineteenth century, where George Green (1838) and Lord Kelvin (1856) made significant contributions. Anisotropy in seismology was first studied by the Polish geophysicist Maurycy Rudzki (1862–1916). In this case the waves are governed by the elastodynamic wave equations, a set of coupled linear partial differential equations.

As in Huygens' studies of the Icelandic crystal, the medium in which the waves are propagating needs to be characterized. The wave anisotropy is caused by material properties being aligned in preferential directions. For instance, in seismic wave propagation in the Earth, the medium is defined by the geology, where fine layering and fractures can cause anisotropy. A whole range of different types of anisotropy can be identified, which goes by names such as triclinic, orthorhombic, hexagonal anisotropy (Helbig, 1994; Červený, 2001).

As we have seen in Chap. 4, in Huygens' studies the anisotropy could be described by an ordinary ray (labeled *o*) and an extraordinary ray (labeled *e*). In modern anisotropy, the wave structure is considerably richer. General anisotropy is described by a tensor with 21 independent elements, which need to be specified in every point of the three-dimensional space.

Despite this complexity it has been established that elastodynamic and electromagnetic anisotropy are governed by equivalent wave equations (Ikelle, 2012). In fact, due to intrinsic symmetries, the electromagnetic case appears to be simpler than the elastodynamic case. In its most general form, elastodynamic wave propagation is governed by 18 coupled equations, but electromagnetic wave propagation only by 12 equations. Mathematically, we can consider the electromagnetic case as a special case of the elastodynamic case. As a result, the same solutions techniques can be used for the modeling of waves.

In particular, Huygens' principle can be invoked. In all cases described above, we have a wave equation of the type presented earlier in this Chapter (7.18): $Lu(x) = f(x)$. The main task is to construct the Green's functions (7.19). This can be a daunting task, and admittedly we step here over some sometimes formidable difficulties. However, once done, the Green's functions can be harnessed in Kirchhoff's integral to give the solution of the wave equation and hence model the wave propagation. Kirchhoff's integral is the complete wave-based formulation of Huygens' principle.

Huygens' principle does not explicitly consider a medium. All that Huygens' principle needs is a wave equation to solve.

Hadamard's Syllogism

In the late 19th and early twentieth centuries, physicists and mathematicians took a keen interest in understanding the mathematical foundations of wave propagation. The fundamental questions included how waves propagate through different media and how initial and boundary conditions would affect the evolution of wave solutions. Although the work by Fresnel, Green, Helmholtz and Kirchhoff had resulted in a

Fig. 7.13 Jacques Hadamard and Richard Feynman

complete and closed-form of Huygens' principle, its original formulation by Huygens was merely qualitative and there was still a need for a more rigorous framework.

The French mathematician Jacques Hadamard (1865–1963, see Fig. 7.13) contributed to this effort by focusing on proving under which conditions Huygens' principle holds for solutions of wave equations. In his lectures on Cauchy's problem in linear partial differential equations at Yale University in 1923, he analyzed Huygens' principle in the form of a syllogism.

A syllogism (from Greek "logical conclusion") is a construct in mathematical logic dating back to Aristotle, consisting of two premises—a major premise and minor premise—followed by a conclusion. A well-known syllogism is "All men are mortal" (major premise), "Socrates is a man" (minor premise) and "Socrates is mortal" (conclusion). Hadamard's syllogism for Huygens' principle, then, runs as follows (Hadamard, 1924):

A. Major premise: To deduce from a phenomenon known at time t_0 the effect produced at a later time t_2, one can start by calculating the effect at an intermediate time t_1 then from this one to deduce the effect in t_2.
B. Minor premise: If the initial disturbance at time t_0 is located in the vicinity of a point O, its effect at time t_1 will be zero everywhere, except in the vicinity of a sphere S_1 with center O and radius $c(t_1–t_0)$ denoting by c the speed of propagation.
C. Conclusion: The initial disturbance can, from the point of view of its effect at the final instant t_2, be replaced by a system of disturbances taking place at the intermediate instant t_1 and suitably distributed over the surface of the sphere S_1.

In summary, the premises A and B imply the conclusion C. Here, the major premise A may be referred to as a step-by-step description of the propagation from one wavefront to a later wavefront (noting $t_0 < t_1 < t_2$). This statement seems obvious (Baker and Copson, 1939), and Hadamard himself characterized it as a truism, but it embodies the principles of determinism and causality, common in classical physics.

The "conclusion" C formulates the computational aspect of Huygens' principle, which was the main focus of the work of Fresnel, Helmholtz and Kirchhoff.

The minor premise B turns out to be the tricky one, as it describes the locality of the wave propagation. The monochromatic (single-frequency) waves in Helmholtz' theorem do not obey the locality principle. Kirchhoff's generalization to polychromatic waves (waves of arbitrary wave form) do satisfy the locality principle, but this depends on the dimensionality of the space under consideration.

Indeed, Hadamard formulated and proved a theorem which says that the minor premise does not hold for linear partial differential equations with an even number of spatial variables. As noted at the end of Chap. 4, this implies that Huygens' principle is valid only in odd-dimensional spaces. Two-dimensional wave propagation, such as surface waves on a water surface or surface waves in seismology do not follow Huygens' principle in its simplest form, as they leave a trail behind them which violates the locality principle (premise B). Hadamard's "Method of Descent" offers a solution by adding an auxiliary spatial parameter and regarding the even-dimensional wave problem as a specialization of an odd-dimensional one for which Huygens' principle is valid (Baker and Copson, 1939).

Hadamard went on to ask for which equations Huygens' principle is true and conjectured that it is an exclusive principle for wave equations. Are there any other equations than the wave equation, in other words are there any other physical processes than waves, which are governed by Huygens' principle? Hadamard believed not. However, this conjecture was refuted by the German mathematician Karl-Ludwig Stellmacher (1909–2001) who constructed counterexamples (Günther, 1988).

Huygens' Principle Today

Huygens' principle celebrated its 300th birthday in 1990 with a symposium in his hometown The Hague. We quote here a number of observations from the proceedings. Following the work of mathematicians Hadamard (1923, see Fig. 7.13) and Riesz (1948), Huygens' principle plays a central role in linear partial differential equations, oscillatory integrals and Fourier integral operators, even more so than Hadamard and Riesz themselves seem to have realized (Duistermaat, 1990). And after the discovery of the quantum nature of light and the recognition of its dual wave-particle character in the early twentieth century, Huygens' principle has found an application in quantum mechanics, as was testified by Feynman (1948, see Fig. 7.13):

> Actually Huygens' principle is not correct in optics. It is replaced by Kirchhoff's modification which requires that both the amplitude and its derivative must be known on the adjacent surface. This is a consequence of the fact that the wave equation in optics is second order in time. The wave equation in quantum mechanics is first order in the time; therefore, Huygens' principle is correct for matter waves, action replacing time.

The historical implications of the *Traité de la Lumière* and Huygens' principle are not to be underestimated. It has been, and still is, the subject of a vast amount of publications. In the wave-particle debate of light propagation, Huygens stood firmly on the side of wave theory, and his work paved the way for the almost total acceptance of the wave theory of light in the nineteenth century. As stated by Ernst Mach in 1913 and quoted by Emil Wolf (1990):

> In Huygens' Traité de la Lumière appears the first attempt to correlate and render consistent the different properties of light which had from time to time been discovered, and to explain some by others, so as to reduce the number of necessary fundamental properties and the number of fundamental concepts indispensable for a correct comprehension of the underlying processes. The principal fruits of Huygens' labors lay in his demonstration of the possibility of deriving all the essential features of rectilinear propagation, reflection, and simple and double refraction from the rate of propagation of light.

Chapter 8
Huygens and Special Relativity

> Formerly, people thought that if matter disappeared from the Universe, space and time would remain. Relativity declares that space and time would disappear with matter.
> —*Albert Einstein*

By the turn of the 19th to the twentieth century the following contradictory facts had been established:

1. The speed of light was found to be universally constant, regardless of the motion of the light source and regardless of the motion of the observer who observes the light. This was the outcome of the famous Michelson-Morley experiment.
2. Maxwell's equations for electrodynamics, which describe the behavior of electromagnetic waves, including light, had been found to already account for the constancy of the speed of light. Maxwell's equations remain the same in any inertial coordinate frame, if the frames transform according to the rule of Lorentz. This was discovered by Lorentz and we can say that Maxwell's equations are Lorentz invariant.
3. However, Newton's laws of mechanics, which describe the motion of objects, do *not* account for the constancy of the speed of light. In Newtonian mechanics, velocities can be added up without limits, even exceeding the speed of light. In other words, Newton's laws are not Lorentz invariant. Instead, they are Galilei invariant.

To solve this predicament, Albert Einstein devised the special theory of relativity, which reconciles the principles of mechanics with the constancy of the speed of light. The theory introduces concepts such as time dilation, length contraction, and the relativity of simultaneity.

Below we explain these concepts. As we will point out, Christiaan Huygens' work contained elements that can be considered precursors to certain aspects of the theory of relativity.

T. J. Moser and E. A. Robinson, *Walking with Christiaan Huygens*, History of Physics,
https://doi.org/10.1007/978-3-031-46158-3_8

Symmetry of special relativity

Einstein's special theory of relativity, proposed in 1905, is based on two postulates:

1. The laws of physics are identical in all inertial reference frames (that is, frames experiencing no external net force).
2. The speed of light in vacuum is the same for all observers, regardless of the motion of the light source or observer.

The first postulate (the *Relativity Principle*) states that there is no conceivable physical experiment to distinguish one inertial frame from another. The second postulate (the *Light Principle*) is the experimental fact of the constancy of the speed of light, which Einstein boldly elevated to a postulate. This turned out to be a brilliant move. Even although the second postulate is counterintuitive, all experimental facts since then up until now have confirmed the accuracy of Einstein's special theory of relativity.

This theory of Einstein is called "special" because to applies only in special circumstances, the inertial frames of reference discussed below. His more broadly applicable theory, involving non-inertial frames, was developed 10 years later, and is called the general theory of relativity. We will make some references to the general theory in Chap. 9, and focus here on the special theory, and so will occasionally omit the qualifier "special".

Einstein's theory of relativity is often regarded as challenging to comprehend. However, the difficulty is not so much in Einstein's theory, as in the experimental facts that presented themselves before he formulated his theory. By the end of the nineteenth century, the empirical evidence supporting the theory had become increasingly apparent, despite its bewildering nature.

It should be kept in mind, indeed, that the constancy of speed of light was originally only an experimental fact—but one which posed considerable problems for physicists at the end of the nineteenth century to reconcile with Newtonian mechanics, or even resulted in denialism. By analogy, German philosopher Georg Wilhelm Friedrich Hegel (1770–1831) built a remarkable philosophical system, called absolute idealism, which resembled a piece of architecture and with which he claimed to be able to predict long-term historical developments. If an historical event would not fit in his system, he supposedly reacted by saying "the worse for the facts". To bring in Huygens, as we have stated in Chap. 1, he was the complete opposite. Huygens preferred working with conjectures rather than theories. If an experimental fact did not fit in his conjectures, he had not the slightest hesitation about abandoning one for a better one. He could have said "the worse for our theory".

Special relativity deals with an important class of reference frames, the inertial frame. These frames are either non-moving (that is, stationary) or else moving at a constant velocity in a straight line. In other words, an inertial frame is neither accelerating nor decelerating. Rotation is a form of acceleration because the velocity changes direction in rotation. Thus an inertial frame is not rotating. Let us consider two inertial frames. Special relativity is only concerned with the relative velocity v

between the two frames. In other words, special relativity is only concerned with the speed at which two frames are either moving apart (case of positive velocity v) or else moving together (case of negative velocity v).

Let us make this more precise. According to the principle of relativity, all laws of physics apply without change in inertial frames moving at constant speed relative to each other. As a result of this principle, it is impossible to specify the speed of a frame without referring to another frame in which the speed is measured. This is due to the fact that motion is always relative and is determined by the observer's reference point. Therefore, to avoid a circular definition, an inertial frame is defined strictly as one in which Newton's law of inertia holds true. Christiaan Huygens wrote: "*Dicent, an igitur si unicum quodpiam corpus in toto immenso spatio mundano existat, deus ipse id movere non poterit motu progressivo.*" (It is said that if a single body should exist in all infinite space, God himself would not even be able to impart progressive motion to that body, quoted by Schouten, 1920).

Huygens, throughout his work, made occasional references to the principle of relativity. The concept of the *fixed stars* has been very popular during many centuries since antiquity as an absolute reference frame for celestial movements, such as of planets and moons in our Solar system. In one way or another, the fixed stars were used by Ptolemy, Plato, Aristotle, Nicolaus Copernicus, Tycho Brahe, Johannes Kepler, Isaac Newton and many others. Newton wrote, in *Philosophiae Naturalis Principia Mathematica* (1689):

> We may distinguish rest and motion, absolute and relative, one from the other by their properties, causes, and effects. It is a property of rest, that bodies really at rest do rest in respect to one another. And therefore as it is possible, that in the remote regions of the fixed stars, or perhaps far beyond them, there may be some body absolutely at rest.

In *Considérations sur la Forme de la Terre* (1686–1687) (Considerations on the form of the Earth), Huygens discussed the flattening of the Earth (at the poles compared to the equator) and concluded that this is caused by the Earth's rotation about its axis. However, and this is crucial, for Huygens this rotation is *not* described with reference to the fixed stars, but it is the pure and simple rotation of a single body. The same flattening would have occurred if the Earth was the only body in the universe. No absolute frame is needed. The rotation is noticeable by the centrifugal force. Of course, progress in astronomical insight since the seventeenth century has shown that the fixed stars are anything but fixed. First there is the annual parallax, due to the rotation of the Earth around the Sun, which causes the stars to make an apparent annual orbit. Second, the stars and galaxies are in perpetual real motion relative to each other and to our Solar System. This makes it not only unnecessary, but even fundamentally impossible to define an absolute frame. We discuss Huygens' work on the centrifugal force and its implications for the general theory of relativity in Chap. 9, as well as his work on the flattening of the Earth outside the context of relativity in Chap. 12; in this Chapter, we focus on some contributions of Huygens that point to special relativity.

Because of the constancy of the speed of light, Einstein showed that 3D space and 1D time are coupled together, and must be thought of together, in a construct called "space–time". Our common-sense distinction between space and time, as separate concepts, cannot be valid, if the speed of light is universally constant. We discuss this next.

Special relativity is characterized by an invariant quantity in the realm of space–time. It is a quantity called "proper time." For example, elapsed proper time p for an ordinary mechanical clock is given by the number of rotations of the hands of the clock. For example, for a certain event, p might be 50 rotations. Alternatively, we might take a gyroscope, or a freely spinning wheel, and measure the number of rotations. Proper time is independent of any system of coordinates and it is invariant under Lorentz transformations (discussed later in this chapter). It was introduced by the Dutch physicist Hendrik Antoon Lorentz (1853–1928) and further developed by the German physicist Hermann Minkowski (1864–1909). There is no motion associated with proper time. Since we stand on the solid Earth and to us the Earth is not moving. But the Earth spins on its axis once in every 24 h day. The number of days elapsed serves as elapsed proper time on Earth. We will return to proper time shortly.

As discussed in Chap. 3, Huygens was the first person to determine the speed of light. The velocity of light in a vacuum is about 300,000 km per second (we will use this round number below for convenience and reserve the letter c for the speed of light). A light-second is defined as that distance which light travels in one second of time, and so it is equal to about 300 thousand kilometers. A more practical unit in astronomy is the light year. A light year is the distance that light travels in a year. A light year is equal to $300000 \times 365 \times 24 \times 60 \times 60$ kilometers, or approximately 9.5 trillion km ($= 9.5 \times 10^{12}$ km). The star Vega is about 27 light-years away and is one of the stars closest to us.

Natural units are defined in terms of fundamental physical constants. One such fundamental physical quantity is the velocity of light in vacuum, which is experimentally found to be the same constant at all places, in all directions, and at all times. Let us discuss the relationship between a quantity measured in natural units and the same quantity measured in conventional units. Let capital letters denote quantities in conventional units and the corresponding small letters denote the same quantities in natural units. The velocity of light in vacuum is about 300 million meters per second; in natural units it is 1. In conventional metric units, distance is given in meters, which we may denote as capital X. Also in conventional units, time is given in seconds, say capital T, and velocity, capital V, is given in meters per second. The equations used to convert conventional units to natural units are:

$$x = \frac{X}{c} = \frac{X}{300\,000\,000 \text{ m/s}}, \qquad t = T, \quad v = \frac{V}{c} = \frac{V}{300\,000\,000 \text{ m/s}}. \tag{8.1}$$

With these definitions in terms of natural units, both x and t have the same unit, the second, and velocity v is a dimensionless number.

The use of the velocity of light in defining the unit of distance has the virtue of simplifying the connection between space and time. Any message sent by light or any other electromagnetic signal from Vega to Earth will take about 27 years (again we use the round number for convenience). We see that the use of light-years implies that the communication time and the distance are the same number. For example, from Earth to Vega, the time of communication by an electromagnetic signal is 27 years, and the distance is also 27 light-years. For a light signal, time and distance are numerically equal; for example, 27 years and 27 light-years are numerically the same. However, depending upon the convention used for the directions of the t and x axes, the values of t or x might be either positive or negative. Therefore, for a light signal, instead of writing $t = x$, we write

$$t^2 = x^2 . \tag{8.2}$$

Any material body travels at less than the speed of light, so $t > x$ and the proper time–space interval is

$$\text{time} - \text{space interval} = p = \sqrt{t^2 - x^2} > 0 , \tag{8.3}$$

for any event of a moving material body. However, for any electromagnetic signal, such as light, the time–space interval is equal to zero. This equality for light is a central concept in the special theory of relativity.

The fact that in the special theory of relativity time and space are no longer absolute, but speed of light is, has important implications for the geometry of space–time diagrams.

Newton based his theory of classical mechanics on the concepts of absolute time and absolute space. In his time, these concepts were already available: the notion of absolute time can be traced back to Aristotle; the notion of absolute space was considered by various philosophers, including Descartes. See Fig. 8.1, referring to the elapsed time and distance for an event of a moving material body. As before, we will express both time in seconds and distance in natural units. In this way the velocity of light is unity. We can say that Newton puts time on one leg and distance on the other leg of a right triangle. Velocity equals the slope of the hypotenuse and hence can have any value. Newton allows a physical object to have a velocity greater than one (velocities can add up without any limits).

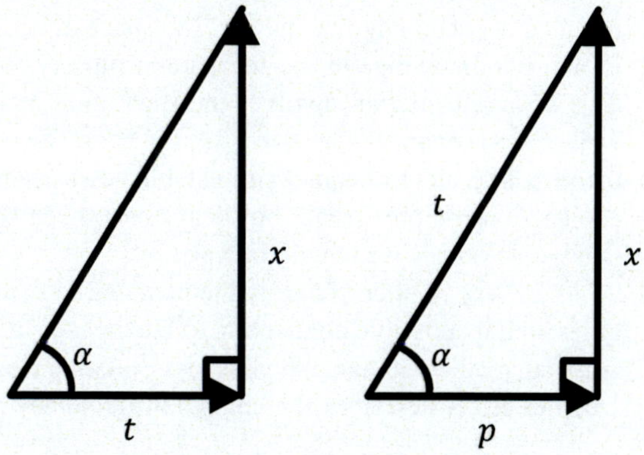

Fig. 8.1 Newton's triangle (left) and Einstein's triangle (right). For Newton, velocity is the ratio of absolute distance and time, and is unbounded (it is the slope of the hypotenuse, and equals tan α). For Einstein, time is put on the hypotenuse and therefore velocity is always less than one (it equals sin α

However, the theory of relativity postulates that the speed of light represents the maximum velocity that can be achieved in nature (velocities can add up only to a number less than or equal to one; in the latter case one of them is the speed of light c). Thus, the magnitude of the velocity of any physical object must be less than one. Such a situation is impossible under Newtonian mechanics. As a result, Einstein demands that the Newtonian concepts of absolute space and absolute time must be abandoned. What can be done? Einstein gave this solution: if one universal coordinate system is impossible, it follows that *each and every inertial frame must have its own coordinate system.*

The important concept is that there is a dual nature between space and time. Newton took time as absolute and space as absolute. Einstein rejects this categorization and instead takes the speed of light as an absolute constant. Einstein supersedes the ideas of absolute time and absolute space by the notion of the combination called "space–time".

How do we come to grips with this strange idea?

Special relativity may be envisaged using the triangle on the right in Fig. 8.1. The t coordinate (the elapsed time of a certain event, a moving material body) is the hypotenuse of a right triangle and the x coordinate (the distance covered in the same event) is the leg of the same right triangle (recall that in natural coordinates, x has the same dimension as t). A leg of a right triangle is always less than the hypotenuse of that triangle. As a result, the x coordinate is always less than the t coordinate. Therefore, the velocity, which is the ratio of distance over time, is always less than one. Recall we are using natural units, in which x has the same dimension as t (seconds) and the velocity of light is one. Einstein achieves his goal. The velocity of any material object is necessarily less than one. Einstein spares himself from the burden of Newton, in which objects can have a velocity greater than the speed of light.

It seems far-fetched, but relativity theory is nothing more than the symmetry of space–time. In space–time, the separation between two events is measured by the time–space interval between the two events, which takes into account not only the spatial separation between the events, but also their temporal separation. It is this dual nature of Einstein's system in regard to space and time that we must reconcile.

The symmetry of special relativity involves a totally new geometric concept of distance between any two events. The space–time interval p must be invariant, that is, it must remain the same as measured from any observer's inertial reference frame. This invariant interval involves the separation in time between any two events as well as their separation in space (see Eq. 8.3). The concept of space–time, along with theories like electromagnetism, establishes a framework for providing consistent explanations of observable physical phenomena within different reference frames that would otherwise remain inexplicable.

A key way to understanding the meaning of relativity theory, its symmetry properties, and its possible connection to Huygens is the Doppler effect, named after the Austrian mathematician and physicist Christian Andreas Doppler (1803–1853). The Doppler effect describes the change in frequency and wavelength of waves due to the relative motion between the source of the waves and the observer. There is a classical Doppler effect which uses Newton's absolute time and space and a relativistic Doppler effect which uses Einstein's relative time and space. The classical Doppler effect is readily observable in everyday situations, such as when a car approaches and passes by at high speed. When the car is approaching the observer, the frequency of the sound waves generated by the car appears higher than when the car is at rest, resulting in a higher pitch. However, as the car moves away from the observer, the frequency of the sound waves appears lower than when the car is at rest, resulting in a lower pitch.

The relativistic Doppler effect can be described by the following equation (Brown, 2018):

$$\frac{f_r}{f_s} = \frac{1 - v_r/c_s}{1 + v_s/c_s}\sqrt{\frac{1 - (v_s/c)^2}{1 - (v_r/c)^2}} \,. \tag{8.4}$$

Here v_s represents the speed of a source moving to the left (relative to the medium in which the wave travels), f_s the frequency of the waves that the source emits, c_s the propagation speed of the signal (a property of the medium), c the speed of light in vacuum, v_r the speed of a receiver moving to the right (relative to the same medium) and f_r the frequency as recorded by the receiver. Equation 8.4 represents the frequency shift of a wave signal (for instance, sound or light) due to the motions of the source and receiver, taking into account relativistic effects.

From Eq. 8.4 two important special cases can be derived. First, for source and receiver speeds much less than the speed of light ($v_{s,r} \ll c$) the formula simplifies to:

$$\begin{aligned} \frac{f_r}{f_s} &= \frac{1 - v_r/c_s}{1 + v_s/c_s} \quad , \text{(a)} \\ \frac{f_r}{f_s} &= \frac{1}{1 + v_s/c_s} \quad , \text{(b)} \\ \frac{f_r}{f_s} &= 1 - v_r/c_s \quad . \text{(c)} \end{aligned} \tag{8.5}$$

This case represents the classical Doppler effect for either sound or light. Equation 8.5a) represents the case where both the source and the receiver are moving (with respect to the medium), for the receiver at rest ($v_r = 0$) we have Eq. (8.5b), for the source at rest ($v_s = 0$) we have Eq. (8.5c).

Second, if the signal speed approaches the speed of light ($c_s \to c$) the formula simplifies to

$$\frac{f_r}{f_s} = k = \sqrt{\frac{1 - v/c}{1 + v/c}}, \tag{8.6}$$

where we use the symbol k for later reference in this chapter. In this case, the relativistic Doppler effect for light signals, v represents the relative speed between the source and receiver. There is no longer an absolute speed corresponding to a hypothetical universal frame. There is also no longer any reference to a medium.

Based on these expressions, we can consider the relativistic Doppler shift to be symmetric between source and receiver because it depends only on their relative mutual speed v. If the source emits a signal at frequency f_s, it will be perceived by the receiver as f_r. If they switch roles and the receiver emits a signal at frequency f_s, it will be perceived by the source as f_r. The relativistic Doppler effect takes into account the time dilation and length contraction effects predicted by the special theory of relativity, which result in symmetric changes in the observed frequency and wavelength.

On the other hand, the classical Doppler effect is not symmetric because it depends on the respective speeds of the source and receiver with respect to the medium frame. If the source and receiver are moving at different speeds ($v_s \neq v_r$), for instance if one is moving and the other is not, then switching of source and receiver roles will result in different Doppler shifts f_r/f_s. The medium as a third party ruins the symmetry. This lack of symmetry is due to the classical assumption of absolute time and space, which is inconsistent with the principles of relativity.

The time dilation between two observers, moving at a constant relative speed (along the same line), denoted by the symbol k, can be understood by considering the time of transmission of an impulsive light signal between the two observers. It can also be understood by considering a monochromatic (single-frequency) light signal. The relativistic Doppler factor is then equivalent to the receiving period at one location divided by the sending period at another location. The period of the signal in time units is the inverse of the frequency. Therefore, we express the relativistic Doppler factor by the letter k in Eq. 8.6.

To illustrate the implication of this, consider a remote sensing experiment, with one observer (A) positioned on an inertial frame at rest, and another observer (B) positioned on an inertial frame which coincides with A's frame at time 0 and then moves away along a straight line at a constant speed v. For easy arithmetic, we assume that $v = 0.6$ ($= 3/5$) (in natural units, so $c = 1$). Observer A sends a sinusoidal light signal starting at time 4 to observer B, who, upon receiving it, reflects it back to observer A, who receives it starting at time 16. We ask, what is the Doppler shift in this example? And we answer that question from both viewpoints, classical Newton and relativistic Einstein.

From Newton's viewpoint, which corresponds to our everyday common sense, there is only one universal time. See Fig. 8.2 (top). Obviously, the reflection time is halfway between the sending time and the receiving time, that is, at $t = (4+16)/2 = 10$, the arithmetic mean of the sending and receiving times. The classical Doppler shift for observer B (with the velocity assumed above), follows from Eq. 8.5c and is $f_r/f_s = 0.4$ ($= 2/5$). This signal is then reflected back to observer A, where the roles of source and receiver are switched, so that the frequency shift on the way back is given by Eq. 8.5b: $f_r/f_s = 1/1.6$ ($= 5/8$). Therefore the frequency eventually received by A is shifted by $0.4 \times 1/1.6 = 0.25$ ($= 1/4$).

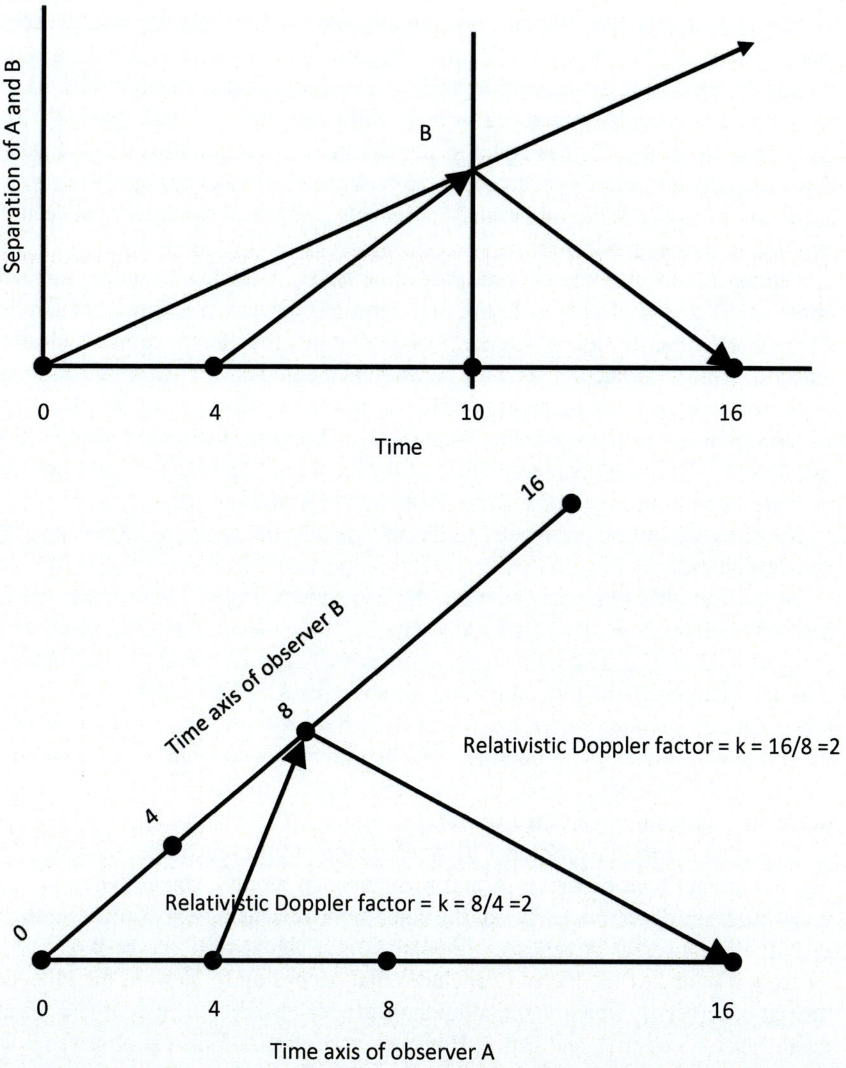

Fig. 8.2 Top: Remote sensing experiment according to Newton: there is one universal time axis. Bottom: same experiment according to Einstein: there are two time axes, one for each observer, and the two relativistic Doppler factors are the same, as required

However, according to Einstein, this analysis is over-simplified since it does not recognize the relativistic dilation of time caused by the relative motion of the two observers. Hence, Fig. 8.2 (top) must be replaced with Fig. 8.2 (bottom), which shows a different time-axis for each observer. Of course, each observer has his own space-axis, not shown. Upon sending the signal from A to B, the time is dilated by the relativistic Doppler factor k (defined by Eq. 8.6), and so the frequency is shifted by the factor $1/k$. With the assumption above for velocity, the formula above yields $k = 2$, so the relativistic Doppler shift in frequency at B is 0.5. This signal is then reflected back to observer A, with a further identical frequency-shift, so that the frequency eventually received by A is shifted by $0.5^2 = 0.25$ $(= 1/4)$. This relativistic frequency shift is equal to the classical frequency shift derived above. However, as discussed above, according to Newton the frequency shifts from A to B and from B back to A are different. This reflects the lack of symmetry between the two frames in the classical sense.

Further, the local reflecting time (for observer B) is the local sending time (for observer A) $\times k = 4 \times k = 8$ (the time required for transmission is included in this time-dilation, for this experiment). Note that this is different from the classical reflecting time (10) calculated by the Newtonian method above. After reflection, the local receiving time (for observer A) is his local sending time $\times k \times k = 4 \times k \times k =$ 16, as stipulated at the beginning. According to Einstein, the local reflection time (for observer B) is the geometric mean of the local sending time (= 4, for observer A) and receiving time (= 16, also for observer A): $8 = \sqrt{4 \times 16}$.

Needless to say, experiments verify the relativistic analysis, rather than the classical analysis.

To illustrate the symmetry between the two frames, Fig. 8.3 shows that what is true for observer A is also true for observer B. If we switch the roles of observers A and B, and let B send a light signal to observer A at time 4 and receive it back at time 16. Then again the reflecting time for observer A is 8. In other words, there is full symmetry between observers A and B. In hindsight, we can say that choosing the frame of observer A to be at rest and B moving away from it, has a meaning only in the Newtonian context. In relativistic context, the only thing that counts is the relative speed between the two frames.

To further clarify the difference between midpoint time and proper time, consider Fig. 8.4. Again both observers A and B move apart along a single line at velocity $v = 0.6$. Both observers carry out the same remove sensing experiment as shown in Fig. 8.2. Observer A sends an impulsive light signal to observer B at $t = 4$, receives it back at $t = 16$, and concludes that according to Newton the reflection time at observer B, which is the midpoint time for observer A, was at $t = 10$ and the distance between A and B is 6. However, according to Einstein observer B will receive A's signal at $t = 8$. This is the proper time for observer B.

Vice versa, the same applies with the roles of observers A and B switched. The two observers are completely equivalent. Each observer, by remote detection, finds that the other observer is at midpoint time 10 and distance 6. However, the other observer is at proper time 8 and distance zero in his own coordinate system. In other words, the midpoint time 10 is on the time axis of the one doing the remote detection, and the corresponding proper time 8 is on the time axis of the other. The reason why midpoint time 10 is not the same as the corresponding proper time 8 is that two different coordinate systems are used.

In Newtonian theory, the difference between the sending and receiving time is split to obtain the arithmetic mean, which is called the midpoint time. In relativity theory the difference is proportioned to obtain the geometric mean, which is the

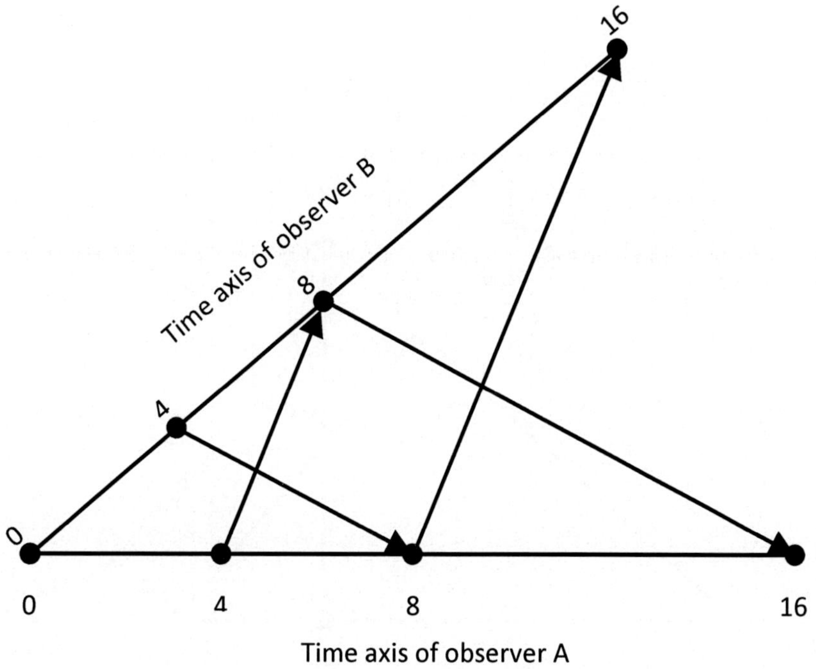

Fig. 8.3 Symmetry for observer A and observer B

proper time. Because the arithmetic mean of two positive numbers is always greater than the geometric mean, it follows that the midpoint time is stretched or dilated with respect to the proper time. This phenomenon is the dilation of time.

The definitions proposed by Newton have proven adequate at the low velocities at which we conduct our daily business. However, high velocities (that is, velocities close to the speed of light) occur in particle physics and at astronomic scales. Global Positioning System (GPS) and satellite communication rely on relativity for accurate time measuring. In such circumstances Newton's theory is not sufficient. The special theory of relativity uses the intriguing fact that the velocity of light is constant to unite the concepts of space and time.

We have to remember that what Einstein accomplished is certainly very extraordinary. Everywhere you look in nature, you might see symmetry. If you look at plants and animals, you will find that they have symmetrical body shapes and patterns. If you divide a leaf in half along the stem, you may find that one half has the same shape as the other half. There is symmetry everywhere, so why not here?

What do we mean by symmetry? Symmetry in mathematics refers to objects remaining unchanged under transformations like reflection, rotation, or translation. In physics, symmetry encompasses the idea that fundamental laws of nature remain unchanged under specific transformations or operations. It includes spatial, temporal, and combined transformations. Symmetries in physics correspond to the invariance

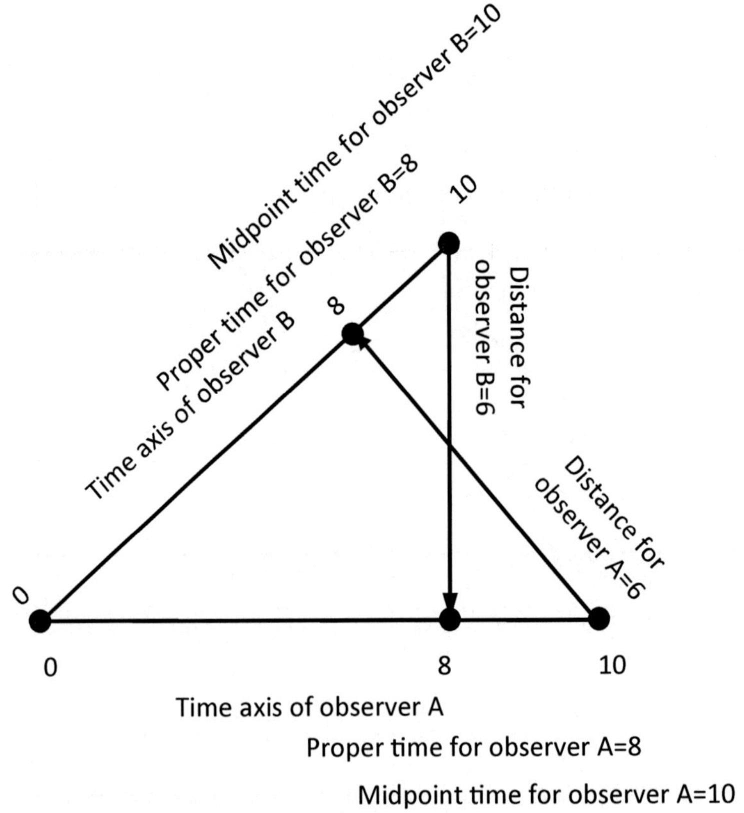

Fig. 8.4 Proper time and midpoint time

of physical laws under these transformations. The Lorentz transformation discussed later in this chapter is one important example.

Before Einstein, one would measure the dimensions of the table in front of us in the familiar three-dimensional coordinates of length, breadth, and height. One would measure the passage of time with the entirely separate and independent coordinate of Newtonian time. But now Einstein has changed everything. Instead of space and time we have space–time. No longer is the time coordinate independent of the spatial coordinates. They are all interlinked and we have seen how the relativistic Doppler factor is used to find the relationship.

Frames of Reference

We look up to the planet Jupiter and we see it moving in the sky. We (Earth dwellers) can measure the time t (in hours) that it takes Jupiter to travel a distance x (in km). An imaginary resident on Jupiter is called a Jovian. To a Jovian, planet Jupiter does not move at all. The number of rotations of the moon Io around Jupiter gives proper time p on Jupiter. The question is: How are t, x, and p related? In Fig. 8.5, the variable t corresponds to the time on the time axis for the observer on Earth, the variable x corresponds to the distance on the distance axis for the observer on Earth, and the variable p corresponds to the proper time for the observer on Jupiter. As seen in Fig. 8.5 these three variables form a right triangle, with sides x and p and with hypotenuse t.

In Fig. 8.5 the Jupiter frame is called the *target frame*. The target is a *motionless body* in that frame. Because the target does not move, we have no need for the distance axis. As a result, we will only plot the time axis. The time axis in the target frame is shown as line OP in Fig. 8.5. The time p plotted on this axis is the *proper time* in the target frame. We will name the Earth frame the *observer frame*. From the point of view of an observer on the Earth frame, the target is certainly moving. The coordinates of the target in the observer frame are the distance $x = PT$ and the time $t = OT$. The fundamental structure of special relativity is expressed in terms of distance$x$, time t, and proper time p.

Note for the sake of completeness that the situation outlined in Fig. 8.5 is three-dimensional, with Jupiter moving at a constant speed relative to the Earth (approximately over short distances; over longer distances the elliptical planetary motion and general relativity comes into play). This situation is equivalent to the one-dimensional situation discussed in Figs. 8.2–8.4, except that the reference frames with origins at Jupiter and the Earth, respectively, never coincided.

The speed of light in vacuum, usually denoted c, is a universal physical constant that is important in many areas of physics. Its exact value is defined as 299,792,458 meters per second (approximately 300,000 km/s or 300 m/μs or 186,000

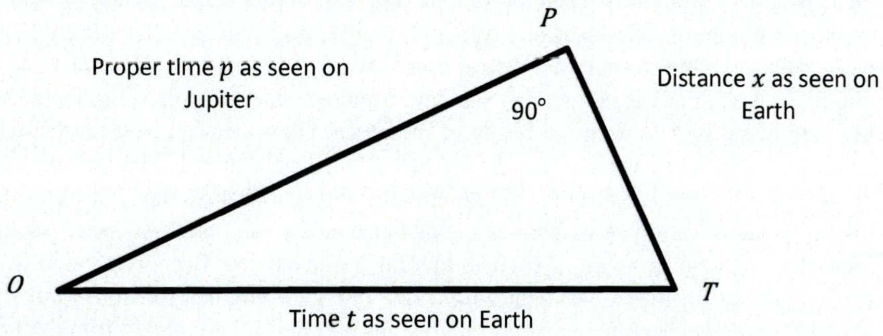

Fig. 8.5 The problem is to convert the proper time p (number of rotations of Io) into Earth time (number of days)

miles/s). It is exact because by international agreement a meter is defined as the length of the path taken by light in vacuum during a time interval of 1/299792458 s (definition by the 17th General Conference on Weights and Measures, 1983). According to special relativity, c is the upper limit for the speed at which conventional matter and information can travel. Although this speed is most commonly associated with light, it is also the speed at which all massless particles and field perturbations travel in vacuum, including electromagnetic radiation and gravitational waves. Such particles and waves travel at c, regardless of the motion of the source or the inertial reference frame of the observer. Particles with a nonzero rest mass can approach c, but they can never actually reach it. In the special and general theories of relativity, c interrelates space and time, and also appears in the famous equation of mass–energy equivalence $E = mc^2$.

The time–space inner product of the vectors (t_1, x_1) and (t_2, x_2) is defined as:

$$(t_1 t_2 - x_1 x_2) \,. \tag{8.7}$$

The inner product of the vector (t, x) with itself is

$$(t^2 - x^2) \,. \tag{8.8}$$

We can characterize three kinds of vectors according to the table:

Vector (t, x) is time-like if $t^2 - x^2$ is positive
Vector (t, x) is light-like if $t^2 - x^2$ is zero
Vector (t, x) is space-like if $t^2 - x^2$ is negative

Time-like straight lines can be called *inertia lines*, as they represent the paths of unaccelerated bodies. Light-like lines are simply called *light-lines*, and they are the paths of photons, neutrinos, and gravitons. Space-like straight lines are sometimes called *separation lines*, since particles that occupy distinct locations on a space-like straight line must be separate and distinct particles. The reason is that two locations on a space-like straight line cannot both lie on the path of a single particle.

Now we come to the most difficult aspect of relativity theory, the one that can never fully be understood in a physical sense. We consider one quantum of light, a photon, traveling at speed $c = 1$ in vacuum. Suppose that it travels along the light-like line given by $t = x$ in our frame of reference. Now we want to try to explain what we mean when we say that a photon has no inertial frame of reference. Let us try to construct such a frame mathematically, and see what results. The light-like line in question would serve both as the distance axis x' and the time axis t' of the inertial frame of the photon, if such an inertial frame existed. Because $t^2 - x^2 = 0$ for a light-like line, it follows that a light-like line is orthogonal to itself. Thus the x' and t' axes, which coincide, are orthogonal as required for an inertial frame. Thus mathematically the frame of reference for a photon consists of two orthogonal axes x' and t' which coincide with the light line. Because this frame is one-dimensional, it can only reach events that lie along the given light-like line of the given two-dimensional x, t frame. Thus this constructed photon frame is not a bona-fide inertial frame, and this is what we mean when we say that there can be no inertial frame for a photon.

Because the x' and t' axes of the constructed photon frame coincide, we can say that all distances exist at the same time, and also that all times exist at the same distance. That is, a photon is at *all places at the same time*, and the same photon is at *all times at the same place*. In other words, when we perceive that a photon has travelled a distance x in time t, the photon perceives that it has spent no proper time for this journey. Thus for a photon all locations in its path are simultaneous, so the photon covers an infinite distance (to us) in zero time (to it). We may say that a photon is infinite because its presence in space goes from negative infinity to infinity. This is the space aspect of a photon.

Now let us look at the time aspect. When we perceive that a photon has travelled a distance x in time t, the photon perceives that it has covered no distance for this journey. Thus for a photon all times in its path are at the same location, so the photon spends an eternity of time (to us) and yet remains at the same place (to it). We may say that a photon is eternal since its presence in time goes from negative eternity to eternity. This is the time aspect of a photon.

In 1637 Descartes characterized light as a pressure wave transmitted at infinite speed. As a consequence, in its own frame light endures forever and is present everywhere. Let us now think about a photon of light, traveling at speed c. For a photon, all times t (for us) are the same time for it and all places x (for us) are the same place for it. Time does not pass for a photon, so a photon is ageless. Space does not extend for a photon, so a photon is space-less. In its own perception, a photon endures forever and is present everywhere. Although his was a philosophical statement, whereas relativity theory is based on the experimental fact of the constancy of the speed of light, according to the space aspect of relativity theory, Descartes was right.

Time and space are frozen (that is, null and void) for a photon which to us appears to be traveling at speed c in vacuum. By its clock, the photon does not age. It takes no time (to it) to travel any distance (to us). By its measuring stick in meters, a photon does not move. It requires no distance (to it) to travel any length of time (to us). A photon (to it) is at the same time at all places, and at the same place at all times. Time and space (in our human sense) do not exist for a photon. A photon is infinite, eternal, perfect. A photon is also unique in its own perception (being infinite and eternal). It "thinks" it is the only photon in the universe and eternity.

Light is a unique type of wave. By the special theory of relativity, photons do not experience time at all. From the photons' viewpoint, photons never have, nor ever will, move. In our reference frame, we observe a photon being emitted from an atom. The photon might exist in vacuum for billions of years before it is absorbed by another atom. However, in the photon's reference frame, there is zero time elapsed between when it is emitted and when it is absorbed. A photon does not experience distance either. From the perspective of a photon, there is no such thing as time and there is no such thing as distance. It may be said that a photon does not travel at all. Recall that Zeno of Elea imagined that he had abolished motion. Although his was a philosophical statement, whereas relativity theory is based upon the experimental fact of the constancy of the speed of light, according to the time aspect of relativity theory, Zeno was right.

Bacon, Galileo, Roemer and the Movement of Light

Roger Bacon (c1214–c1294, see Fig. 8.6) was a medieval English philosopher and Franciscan friar who placed considerable emphasis on the study of nature through empiricism. He was an early advocate of the modern scientific method. Bacon revealed the importance of empirical testing when he obtained results that were different from what Aristotle would have predicted. Bacon wrote about optical magnification and even applied its use to observing the sky. He wrote that by using lenses, "the Sun, Moon, and Stars may be made to descend hither in appearance... which persons unacquainted with such things would refuse to believe." For this he was labeled a magician and imprisoned.

Bacon claimed that the movement of light required a finite, though imperceptible, interval of time. However, most writers up to the seventeenth century held to the mistaken doctrine of instantaneous transmission. Galileo tried in 1607 to measure the velocity of light with the aid of lantern signals, but without success, for light travels Earthly distances in extremely short fractions of time, not feasibly measured by Galileo's method.

Fig. 8.6 Roger Bacon (c1214–c1294)

In 1612, having determined the orbital periods of Jupiter's moons, Galileo proposed that with sufficiently accurate knowledge of the orbits of the moons, one could use their positions as a universal clock, and this would make possible the determination of longitude. In other words, Galileo proposed a cosmic clock based on the times of the eclipses of the moons of Jupiter. Galileo proposed this method to the Spanish crown but implementation of the cosmic clock proved to be unworkable, because of the inaccuracies of Galileo's timetables and the difficulty of observing the eclipses on a ship. However, with refinements, it was believed that the method could be made to work on land.

Ole Christensen Roemer (also spelled and called Ole Rømer or Olaf Roemer, Fig. 8.7) was born on 25 September 1644 in Aarhus, Denmark. He matriculated at the University of Copenhagen. His mentor at the University was Erasmus Bartholin, who had been given the task of preparing the astronomical observations of Tycho Brahe for publication. In 1668, while Roemer was living in Bartholin's house, Bartholin discovered the double refraction of a light ray by Iceland spar (calcite). Huygens spent over a year experimenting with Iceland spar after Bartholin sent him some samples. A crystal of Iceland spar has two remarkable properties. First, it is a natural polarizing filter, separating light of different polarizations. Second, because of its natural polarization, Iceland spar is birefringent, meaning light rays entering the crystal become polarized, split, and take two paths to exit the crystal—creating a double image of an object seen through the crystal. There is evidence that the Vikings used the polarizing effect of Iceland spar to navigate the North Atlantic. The constant fog and mist in the North Atlantic often made navigation by the stars or by the Sun impossible. The Vikings called Iceland spar a 'sunstone' because the polarizing effect can be used to find the direction of the Sun, even, so it seems, in dense fog and overcast conditions. Christiaan Huygens researched the crystal very thoroughly and put forward the theory in *Traité de la Lumière*.

Giovanni Domenico (Jean-Dominique) Cassini (1625–1712) was an Italian astronomer who was invited to France to help set up the Paris astronomical observatory. It opened in 1671 and Cassini remained there as director until his death in 1712. Cassini had observed the moons of Jupiter between 1666 and 1668, and discovered incongruities in his measurements that, at first, he attributed to light having a finite speed. In 1672 Roemer went to Paris and continued observing the moons of Jupiter as Cassini's assistant. Roemer observed that times between eclipses were shorter as Earth approached Jupiter, and were longer as Earth moved away from Jupiter. Cassini made an announcement to the Academy of Sciences on 22 August 1676:

> Light seems to take about ten to eleven minutes [to cross] a distance equal to the half-diameter of the terrestrial orbit.

Because the Earth revolves around the Sun in an elliptic orbit, the half-diameter of the terrestrial orbit varies. The mean Earth-Sun distance is now defined as the astronomical unit (au) and exactly as 149,597,870,700 m, which can be rounded to 150,000,000 km. Thus the diameter of Earth's orbit is about 300,000,000 km. Cassini's announcement can be reworded in this way:

Fig. 8.7 Ole Roemer (1644–1710)

Light takes about 20 to 22 minutes to cross the distance of one diameter (300,000,000 km) of Earth's orbit.

In other words, light takes time to move, so light is not instantaneous. This was a momentous result. The next announcement was made by an anonymous person who wrote a short paper, *Démonstration touchant le mouvement de la lumière trouvé par M. Roemer de l'Académie des sciences*, published 7 December 1676 in the *Journal des Scavans*. This paper modifies Cassini's announcement by the statement that light takes about 22 min (instead of 20 to 22 min) to cross the diameter of Earth's orbit. Actually the modern correct value is approximately 16.67 min, which is 1000 s. Cassini and Roemer were observational astronomers. They had discovered that light is not instantaneous and they went back to their work of taking observations. Fortunately, Huygens went on from there.

Huygens very much wanted to know the numerical value of the velocity of light. For years, he had claimed that the velocity of light is not infinite but finite. Apparently Huygens never saw Cassini's announcement of 22 August 1676, but he did see the anonymous summary of Roemer's work that was published on 7 December 1676. Huygens now had the numerical value of the time it takes to cross the diameter of the Earth's orbit: 22 min or 1320 s. Huygens knew the estimated mean diameter of the Earth's orbit: 11,000 Earth diameters or 280 million km. He based this on an indication of the Sun's parallax of 9.5 arc seconds, which Cassini had received in 1673 from an observation of Mars. In 1678 Huygens carried out the division and obtained a numerical value of the speed of light of 212,000 km/s in today's units.

With modern accurate values for the mean diameter (300 million km) and the travel time delay for light (1000 s), he would have obtained:

$$c = \frac{\text{diameter}}{\text{time to cross diameter}} = \frac{300,000,000\text{km}}{1000\text{s}} = 300,000\text{km/s}\,, \tag{8.9}$$

which is the modern accurate value for the speed of light. However Huygens did not stop there. In *Traité de la Lumière*, Huygens gives a most remarkable derivation which is discussed in the following section.

Huygens and the Relativity Equations

The left side of Fig. 8.8 shows the Sun at A, the annual orbit of Earth as $BCDE$ and Jupiter at F. The right side of Fig. 8.8 shows the path $KBCL$ of Earth and Jupiter moving apart simplified to the straight-line outward path KL, and also the path $LDEK$ of Earth and Jupiter moving together simplified to the straight-line inward path LK. More specifically, Huygens simplifies the mathematics by making the following assumptions. Jupiter and Earth both revolve around the Sun. Since one Jupiter year is so much longer than one Earth year, Huygens assumes that Jupiter is not moving. But Huygens also assumes that the speed of light measured at B and E is equal and independent of the Earth's speed. While he made these assumptions for the sake of simplicity, they in fact already hinted at the invariance of the speed of light irrespective of the moving frame later proposed by Maxwell and Einstein. He would write: "Non est mathematicè difficilis materia, sed physicè aut hyperphysicè." ("Not the mathematics is difficult, but the physics or the hyperphysics."). Huygens also replaces the elliptic orbit of the Earth by two straight-line paths as shown on the right side of Fig. 8.8. Implementing these approximations, Huygens' assumptions are:

(1) The Earth travels in a straight line away from Jupiter (from K to L) for one half-year on the outward path
(2) The Earth then travels back toward Jupiter (from L to K) for the remaining half-year on the inward path

A clock ticks at equal intervals of time: Jupiter's moon Io acts like a clock on Jupiter. On Fig. 8.8, an eclipse GH occurs when Io is hidden behind Jupiter. An eclipse starts when Io is at G and ends when Io is at H. An orbit is the time between eclipses. In other words an orbit occurs when Io goes once around Jupiter from H to H, along $HNGH$.

In his book *Traité de la Lumière*, Christiaan Huygens gave a complete mathematical analysis of the problem. Huygens works with astronomical bodies and signals of light. A half-year on Earth is

$$t = \text{half - year} = \frac{365}{2} = 182.5\text{days}\,. \tag{8.10}$$

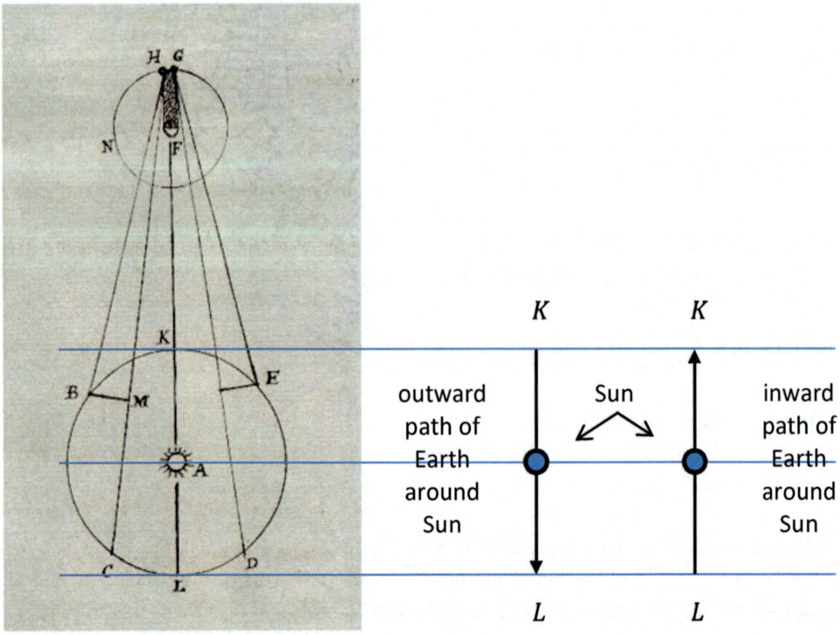

Fig. 8.8 Left side: Figure from Chap. 1 of Christiaan Huygens' *Traité de la Lumière*. Right side: Simplified version [with permission from Leiden University Library, HUG 31, ff. 255r.-312v.]

As seen from Earth, the moon Io makes about 206 orbits of Jupiter in a full Earth year (again we use this round number in the following). It was expected on the outward passage one would see 103 orbits in time t, and on the inward passage one would see 103 orbits in the same time t. However, on the outward passage Roemer saw the 103 orbits in the receiving time given by

$$r = t + x = 182.5\text{days} + 1000\text{seconds} \,. \tag{8.11}$$

Roemer concluded the time lag of 1000 s is in fact the additional time that it takes for light from Jupiter's moon to travel the extra distance across the diameter of the Earth's orbit. In other words, the light has to go the additional distance KL which is x=1000 s (natural units). If the velocity of light were infinite, the time lag would have been zero. Because the time lag was 1000 s, Roemer came to the historic conclusion that light is not instantaneous.

Let us now explain what Huygens did. Looking backward after the passage of nearly four centuries, we do have the advantage of hindsight. Huygens did the simplest possible thing. He had no way of determining the absolute velocity of either Jupiter or Earth to some fixed reference frame, like that of the fixed stars. As stated earlier in this Chapter, the fixed stars compose the background of astronomical objects that appear to not move relative to each other in the night sky compared to the foreground of Solar System objects that do. Because he had no other choice, Huygens was forced to consider only the relative velocity of Earth and Jupiter. It mattered not whether Earth was moving, or Jupiter was moving, or both were moving.

To show that such a situation describes the essence of Einstein's special relativity, we now resort to some simple mathematics. See Fig. 8.9. Earth and Jupiter are moving apart on the outward path KL. It follows that the "*receiving*" *time* r (related to the outward passage) is found by adding distance x to the time t. In his mind's eye, Huygens looked at a diagram such as given in Fig. 8.9. He immediately sees that the over-shoot x in the outward passage LM is exactly the same as the under-shoot x in the inward passage MK. It follows that the "*sending*" *time* s (related to the inward passage) is found by subtracting distance x from the time t. In other words, Huygens unconsciously saw the fundamental symmetry in Einstein's Special Relativity.

The *sending time* s is found by subtracting distance x from the time t; that is,

$$s = t - x \,. \tag{8.12}$$

In words, sending time is equal to 182.5 days minus 1000 s.

The *receiving time* r is found by adding distance x to the time t; that is,

$$r = t + x \,. \tag{8.13}$$

In words, receiving time is equal to 182.5 days plus 1000 s.

In essence, Huygens in his reasoning introduces two projection factors; one for sending and the other for receiving. On Earth, the sending time and the receiving time can be recorded. The problem is to find the proper time on planet Jupiter. Huygens

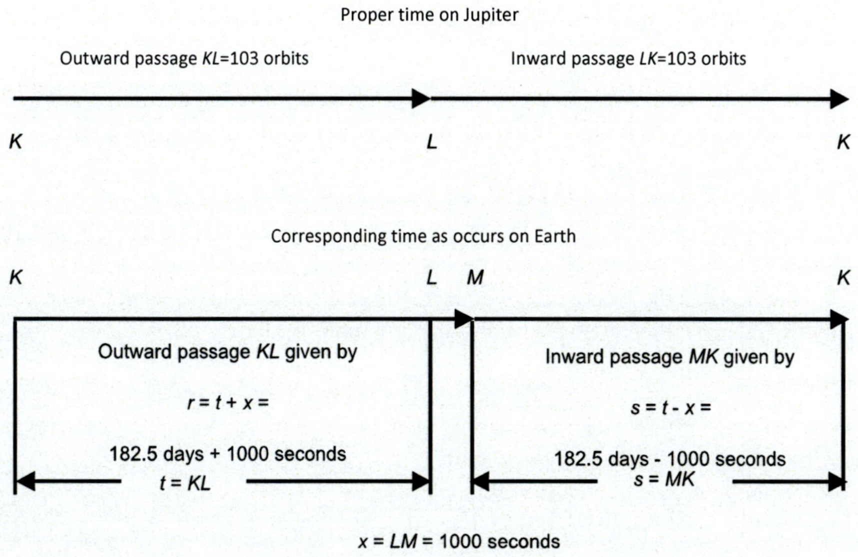

Fig. 8.9 Equal intervals of proper time on Jupiter correspond to unequal intervals on Earth

solves this problem by using the mathematics of projection. The shadow of a tree is the projection of its height. The projection factor k is

$$k \times \text{height} = \text{shadow} . \tag{8.14}$$

The projection factor represents a critical element in the reasoning of Christiaan Huygens. We know that p is the proper time on Jupiter given by 103 orbits of its moon Io. The sending time s is projected onto the proper time p. In symbols, the relationship is

$$k_S s = p, \text{ so } k_S = \frac{p}{s} . \tag{8.15}$$

In turn the proper time p is projected onto the receiving time r. The relationship is

$$k_R p = r, \text{ so } k_R = \frac{r}{p} . \tag{8.16}$$

Assuming symmetry, Huygens made the two projection factors the same. In other words, Huygens defines k as

$$k = k_S = k_R . \tag{8.17}$$

The equality of these two projection factors is the essence of special relativity. The projection factor k, a concept that Huygens created, is therefore known as the Optical Doppler factor or the Relativistic Doppler factor. The equality of the two projection factors gives Huygens' projection-factor equation (in natural units):

$$k = \frac{p}{s} = \frac{r}{p} \text{ where } r = t + x, s = t - x . \tag{8.18}$$

This equation yields

$$p^2 = rs \text{ or } p = \sqrt{rs} . \tag{8.19}$$

This says that the proper time is equal to the geometric mean (as mentioned earlier in this Chapter) of receiving time and sending time. We have

$$p^2 = rs = (t + x)(t - x) = t^2 - x^2 . \tag{8.20}$$

By taking the square root, we obtain

$$p = \sqrt{rs} = \sqrt{(t + x)(t - x)} = \sqrt{t^2 - x^2} . \tag{8.21}$$

This equation says that the proper time p is equal to the time–space interval $\sqrt{t^2 - x^2}$. The speed of Jupiter (as seen on Earth) is

$$v = \frac{\text{distance}}{\text{time}} = \frac{x}{t} . \tag{8.22}$$

The proper time p becomes

$$p = \sqrt{t^2 - v^2 t^2} = t\sqrt{1 - v^2} . \tag{8.23}$$

Because $x < t$ the speed v is less than one, which means that $\sqrt{1 - v^2}$ is less than one. As a result, the observed time t on Earth is greater than the proper time p on Jupiter; that is

$$t > p . \tag{8.24}$$

Time dilation is the difference in the elapsed time measured by two clocks due to them having a velocity relative to each other. A person on Jupiter will measure the elapsed time as the proper time p. An observer on Earth will look up to Jupiter and measure the elapsed time t on the moving clock on Jupiter as time t. Because t is greater than p, we say time t is the dilated value of p.

Lorentz Transformation

There are three concepts that play a central role in relativity theory, namely invariance, time-distance, and causality. The concept of invariance is basic to the science of physics. We say that the laws of physics are invariant with regards to certain transformations. Let us explain. Although the concept of invariance is simple, it is extremely powerful. If one stands at one place, or if one stands at another place, things may look different, but the underlying laws of physics are still the same. Translational invariance says that the laws of physics do not change if we move from one place to another. What we see in different places might be different, but the laws of physics are not. Similarly, rotational invariance says the laws of physics do not change if we spin around and see the world while facing different directions. What we see in different directions might be different, but the laws of physics are not. Invariance principles are connected to the conservation of physical quantities. The law that states that momentum is conserved derives from translational invariance. The law that states that angular momentum is conserved derives from rotational invariance.

Now let us look at time and distance. The ancients believed in geometry. In the seventeenth century when Descartes introduced algebra into geometry, the new discipline of analytic geometry opened up. Algebraic equations provide a useful method to write down relationships between different physical quantities. Not least among these relationships is the equation for speed. By instinct people understand the difference between fast and slow, but things changed when the notion became formalized in an algebraic equation, namely "speed equals distance divided by time," or $v = x/t$. In this equation v is the speed, x is the distance, and t is the time. For example, if you drive 120 km in two hours, your (average) speed is 120 divided by 2. The answer is the (average) speed is 60 km/h. At this point we are already at the heart of relativity theory. Let us explain.

We ask whether this equation invariably furnishes a description of nature with which everyone would agree. In other words, does the equation deal with invariant quantities? As it turns out, neither distance nor time are invariant. Both depend upon the perspective of the viewer.

For example, an airplane has an air-speed and a ground-speed. Both are the same if there is no wind. Let us exclude the case of no wind. If the plane is traveling into the wind, the ground-speed is less than the air-speed. If the plane is traveling with the wind, the reverse is true. These differences are apparent to anybody flying across the Atlantic; since the prevailing winds are from the west, east-bound flights have a greater ground-speed than do west-bound flights, although the air-speed is the same for both. Air-speed and ground-speed are different because the distance as measured in the air is not the same as the distance as measured on the ground. Everyone accepts that distance is not an invariant quantity.

Commonsense says that time as measured in the air is the same as the time as measured on the ground. Einstein says no! More precisely, Einstein says what is true for the distance is also true for the time. In other words, neither distance nor time are invariant with respect to the choice of the observation point. Time as measured on the airplane is not the same as the time as measured on the ground, just as with distance (although this time-difference is small for trans-Atlantic travel, much smaller than "jet-lag"). This is the essence of the special theory of relativity.

The theory of relativity violates our commonplace understanding of time. What do we do? Space and time must be merged into one entity called space–time. What is the next step? Let us look at causality. The order of cause and effect cannot be reversed. The effect cannot precede the cause. If this were not true, we would be faced with irresolvable contradictions. For example, consider one period of a periodic wave. It is the time difference between two successive crests. It follows that we will encounter one crest first and the other crest later. In this way we can say that first precedes second.

We can agree with Einstein that time is not invariant, but still, we must insist that causality holds. In common usage, an "event" is an occurrence, especially one that is particularly significant, interesting, exciting, or unusual. But in the theory of relativity, an "event" is an occurrence precisely defined at a single point in space–time. Some events occur so far apart from each other in space and in time that they cannot have any influence on each other. In such cases the designation of one as "cause" and the other as "effect" can be reversed. Let that be; we are not concerned with such distant events. We are interested in events that can have an influence upon each other, and so can be ordered. Causality must be preserved. The insistence on a causal universe constrains the possibilities of the melding of space and time. (Note that, at the quantum level, causality may not be asserted so confidently, but that regime is excluded from this discussion.)

We must use a common unit of measurement for both space and time. Recall that originally, in 1791, the meter was defined as the distance equal to one ten-millionth part of the quadrant of the Earth. Later, a platinum bar with a rectangular cross section and polished parallel ends was made to embody the meter. The meter was defined as the distance between the ends, when the bar is held at a specified temperature. However, now the meter is defined in terms of time. The definition states that the meter is the length of the path traveled by light in vacuum during a time interval of 1/299,792,458 of a second. From the definition we see that the calibrating speed is the speed of light, namely $c = 299{,}792{,}458$ m/s. In this way, both space and time are measured in seconds.

Coming back to the statements made in the introduction of this Chapter, we are now in a position to find an invariant quantity in the realm of space–time. An inertial frame of reference is a frame of reference that is not undergoing acceleration. We may ask, "acceleration relative to what?" So, a better definition of an inertial frame of reference is one wherein a physical object with zero net force acting on it moves with a constant velocity (which can be zero). An inertial frame of reference can be defined in analytical terms as a frame of reference that describes time and space homogeneously, isotropically, and in a time-independent manner. Conceptually, the physics of a system in an inertial frame has no forces external to the system.

Suppose that we have two inertial frames of reference that are moving apart at constant velocity along a straight line. Call one frame (1) the observer and the other frame (2) the target (for instance, to stay with Huygens, think of the Earth as the observer and Jupiter as the target, as discussed in the previous section). The target frame has space coordinate x_0 and time t_0 in terms of the observer frame. We could say that the frame of the target is moving away from the frame of the observer. Alternatively, we could say that the frame of the observer is moving away from the frame of the target. It is only the relative motion that counts in relativity theory.

Let distance and time of an event on one frame be x_1 and t_1. Let distance and time for the same event on the other frame be x_2 and t_2. Assume that the proper time p is the same for both frames; that is,

$$p = \sqrt{t_1^2 - x_1^2} \text{ and } p = \sqrt{t_2^2 - x_2^2}\,. \tag{8.25}$$

In this scheme space x and time t are no longer absolutes. They have been sacrificed in favor of an absolute space–time quantity p. The first frame measures x_1 and t_1 and the second frame measures x_2 and t_2 but they both get the same space–time invariant p.

We are now ready to find the relationship between two inertial frames: one with projection factor k_1 and the other with projection factor k_2. We wish to find a projection factor k_0 that transforms projection factor k_1 into k_2. The sending time (related to the inward passage), the receiving time (related to the outward passage), and the proper time of k_0 are respectively s_0, r_0, and p_0. We impose the requirement that

$p_0 = 1$ and that the two given inertial frames have the same proper time p; that is, $p_1 = p_2 = p$. The projection equation which accomplishes the transformation above is

$$k_2 = k_0 k_1 \,. \tag{8.26}$$

This projection equation may be written as

$$\frac{r_2}{p_2} = \frac{r_0}{p_0}\frac{r_1}{p_1} \text{ or } \frac{r_2}{p} = \frac{r_0}{1}\frac{r_1}{p} \,. \tag{8.27}$$

The same projection equation may also be written as

$$\frac{p_2}{s_2} = \frac{p_0}{s_0}\frac{p_1}{s_1} \text{ or } \frac{p}{s_2} = \frac{1}{s_0}\frac{p}{s_1} \,. \tag{8.28}$$

The above two Eqs. (8.27) and (8.28) become

$$r_2 = r_0 r_1 \text{ and } s_2 = s_0 s_1 \,. \tag{8.29}$$

which are respectively

$$t_2 + x_2 = (t_0 + x_0)(t_1 + x_1)$$

$$t_2 - x_2 = (t_0 - x_0)(t_1 - x_1) \,. \tag{8.30}$$

If we expand each equation, we obtain

$$t_2 + x_2 = t_0 t_1 + t_0 x_1 + x_0 t_1 + x_0 x_1$$

$$t_2 - x_2 = t_0 t_1 - t_0 x_1 - x_0 t_1 + x_0 x_1 \,. \tag{8.31}$$

These two Eqs. (8.31) may be combined to yield:

$$t_2 = t_0 t_1 + x_0 x_1$$

$$x_2 = t_0 x_1 + x_0 t_1 \,. \tag{8.32}$$

These equations are called the *Lorentz equations* or *the Lorentz transformation* (first formulated in final form in 1899 by Lorentz). If we factor out the quantity t_0, we obtain

$$t_2 = t_0\left(t_1 + \frac{x_0}{t_0}x_1\right)$$

$$x_2 = t_0\left(x_1 + \frac{x_0}{t_0}t_1\right). \tag{8.33}$$

Let us now introduce the dilation factor γ (which quantifies the stretching of time and distance, its form is given below) and the relative velocity given by,

$$v_0 = \frac{x_0}{t_0} \tag{8.34}$$

The Lorentz equations may then be written as

$$t_2 = \gamma(t_1 + v_0 x_1)$$

$$x_2 = \gamma(x_1 + v_0 t_1). \tag{8.35}$$

The Lorentz equations as written give the time t_2 and the distance x_2 in terms of time t_1 and the distance x_1. Let us also define the velocities

$$v_1 = \frac{x_1}{t_1} \text{ and } v_2 = \frac{x_2}{t_2}. \tag{8.36}$$

If we divide the lower Lorentz equation by the upper Lorentz equation, we obtain the celebrated *Einstein velocity addition formula* (1905) given by

$$v_2 = \frac{v_0 + v_1}{1 + v_0 v_1}. \tag{8.37}$$

The Einstein velocity addition formula restricts velocity v_2 to have an absolute value less than the velocity of light, which by these conventions is $c = 1$. If one of the velocities is the speed of light, say $v_0 = 1$, then so is v_2, regardless the other velocity v_1. If both speeds v_0 and v_1 are subluminal (smaller than the speed of light), then so is v_2.

In contrast, the Galilean addition formula

$$v_2 = v_0 + v_1 \tag{8.38}$$

(which we use in everyday life) does not impose any such bound. We can see from these equations that if the velocities v_0 and v_1 are very small compared to the speed of light (that is, if the velocities are $\ll 1$), then Eq. (8.37) becomes equivalent to (8.38). This is why we do not commonly experience these relativistic effects in everyday life.

The above form of the Lorentz equations can be inverted so as to give the time t_1 and the distance x_1 in terms of time t_2 and the distance x_2. In the given Lorentz equations we merely interchange the position of the subscripts 1 and 2, and change v_0 to $-v_0$. The result is the inverted form of the Lorentz equations given by

$$t_1 = \gamma(t_2 - v_0 x_2)$$

$$x_1 = \gamma(x_2 - v_0 t_2)\ . \tag{8.39}$$

Finally, to ensure that the transform pair (Eqs. (8.35) and (8.39)) satisfies Einstein's postulates, it can be proven that the dilation factor γ is given by:

$$\gamma = \frac{1}{\sqrt{(1 - v_0^2)}}\ . \tag{8.40}$$

With the success of Maxwell's electromagnetic theory, Einstein (1905) realized that the finite speed of light (which is an intrinsic part of electromagnetic theory) must be applied to mechanics as well. Maxwell's theory required no change as it was developed on the basis of finite signal speed. However, Newton's laws of motion had to be rewritten. Einstein proposed the two postulates, stated at the start of this Chapter, upon which relativity theory is based. The first postulate is that all the laws of physics are the same in every inertial reference frame. The second postulate is that light is always propagated in empty space with a definite velocity c which is independent of the state of motion of the emitting body. The second postulate is equivalent to the condition that the velocity c of light in vacuum is the ultimate signal speed.

On this basis, Einstein changed Newton's laws. As noted by American physicist Richard Feynman (1918–1988) the only change that was needed was that the mass (as measured in a moving inertial frame) needed a relativistic correction factor: $m = \gamma m_0$, where m_0 is the rest mass of a body that is not moving, γ is again the dilation factor introduced above (Eq. 8.40) and m the relativistic mass of the same body (which now depends on velocity v and becomes infinite for v approaching the light speed). One result was Einstein's best known equation $E = mc^2$. It says that energy is equal to mass times the square of the velocity of light. This equation transformed physics.

This analysis applies to inertial frames, without acceleration or external forces. Of course, our common experience does not occur in such inertial frames. For example, as you read this, you are spinning along with the Earth, about its axis, and you are orbiting the Sun, along with the Earth, and you are orbiting about the center of the galaxy along with the Solar System. These rotations are accelerations. And they are accompanied by pervasive gravitational forces.

So, your experience is not within an inertial frame, as strictly defined above. But, those rotations and gravitational forces are canceled, as you sit in your chair, by contact with the ground. So, in this sense, your frame is indeed an inertial frame, since it has no net acceleration or external forces.

Electromagnetic Waves and Mechanical Waves

"Slow light" occurs when a propagating pulse of light is substantially slowed by the interaction with the medium in which the propagation takes place. For instance, sunlight in the atmosphere bends and slows down and therefore we still see the Sun even although it has set already minutes before. Classical physics' most basic description of light is that of an electromagnetic wave in a vacuum. According to Maxwell's equations, these waves move at a certain speed, represented by the letter c. It is usual to refer to this well-known physical constant as the speed of light. A common belief that "the speed of light is always the same" has its roots in the special relativity postulate, which states that the speed of light is constant in all inertial reference frames. However, light is frequently more than just an electromagnetic field disruption.

Light traveling within a medium is not just a disturbance solely of the electromagnetic field, but rather a disturbance of the field and also the positions and velocities of the charged particles (electrons) within the material. The motion of the electrons is determined by the field (due to the Lorentz force) but the field is determined by the positions and velocities of the electrons (due to Gauss' law and Ampère's law). The behavior of a disturbance of this combined electromagnetic-charge density field (that is, light) is still determined by Maxwell's equations, but the solutions are complicated because of the intimate link between the medium and the field.

Understanding the behavior of light in a material is simplified by restricting the types of disturbances studied to sinusoidal functions of time; more complicated disturbances can be described as weighted sums of these sinusoidal functions. For these types of disturbances, Maxwell's equations are commonly written with terms which account for the interaction with the medium. The "phase velocity" describes how these special disturbances propagate through a material at a speed lower than c. The refractive index n of the material is defined as the ratio of c over the phase velocity of the wave. The index of refraction of a given material is not constant, but it is affected by temperature, pressure, and the frequency of the (sinusoidal) light wave. This results in the phenomenon known as dispersion.

Electromagnetic waves are waves that require no medium to travel whereas mechanical waves need a medium for their transmission. In other words, electromagnetic waves can travel in a vacuum whereas mechanical waves do not. The mechanical waves need a medium such as air, water, metal, or rock in which to travel. Ripples in water are an example of mechanical waves. Light and radio signals are examples of electromagnetic waves. All electromagnetic waves travel at the speed c of light in vacuum. All mechanical waves (such as sound) travel at less than the speed c of light in vacuum.

As we have just seen, light traveling within a medium is no longer a disturbance solely of the electromagnetic field, but rather a disturbance of the field and the positions and velocities of the charged particles within the material. When light travels through a medium such as glass, its velocity is reduced to a value less that c. As a result, it must be treated similarly to a mechanical wave.

Universal Constant Speed of Light

The apparently unrelated laws and phenomena of electricity and of magnetism were integrated by James Clerk Maxwell. In 1861, Maxwell published the beginnings of his famous equations, known as "Maxwell's equations." Maxwell's equations established the electric and magnetic properties of empty space and also of actual media. Maxwell found that the speed of a massless electromagnetic wave in vacuum was very close to the measured speed of light, and proposed the two might match exactly. Maxwell was right. Light is an electromagnetic wave, and the speed of light could be determined based on electric and magnetic constants. These results were taken by Einstein in 1905 as part of his theory of special relativity. In fact, the constancy of the speed of light is considered the cornerstone of special relativity. Unlike space and time, the speed of light is a constant (designated by the letter c) independent of the observer. The constant speed of light underpins much of what we understand about the Universe. It is the c that is in the famous equation $E = mc^2$.

Seeing is believing. In other words, actual physical evidence is convincing. For most situations, physical evidence supersedes mathematical achievements, even accomplished with great skill. A person wants to see experimental results that show that speed of light is a constant. Maxwell understood this desire. Let us see what happened. In the special theory of relativity, light is the privileged entity. There is no inertial frame in which light is at rest. Einstein takes this property as the *Light Principle*; namely, light travels in every inertial frame at the ultimate signal velocity c (equal to unity in natural units), no more and no less. Each inertial frame observes any other inertial frame from the same point of view. Einstein takes this property as the *Relativity Principle*, namely, all the laws of physics are the same in every inertial frame.

The water waves which Huygens observed on the canals of Holland are propagated as disturbances in the water. With this model, Huygens postulated "*ethereal matter*" spreading throughout space in which light is propagated. Huygens designated his ethereal matter as *luminiferous ether* (that is, *light-transmitting ether)*. The ether concept goes back to Aristotle's perfect ether; namely, that quintessence which composed the heavens. On the other hand, Newton had argued that light is a stream of little particles, and under the weight of Newton's authority, Huygens' wave theory of light fell into oblivion. However, in the early nineteenth century, Huygens' wave theory together with his *luminiferous ether* were resurrected by Thomas Young and Augustin Fresnel. Since all other kinds of waves, including sound waves, can be transmitted only in a physical medium, it seemed natural to suppose that light waves would also require some transmitting medium.

Since the time of Maxwell we know that light travels through a vacuum at speed c. We also know that light also travels through air, water and glass but at different speeds all less than c. The ratio between c and the speed v at which light travels in a material is called the *refractive index* n of the material ($n = c/v$). These lower-than-c speeds are the result of absorption and re-radiation delay between atoms. When light is absorbed, it ceases to be light. For this reason, we can say that light only travels in the vacuum between atoms. A generalized form of Maxwell's equations describes this situation via electrical and magnetic properties of the medium, resulting in an effective speed of light in the medium, which is less than the speed of light in vacuum c.

In conclusion, electromagnetic waves can only travel in vacuum, whereas mechanical waves can only travel in material. In fact, Robert Boyle (1627–1691) had shown that sound waves do not travel in vacuum. Huygens knew this property of mechanical waves, but he had no way to know this property of light waves. Huygens experimentally demonstrated that light travels in a vacuum. Following Descartes, Huygens did not accept action at a distance. Action-at-a-distance forces are those types of forces that result even when the two interacting objects are not in physical contact with each other, yet are able to exert a push or pull despite their physical separation. Examples of action-at-a-distance forces include gravitational forces.

Sound can be heard by the ear because mechanical energy produced by the source is transferred to the ear through the movement of particles. When a force is exerted on a particle, it moves from its rest position and exerts a force on the adjacent particles. These adjacent particles are moved from their rest position, push against the next, and this continues throughout the medium. This transfer of energy from one particle to the next is how sound travels through a medium. A mechanical wave distributes energy through a medium by the transfer of energy from one particle to the next.

Why did Huygens' rejection of action at distance force him to postulate the ether instead of vacuum? As we have just seen, in a mechanical wave the force from a moving particle is exerted on the adjacent particles. Suppose there were action at a distance. Then the force from a moving particle could exert force on remote particles separated from the moving particle by vacuum. In this way, a light wave could cross the vacuum between Jupiter and Earth. Huygens did not accept this explanation. For this reason, Huygens postulated the existence of luminiferous ether as the medium for the propagation of light. It was invoked to explain the ability of the wave-based light to propagate through empty space (that is, vacuum), something that mechanical waves cannot do.

In Newtonian mechanics, all states of uniform straight-line motion are mechanically equivalent to a rest frame in absolute space. The Galilean transformation between inertial frames holds. Time is absolute. The Galilean velocity addition formula ($v_2 - v_0 + v_1$) allows adding up velocities without upper limit c. Thus, in Newtonian mechanics there is no privileged frame. This is the content of Newton's Principle of Relativity which says that all the laws of mechanics are the same in every inertial frame. Thus, Newtonian mechanics may be described in terms of equal status of inertial frames.

The luminiferous ether of Huygens is radically different from any known physical substance. In Newton's absolute space, all the inertial frames obeyed the same laws. As a result, they were physically equivalent. However, with the addition of the luminiferous ether, things changed. The frame containing the ether at rest was specially privileged. Thus, at the time of Maxwell's work in the 1860s, the theory of light might be described in terms of privileges; namely, one privileged universal frame stood apart from all of the other inertial frames. In theoretical physics in the

1860s, the situation prevailed that light had a special privileged status and mechanics had an ordinary status, both standing side by side. This incongruity represented a fatal crack in the structure that had to be repaired.

The first step in resolving this difficulty would be a search for the ether frame through measurements on the velocity of light. All through the middle years of the nineteenth century, there were many attempts both experimental and theoretical to detect Huygens' luminiferous ether and to determine its properties. These early experiments had failed to come up with any evidence for the existence of luminiferous ether. However, with the introduction in 1864 of James Clerk Maxwell's celebrated equations, new hopes were raised that the Earth's motion through stationary luminiferous ether could be detected by a suitable optical experiment. Maxwell had shown that light could be described as an electromagnetic wave. Light could travel through space, and the regions between the planets and stars seemed empty of any medium to carry the waves. Yet we see the Sun and stars. The accepted explanation in Maxwell's time was that outer space was filled with an extremely fine, imponderable substance, the ether, which is the carrier or medium of these electromagnetic waves. In effect it was the *luminiferous ether of Huygens*. As we have seen, Maxwell had successfully computed the speed of light from the electromagnetic constants by the equation $c = 1/\sqrt{\mu_0 \varepsilon_0}$. But what was this speed to be measured against? Speeds are always relative to measuring posts and instruments which make up the reference frame.

Nineteenth-century physicists expected that, if they measured the speed of light taking the Earth as the frame of reference, they would obtain varying results. The reason for their conclusion was that the Earth is constantly changing its velocity with respect to the fixed stars. Imagine we are measuring the speed of light coming from a fixed star. As mentioned earlier in this Chapter, the term "fixed star" is derived from the common experience that the relative position of the stars remains unchanged, in marked contrast to the "wandering stars", the planets ("planet" comes from Greek for "wandering"). At the moment we make our measurement, the Earth occupies some position in its orbit around the Sun and is moving at its orbital speed of 30 km/s. But six months later, the Earth would be halfway around the Sun and moving in the opposite direction. If the fixed stars were the ether reference frame, then the relative velocity of that frame with the Earth frame would depend upon the Earth's position in orbit and the speed of the Earth about the Sun, which is 30 km/s. This speed in natural units (where $c = 1$) is = 30/300,000 = 1/10,000.

In 1879, James Clerk Maxwell wrote an acknowledgement of some astronomical tables he had received from the American astronomer David Peck Todd. These tables contained many observations of the planet Jupiter. Maxwell was interested in measuring the velocity of light in various reference frames and described in his letter an ingenious method based upon observations of the eclipses of Jupiter's moons. In 1678 Huygens had been the first to determine the velocity of light by studying the time lags observed by Roemer in detecting these eclipses.

Maxwell wanted to use a similar method to determine the dependence of c on the movement of the Earth. Unfortunately, the astronomical data available to Maxwell were not accurate enough for use in Maxwell's method. Maxwell's method was based upon one-way times from sources to distant receivers. In his letter to Todd, Maxwell remarked that this distinguished his approach from methods based upon two-way times; that is, methods for which the source and receiver are at the same point, and the signal is returned by reflection. In other words, a two-way experiment is one that uses a beam of light that is returned to its starting point. If the ether would exist the outward and return times would differ, but only to second order in the ratio V/c, where V is the Earth's orbital speed and c the speed of light. The two-way optical experiment would have to be capable of measuring the extremely small but finite quantity represented by the square of the ratio of the Earth's orbital speed V to the speed of light; in symbols the quantity is $(V/c)^2 = (30/300,000)^2 = (1/10,000)^2$, which represents one part in one hundred million. These two-way experiments could be performed on Earth, but with current instrumentation Maxwell remarked that the results would be undetectably small because no known optical device could approach this sensitivity. Maxwell died that same year, and in respect and admiration for him Todd had the letter published in *Nature*, vol. 21, 1880.

Maxwell's letter in *Nature* was read by the American physicist A.A. Michelson (1852–1931). The part of the letter that particularly attracted Michelson's attention was the statement in the final paragraph that "in all terrestrial methods of determining the velocity of light, the light comes back along the same path again, so that the velocity of the Earth with respect to the ether would alter the time of the double passage by a quantity depending on the square of the ratio of the Earth's velocity to that of light, and this is quite too small to be observed." Michelson immediately began thinking about better instrumentation to achieve Maxwell's two-way experiment. In 1881, Michelson had some results. In 1887, he performed a refined version of the experiment working in collaboration with E.W. Morley. The Michelson-Morley experiment remains the main experimental pillar of special relativity, and almost invariably it makes up the lead-off discussion.

Let us give some details on how the Michelson-Morley experiment was set up. Basically, they studied two light beams passing through a beam splitter reflecting on two perpendicular mirrors and then recombining, recorded at a sensor. The phase shift between the two signals would indicate the speed of the Earth through the ether when the setup was rotated, but the result would be zero (or close to zero) if an ether did not exist.

The Michelson-Morley experiment in 1887 played the crucial role in the overthrow of the luminiferous ether theory and in the genesis of the theory of relativity. The experiment was designed to determine the absolute motion of the Earth through the privileged inertial frame, namely, the luminiferous ether frame. The result of the Michelson-Morley experiment was that the luminiferous ether frame, if it exists, is unobservable. In this way, the luminiferous ether is deposed from its status of a privileged frame. With no privileged frame, it follows that the Relativity Principle applies not only to mechanics but to all laws of physics. Thus, the privileged status of light in classical physics is replaced by a situation in which all inertial frames are of equal status. However, light, now without its privileged frame of the ether, finds refuge in no frame at all, the vacuum. Belonging to no frame, light has the same velocity in all inertial frames. This is the Light Principle, which re-establishes the privilege of light but in a different guise. By merely changing the word "ether" to the word "vacuum," all of Huygens' results in the seventeenth century are valid today.

Concluding Remarks: Zeno, Huygens and Einstein

Einstein's special theory of relativity unites the concepts of space and time. The theory is replete with puzzling propositions reminiscent of the assertions of Zeno. Zeno in fact offered various groups of paradoxes. In particular, Zeno's paradox of Achilles and the tortoise is a clear example of an attempt by antique science to explain the concept of infinite sequences. An infinite sequence of terms may or may not converge to a finite limit. In this context, a limit is the value that a sequence of terms approaches as the number of terms increases. Limits are essential to calculus (and mathematical analysis in general) and are used to define continuity, derivatives, and integrals. Zeno's paradox of Achilles and the tortoise can serve to demonstrate infinitesimal calculus. This one was finally solved by Cauchy's formula for the sum of a geometric series. The conclusion is, of course, that Achilles does overtake the tortoise eventually.

Zeno's paradox of the Arrow is an argument against the reality of motion. Zeno's argument purports to show how an arrow in flight does not move. The paradox of the Arrow can serve to demonstrate the relativity of motion. In fact, in Einstein's relativity theory, a frame of reference can always be found in which an arrow in flight does not move: it is the frame of the arrow itself.

A postulate is defined as a statement which is taken to be true without proof. The word "postulate" comes from Latin, its Greek synonym is "axiom". Postulates form the basic structure from which lemmas and theorems are derived. The whole of Euclidean geometry is based on five postulates known as Euclid's postulates. Euclid's fifth postulate (also known as the parallel postulate, see Fig. 8.10), states that, in two-dimensional plane geometry:

> If a line segment intersects two straight lines forming two interior angles on the same side that sum to less than two right angles, then the two lines, if extended indefinitely, meet on that same side.

It follows that if the two angles sum to exactly 180° (for example, two 90° angles), then the two straight lines are parallel, and do not meet (hence the name of the postulate). In geometry, this postulate was long considered to be obvious or inevitable, but proofs were elusive. Eventually it was discovered that overturning the postulate required different, geometries. For example, in two-dimensional spherical geometry, like the surface of the Earth, the postulate is not valid, since two meridians, both intersecting the equator at 90°, do intersect at the pole. A geometry where the parallel postulate does not hold is known as a non-Euclidean geometry (which as a matter of fact plays a role in the Minkowski space representation of the Special Theory of Relativity).

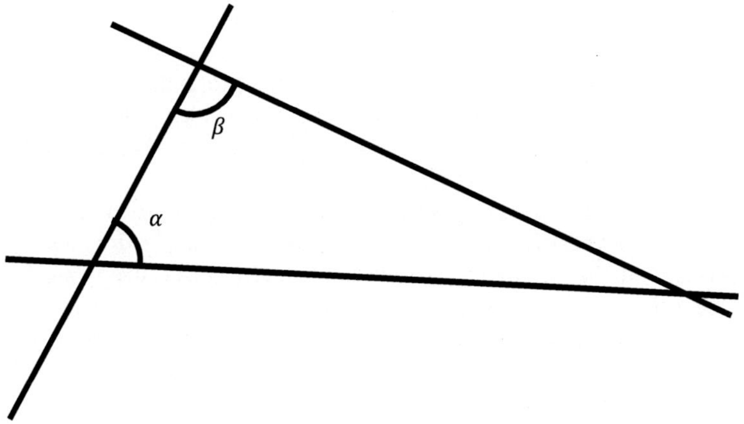

Fig. 8.10 Euclid's fifth postulate

An inertial frame of reference can be defined in this way. Consider a body with zero net force acting upon it. Such a body is not accelerating in an inertial frame of reference; that is, the body must be at rest or else traveling at a constant speed in a straight line. For example, a passenger on a train going at a constant velocity feels that he is at rest while the ground outside is moving away from him. On the other hand, a person on the ground seeing a train going at a constant velocity feels that he is at rest while the train is moving away from him.

In physics, Albert Einstein's 1905 special theory of relativity is derived from first principles now called the postulates of special relativity. Einstein's formulation only uses two postulates, though his derivation implies a few more assumptions. Here vacuum refers to the ideal vacuum of free space, which is not the same as a physically obtainable vacuum. To recall, the two postulates of special relativity are.

(1) First postulate (principle of relativity): The laws of physics take the same form in all inertial frames of reference.
(2) Second postulate (invariance of c): As measured in any inertial frame of reference, light is always propagated in vacuum with a definite velocity c that is independent of the state of motion of the emitting body. Or: alternatively the speed of light in vacuum has the same value c in all inertial frames of reference.

The two-postulate basis for special relativity is the one historically used by Einstein, and it remains the starting point today. As Einstein later admitted, the theory of special relativity implicitly relies on some additional assumptions, such as spatial homogeneity, isotropy, and independence of history (Pais, 1982).

The postulate of relativity is easy to accept. This postulate had been generally accepted since the work of Galileo. The great difficulty lies in the second postulate, the invariance of the speed of light, which is very counterintuitive. It is common understanding that the constancy of the speed of light is (only) an experimental fact. The situation is similar to that of Euclid's fifth postulate. Einstein devised the postulate of Light as an essential aspect of his special theory of relativity as given in 1905.

The universal constancy of the speed of light gives an argument in favor of empiricism, as opposed to the rationalism of Descartes'. It is an empirical fact, which, no matter how counterintuitive, we have to come to grips with. Huygens would have accepted this.

Relative to the Earth's frame of reference, sunlight originates from a moving source (the Sun) whereas candle light within a room originates from a non-moving source (the candle) on Earth. In order for the Huygens wavelet theory and Huygens' principle to work for light propagation in vacuum, the wavelets emanating from a given wavefront of light have to have the same spherical shape whatever the source of the light, moving or not. The radius of the sphere depends upon the velocity of the light. The only way this condition can be met is that the velocity of light is the same constant whatever the frame of origin. This condition is assured by the second postulate, which states that light is always propagated in vacuum with a definite velocity c that is independent of the state of motion of the emitting body. The same applies to all electromagnetic waves.

Spacetime is any model that links the three dimensions and the one dimension of time into a single four-dimensional model. Until the twentieth century, it was assumed that the three-dimensional universe was independent of one-dimensional time. It is convenient that all four be measured in the same unit, either in meters or in seconds. As a result, a conversion factor is needed to change distance into time or, inversely, time into distance. The apparent such conversion factor is c, the speed of light.

The Michelson-Morley experiments of 1887, in a failed attempt to prove that light travels through a medium known as ether, had unexpectedly demonstrated that light travels at the same speed in a vacuum regardless of whether it was measured in the direction of the Earth's motion or at right angles to it.

Thus, whether a source of light is moving towards you or away from you (in a vacuum), the light still travels at the same constant c. This behavior, which is contrary to classical physics, indicates that the speed of light is constant and does not depend on the speed of its source or its observer. In 1905, Einstein realized that the idea of ether as a medium for light to travel in was totally unnecessary. This conclusion had actually been reached earlier, by the French mathematician Jules Henri Poincaré (1854–1912) from a more mathematical point of view, in the years leading up to 1905.

Einstein realized that if one could catch up to a beam of light, as if you could travel on it (as on horseback), one would see a stationary electromagnetic wave, which is impossible. Einstein hypothesized, therefore, that the finite speed of light actually plays the role of "infinite" speed in our universe in the sense that nothing can ever travel faster than light. The speed of light is independent of the speed of the observer as well as of the speed of the source of the light. Everyone in the universe, no matter how fast they were moving, would always measure the speed of light at exactly the same value c. Hence, one could not "travel on a beam flight", and so the dilemma is resolved.

In all Earth-bound measurements, the speed of light has remained constant at c. However, the theoretical consistency of the speed of light cannot be established. In fact, it is conceivable that the speed of light was different in the past, and that light traveled much faster in the early universe than it does now. It is also conceivable that the speed of light differs in distant parts of the universe beyond our comprehension. And it is conceivable that substituting different postulates for the postulate of constant speed of light would result in valid, albeit different, relativity theories. It should be noted that Einstein did not prove that the speed of light is constant. Indeed, it is difficult to find acknowledgement that the constancy of the speed of light is only based on experiment and not on theory. Instead, Einstein took it as a postulate together with the postulate of relativity from which he derived the rest of the special theory of relativity.

Whether Huygens could have foreseen the emergence of the special theory of relativity remains of course highly speculative. But as we have seen, he had a keen sense for the relativity of space and time. In his derivation of the speed of light he (consciously or unconsciously?) applied the symmetry of what we now call the relativistic Doppler factor. And although Huygens' concept of the luminiferous ether still provided a preferential frame and therefore violated Einstein's second postulate, we may refer back to Maxwell's famous statement in Encyclopedia Britannica (quoted in Chap. 3): "The only ether which has survived is that which was invented by Huygens to explain the propagation of light." Huygens' ether was an extraordinary concept and it is indeed not too far-fetched to state that by merely changing the word "ether" to the word "vacuum," all of Huygens' results in the seventeenth century are valid today.

Chapter 9
Huygens and Centrifugal Force

> Si mobile circumferentiam percurrat quo tempore bini recursus peraguntur penduli cujus longitudo sit semidiametro aequalis erit mobilis vis centrifuga gravitati aequalis (If a mobile body traverses the circumference at the same time as the two recursions of a pendulum whose length is equal to the semidiameter, the body's centrifugal force will be equal to gravity)
> — *Christiaan Huygens, letter to Henry Oldenburg, 4 September 1669*

Galileo and Motion

Christiaan Huygens created the term "centrifugal force" in 1659; for us to understand this conceptual advance, we first recount some ancient history, and then some more recent history. The ancient Greek philosopher Aristotle (c. 384 BC–322 BC) was one of the greatest thinkers of all time. He held that objects on the Earth stopped moving once applied forces were removed. Aristotle held that objects at rest remained at rest unless a force acted on them, but that objects in motion did not remain in motion unless a force acted constantly on them. The viewpoint of Aristotle is consistent with common experience. Most objects in a state of motion do not remain in that state of motion. For example, a block of wood pushed at constant speed across a table quickly comes to rest when we stop pushing. But Aristotle failed to account for a "hidden" force, the frictional force between the surface and the object. Thus, there are two opposing forces, the force associated with the push, and the friction force that acts in the opposite direction.

Galileo Galilei, born in 1564, threw aside Aristotle's thinking on motion, which had remained unquestioned up to Galileo's time. Instead, Galileo formulated the concept of inertia: namely, that an object in a state of motion possesses an inertia that causes it to remain in that state of motion unless an external force acts on it. Galileo

T. J. Moser and E. A. Robinson, *Walking with Christiaan Huygens*, History of Physics,
https://doi.org/10.1007/978-3-031-46158-3_9

understood that as the frictional forces were decreased (for example, by placing oil on the table) the object would move further and further before eventually stopping. From this he derived a basic form of the law of inertia: if the frictional forces could be reduced to exactly zero (which is actually not possible in a realistic experiment, but it can be approximated to high precision) an object pushed at a given speed across a frictionless surface of infinite extent will continue at that speed forever after we stop pushing.

Galileo's most famous experiment involved falling bodies of different weight. According to Aristotle, heavier objects fall faster than lighter ones, and an object that weighs twice as much as another would fall twice as fast. But Galileo tested this hypothesis and found that a cannon ball weighing 10 pounds will not reach the ground ahead of a musket ball weighing less than a pound, provided both are dropped from the same height. Galileo also showed conclusively, by a short and clear argument, that a heavier body does not move more rapidly than a lighter one.

Galileo's simple argument assumed two stones in free fall, one light and one heavy, which are connected by a very thin and weak string. When the two stones are dropped, the string does not break. The light stone could not pull up on the heavy stone, resulting in a slower fall. The two could not work together as a heavier object resulting in a faster fall. Therefore, two objects of different weight fall at the same speed and hit the ground at the same time (and simultaneously knock down an in hindsight surprisingly naïve theory that had been accepted for millennia).

To express this mathematically, let Δt be any increment of time t and let Δx be the corresponding increment of distance x, through which an object moves in time Δt. Galileo's definition for uniform velocity requires that velocity, defined as the ratio $v = \Delta x/\Delta t$ is a constant. The concept of velocity defined in this way is one of the core ideas of differential calculus, which would begin to evolve into its modern form about 50 years later and serve as the springboard for what we now call "classical physics."

The next problem considered by Galileo, accelerated motion, requires some mathematical imagination. Galileo sought an answer by studying bodies falling with acceleration as occurs in nature. In conformity with his definition of uniform velocity, he defined uniform accelerated motion in this way. The ratio $a = \Delta v/\Delta t$ is a constant. In other words, Galileo applied the same idea to acceleration as he did to velocity. Let Δt be any increment of time t and let Δv be the corresponding increment of velocity v of the aforesaid body. Galileo's definition for uniform acceleration is that acceleration, defined as the ratio $a = \Delta v/\Delta t$, is a constant.

It was too difficult to measure the terminal speed of the body. However Galileo was able to measure the total distance x travelled and the total travel time t. He wanted to find, through mathematics, how the distance increased with the total time of objects falling under the assumption of uniform acceleration. People already knew how to find the total distance from the total time for motion at constant speed. Galileo successfully derived an equation that relates distance to time of fall for motion at constant acceleration.

For instructive purposes, let us first use calculus and then explain how Galileo solved the problem without calculus.

For motion at *constant velocity*, we have:

t = time, x = distance, v = speed = $\frac{x}{t}$ = constant
Solution: $x = vt$
Example: Given $v = 4,\ t = 2$
Solution: $x = vt = 4 \times 2 = 8$

For motion at *constant acceleration* (that is, with the velocity changing uniformly, also known as *uniformly accelerated rate*), we have:

t = time, v = velocity, a = acceleration = constant
Solution: $v = at,\ x = \int_0^t at\ \mathrm{d}t = a \int_0^t t\ \mathrm{d}t = a\left(\frac{t^2}{2}\right)$
Example: Given $a = 3,\ t = 2$
Solution: $v = \frac{\mathrm{d}x}{\mathrm{d}t} = at = 3 \times 2 = 6, x = a\left(\frac{t^2}{2}\right) = 3 \times \frac{4 \times 4}{2} = 24$

In the language of differential calculus, velocity is the first derivative of the distance with time, acceleration is the second derivative of distance with time (and in its turn the first derivative of velocity with time). Just as the physical units of velocity can be expressed as meter-per-second, the physical units of acceleration are meters-per-second-ser-second. As we see, the determination of x involves the evaluation of the integral:

$$x = \int_0^t v\ \mathrm{d}t\ . \tag{9.1}$$

Let us show how Galileo evaluated of this integral. The variable x denotes the distance traversed in time t when a body starts at rest and is uniformly accelerated. Because the body starts at rest, the initial velocity is zero. Let the final velocity be denoted by v. Thus the average velocity is:

$$V = \frac{0 + v}{2} = \frac{v}{2}\ . \tag{9.2}$$

Suppose that the same body moves at the uniform speed given by the average velocity V. It would traverse the given distance x in the same time t. Galileo transformed the case of uniform acceleration to an equivalent case of uniform velocity. In the case of uniform velocity V, the distance is equal to velocity multiplied by time, so

$$x = Vt = \frac{v}{2}t\ . \tag{9.3}$$

In the case of uniform acceleration, the velocity is equal to acceleration times time; that is,

$$v = at\ . \tag{9.4}$$

By combining the two Eqs. (9.3) and (9.4), Galileo obtained his law of motion:

$$x = v\frac{t}{2} = (at)\frac{t}{2} = \frac{1}{2}at^2\ . \tag{9.5}$$

Thus Galileo found that the distance covered by a body starting from rest with uniform acceleration is proportional to the square to the time employed in traversing this distance. Isaac Newton used this most celebrated result, about half a century later, to establish his three Laws of Mechanics.

Einstein's General Theory of Relativity

As we have seen in Chap. 8, Albert Einstein in 1905 made a major advance with his theory of special relativity. However, special relativity excludes gravity. Einstein created the general theory of relativity during the years 1907–1915. This theory unified gravity with relativity. General relativity produced a theory of gravitation that superseded Newton's theory of gravitation.

Mass is a property of any physical object. However, there are two aspects of mass which are apparently distinct. Inertial mass is a measure of resistance to the acceleration in the presence of external force. It is the mass which appears in Newton's Second Law of Mechanics: $f = ma$.

The second aspect of mass is gravitational mass, which determines the strength of gravitational force. Gravity is the tendency of massive objects to attract each other by means of action at a distance. Action at a distance is the concept that an object can be moved, changed, or otherwise affected without being physically touched (as in mechanical contact) by another object. A fundamental statement made by Einstein's general theory of relativity is that these two apparently distinct properties of matter are, in fact, the same.

To elaborate: imagine a rocket ship in deep space far away from gravitational fields of planets and moons. See Fig. 9.1, right side. The rocket ship with its engines off is moving at steady speed in a straight line. With no gravity, a passenger will tend to float around inside the ship. He feels totally isolated from the outside world.

Now turn the engines on so that the rocket ship experiences an acceleration of 9.8 meters per second per second (m/s^2). That acceleration is exactly the same as the acceleration due to gravity on the Earth (Fig. 9.1, left). The rocket ship and its contents, including the passenger, accelerate "upwards" at a constant acceleration. The passenger finds himself moving towards what is now the "floor" and needs to

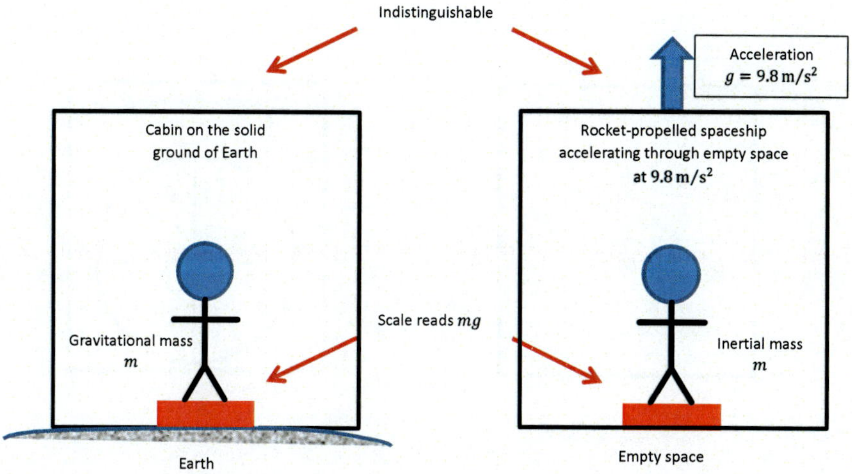

Fig. 9.1 The principle of equivalence states that there is no experiment that will discern the difference between the effect of gravity and the effect of acceleration. In other words, gravitational mass and inertial mass are equivalent. (The numerical value for the acceleration due to gravity is $g = 9.8\text{m}/\text{s}^2$. The mass is denoted by m)

use his leg muscles to stand. If he drops a book, it falls to the floor just as it would fall on Earth.

Einstein states that (from an outside perspective) being inside the uniformly accelerating space ship is physically equivalent to being in a uniform gravitational field. This is the basic postulate of general relativity. Special relativity said that all inertial frames of reference are equivalent. General relativity extends this result to accelerating frames, stating that they are equivalent to inertial frames in which there is a gravitational field. This is called the *Equivalence Principle*. This Principle states that "No experiment whatsoever can be performed that could distinguish between a uniform gravitational field and a uniform acceleration." In other words, gravitational mass and inertial mass are equivalent. Gravitational "force" as experienced locally while standing on a massive body (such as the Earth) is the same as the apparent force experienced by an observer in an accelerated frame of reference.

Huygens' Centrifugal Force

Now let us see how these ideas are related to Huygens' thoughts on centrifugal force. In Fig. 9.2, we have:

(1) The falling weight (that is, an object falling down by gravity),
(2) The whirling weight (that is, the same object whirling in a horizontal circle at the end of a string).

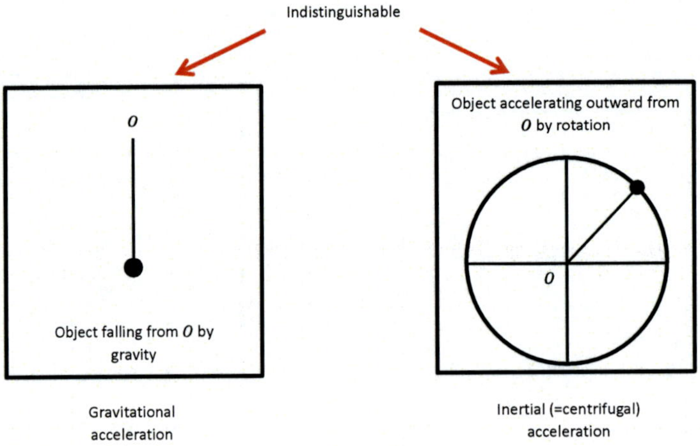

Fig. 9.2 Gravitational acceleration and centrifugal acceleration. Left: The hanging weight. The tension in the string results from the weight's tendency to fall at an uniformly accelerated rate due to gravity. Right: The whirling weight. The tension in the string results from the weight's tendency to fall to the center O at a uniformly accelerated rate due to rotation

Tie an object to one end of a piece of string and whirl the object round in a circle, making the other end of the string the center of the circle. For some rate of whirling, the tension in the string is the same in both cases. Huygens postulated the *principle of equivalence*, namely that the two physical situations (1) and (2) are dynamically equivalent (in the sense that the forces are equal). In case (1) the tension in the string results from the weight's falling at constant acceleration g due to gravity. In case (2) the tension in the string results from the whirling object pulling away from the center of its circular trajectory at the same constant acceleration g. In this instance, the constant acceleration is not due to a change in the magnitude of the velocity but to its direction. In other words, it is not due to gravity but due to rotation.

The object is moving in a circle under the influence of a force transmitted along the string. Now, a taut string can transmit a force only along its length, not transverse to itself. Hence at any given moment the force acting on the object must always be directed along the string, that is, toward the center of the circle.

Christiaan Huygens created the term "centrifugal force" in 1659 and wrote of it in his 1673 *Horologium Oscillatorium.* A full account is given in his book *De Vi Centrifuga* ("*About the centrifugal force*", Complete Works Vol. XVI) which was published posthumously in 1703. Huygens gave the first accurate account of centrifugal force. For this purpose Huygens devised the *Equivalence Principle,* which Einstein, nearly 250 years later, used to formulate general relativity. Using Galileo's law of free fall $x = gt^2/2$ and his own principle of equivalence, Huygens was the first to give a correct derivation of centrifugal force.

See Fig. 9.3. The rotation of an object at the end of a string tends to stretch the string. Let the circle represents the Earth's equator. If gravity were suddenly eliminated when the object is at A, the object would travel with constant speed along the tangent AB. Segment BC is the distance through which the object moving on the circular trajectory (that is, the arc AC) falls away from the straight line trajectory AB. Suppose that it takes time Δt at velocity v for the object to travel from A to C along the circle: that is,

$$v\Delta t = \text{arc}AC \ . \tag{9.6}$$

From Fig. 9.4, we have:

$$\frac{\Delta v}{AC} = \frac{v}{r} \ . \tag{9.7}$$

We make the approximation:

$$AC \approx \text{arc}AC = v\Delta t \ . \tag{9.8}$$

Thus:

$$\frac{\Delta v}{v\Delta t} = \frac{v}{r} \ . \tag{9.9}$$

The acceleration a is defined by:

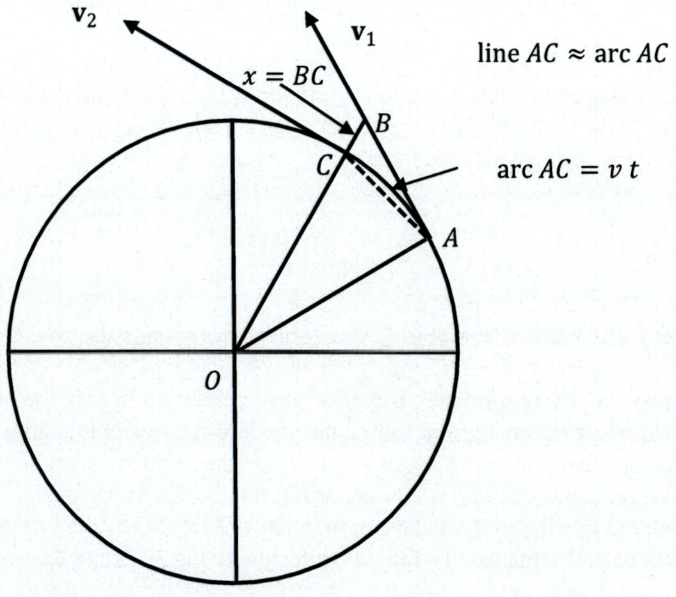

Fig. 9.3 An object rotating at the end of a string

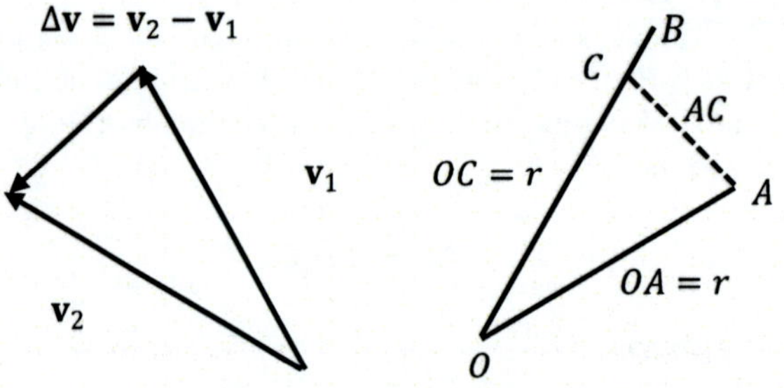

Fig. 9.4 Similar isosceles triangles as shown in Fig. 9.3: the magnitudes of velocity vectors $\mathbf{v}_1$ and $\mathbf{v}_2$ are equal but they make an angle equal to $\angle AOC$. Therefore, base $\Delta v = \|\Delta\mathbf{v}\|$ of left triangle divided by the base AC of the right triangle is equal to side v of left triangle divided by side r of right triangle

$$a = \frac{\Delta v}{\Delta t}\,, \tag{9.10}$$

so

$$\frac{a}{v} = \frac{v}{r} \text{ or } a = \frac{v^2}{r}\,. \tag{9.11}$$

This is the equation for the acceleration a of the object falling from B to C in terms of the velocity of the object in going around the circle and the radius r of the circle.

Huygens argued that because the tension in a plumb line is proportional to the mass of the suspended object, the tension in the string restraining a whirling object must likewise be proportional to the object's mass. Thus the centrifugal force exerted by an object whirling with speed v at the end of a string of length r is given by the formula

$$F = ma = \frac{mv^2}{r}\,, \tag{9.12}$$

where m is the mass of the object. The force F represents the object's effective weight.

The above derivation reflects Huygens' own derivation in two aspects. First, it uses geometrical arguments and formulates the law in terms of the ratios of certain line segments. In this respect Huygens still follows the geometrical tradition. Second, the derivation makes *implicit* use of an infinitesimal calculus argument: in the limit of the point C moving to A we can approximate the arc AC by the line AC.

A modern derivation runs as follows (referring to Fig. 9.3): for BC we have

$$AO^2 + AB^2 = BO^2\,,$$

$$BC = BO - r\,,$$

$$BC = \sqrt{r^2 + (v\Delta t)^2} - r = r\sqrt{1 + \left(\frac{v\Delta t}{r}\right)^2} - r \approx r\left(1 + \frac{1}{2}\left(\frac{v\Delta t}{r}\right)^2\right) - r$$
$$= \frac{1}{2}\frac{v^2}{r}(\Delta t)^2\,. \tag{9.13}$$

For the distance propagated under a uniform acceleration, Galileo obtained $\frac{1}{2}a(\Delta t)^2$ (Eq. 9.5), so a comparison gives $a = \frac{v^2}{r}$. This derivation makes use, in the third step, of a first-order Taylor approximation.

In his derivation Huygens employed basic concepts of the theory of motion that Newton formulated years later in *Principia Mathematica*. Specifically Huygens used:

(1) the relation $F = ma$ between force, mass, and acceleration,
(2) the proportionality between an object's weight F and its mass m, and
(3) the equality of action (the object's weight) and reaction (the tension in the restraining string).

Huygens employed symmetry arguments to derive conservation laws for momentum and for energy. Huygens used what Einstein later called the principle of equivalence to derive the formula for centrifugal force. Two of the greatest achievements in physics are Newton's theory of gravity and Einstein's theory of gravity. Huygens' work provided central elements basic to each of these theories.

Frames of Reference

We want to describe the way an object moves, but we have to ask the question: "moves compared to what"? Galileo's law of inertia says that, if a body is at rest or moving at a constant speed in a straight line, it will remain at rest or keep moving in a straight line at constant speed unless it is acted upon by a force. That "straight line" is drawn in (referred to) a "frame of reference". A frame of reference is usually the world around us. For many practical purposes, the world around us appears to be at rest. Of course, we know that the Earth is spinning around its axis, so while sitting in our chair, we are also spinning at the same rate, although we do not feel it. We return below to this apparent paradox, and the role of centrifugal force.

Now suppose we use some particular portion of the world, such as a railway car, which is moving relative to the rest of the world. Galileo proposed that in all frames of reference which are moving uniformly relative to each other, the laws of nature must be the same. All frames of reference, in which this Law of Inertia holds true, are called inertial frames. The frames in which the Law of Inertia is not applicable are called non-inertial frames. An inertial frame of reference is one in which a body with zero net force acting upon it does not accelerate. In other words, such a body is at rest or moving at a constant velocity. Conceptually, the physics of a system in an inertial frame has no causes external to the system. A non-inertial reference frame is a frame of reference that is undergoing acceleration with respect to an inertial frame. The laws of motion in non-inertial frames vary from frame to frame depending on the acceleration. A rotating frame of reference is an example of a non-inertial frame.

In deriving the centrifugal force Huygens produced the first detailed example of an inertial force. An inertial force is also called a "fictitious force", or a "pseudo-force", or a "d'Alembert force" (named after the French mathematician Jean le Rond d'Alembert 1717–1783). It is a force that appears to act on a mass whose motion is described using a non-inertial frame of reference, such as an accelerating or rotating reference frame. The prime example is the centrifugal force that pushes objects outwards from the center of a rotating reference frame. Another example occurs in an automobile that is accelerating in the forward direction. The passengers perceive that they are acted upon by a force in the backward direction pushing them back into their seats.

An inertial force is due to an object's inertia when the reference frame does not move inertially, and thus begins to accelerate relative to the free object. The inertial force thus does not arise from any physical interaction between two objects, such as contact force or electromagnetic forces. Instead the inertial force is due to the acceleration of the non-inertial reference frame itself. From the viewpoint of the frame, the inertial force is the "force" that makes this acceleration happen. Assuming Newton's second law $F = ma$ (where a is the inertial acceleration), the inertial force F is always proportional to the mass m.

Christiaan Huygens introduced the term centrifugal force in 1659 in his *De Vi Centrifuga.* Newton coined the term centripetal force in 1684 in his *De motu corporum in gyrum.* In Latin, *fugere* means "to flee away", *petere* means "to search for". A centripetal force is a force that causes a body to follow a curved path. See Fig. 9.5. Its direction is towards the momentary center of the circle of curvature and perpendicular to the velocity vector in the inertial system. It is a force by which a body is drawn to a point as to a center. In Newtonian mechanics, gravity provides the centripetal force causing astronomical orbits.

The centripetal force is a "real" force. It attracts the object toward the center and prevents it from flying out. The source of the centripetal force depends on the object in question. For the Moon in orbit, the force comes from gravity. If an object is being whirled around on a string, the centripetal force is provided by tension in the rope. For an automobile moving around a curve, the centripetal force comes from friction between the tires and the road. The centripetal force satisfies the principle of action and reaction, because it corresponds to a counterforce (see Fig. 9.5).

The centrifugal force is not a "real" force. It is a "fictitious" force. The tendency to fly outwards is observed because objects that are moving in a straight line tend to continue moving in a straight line. This behavior is due to inertia, which causes objects to resist to the force that makes them move in a curve. The centrifugal force does not satisfy the principle of action and reaction, because there is no counterforce.

Historically the "real" centripetal force was much easier to grasp than the profound "fictitious" centrifugal force. In his "solar vortex theory" Leibniz conceived of

Fig. 9.5 Centrifugal and centripetal force

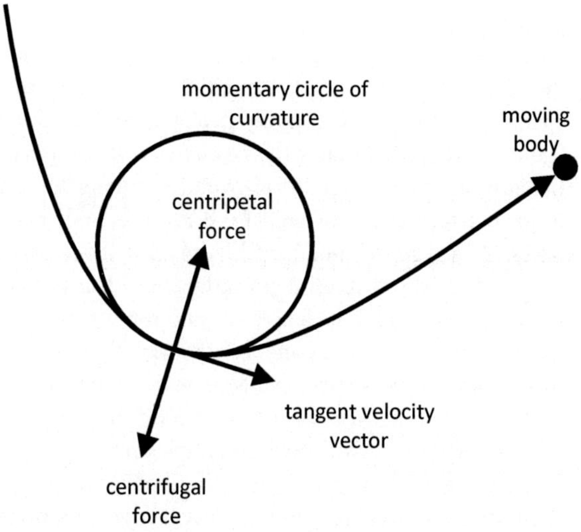

centrifugal force as a real outward force which is induced by the circulation of the body upon which the force acts. Newton in *Principia* crucially limited the description of the dynamics of planetary motion to a frame of reference in which the point of attraction is fixed. In this description, Leibniz's centrifugal force was not needed and was replaced by only continually inward forces toward the fixed point. Like Leibniz, Huygens was a Cartesian and critic of Newton. However, Huygens concluded that Leibniz's invocation of a harmonic vortex was logically redundant, because Leibniz's radial equation of motion follows trivially from Newton's laws. In other words, Leibniz's harmonic vortex as the basis of centrifugal force was dynamically superfluous.

In *Principia*, Newton described the role of centrifugal force upon the height of the oceans near the equator. The effect of centrifugal force in countering gravity, as in this behavior of the tides, has led centrifugal force sometimes to be called "false gravity" or "imitation gravity" or "quasi-gravity".

In 1746, the Swiss mathematician Daniel Bernoulli was the first to put forth the idea that the centrifugal force is "fictitious". Bernoulli demonstrated that the magnitude of the centrifugal force depends on which arbitrary point is chosen to measure circular motion. Later in the eighteenth century Joseph Louis Lagrange explicitly stated that the centrifugal force depends on the rotation of a system of perpendicular axes. In 1835, French mathematician and engineer Gaspard-Gustave de Coriolis (1792–1843) analyzed arbitrary motion in rotating systems, specifically in relation to waterwheels. He described the "compound centrifugal force" which eventually came to be known as the Coriolis Force, which causes the wind to blow clockwise around an anticyclone in the Northern Hemisphere.

It was found that accelerating frames exhibited "fictitious forces" like the centrifugal force. These forces did not behave under transformation (to other frames) like other forces did, thus providing a means of distinguishing various frames. In particular, fictitious forces *did not appear at all* in some frames: those frames differ from that of the fixed stars only by a constant velocity. In short, a frame tied to the "fixed stars" is merely a member of the class of "inertial frames", and absolute space is an unnecessary and logically untenable concept. The preferred, or "inertial frames", were identifiable by the *absence of fictitious forces*. The equations of motion in a non-inertial system differ from the equations in an inertial system by additional terms called inertial forces. This allows us to detect experimentally the non-inertial nature of a system—and it is the only way to do so.

The idea of an inertial frame was used in the special theory of relativity. This theory said that all physical laws (and not just the laws of mechanics) should appear of the same form in all inertial frames. Specifically, Maxwell's equations should apply in all frames. Because Maxwell's equations implied the same speed of light in the vacuum of free space for all inertial frames, inertial frames now were found to be related not by the Galilean transformation, but by the Lorentz transformation. The general theory of relativity eliminated the special position of inertial frames, at the cost of introducing curved space–time. Following an analogy with centrifugal force (sometimes called "artificial gravity" or "false gravity"), gravity itself became a fictitious force, as given in the modern form of Huygens' principle of equivalence. This principle states that there is no experiment that observers can perform in order to distinguish whether any observed acceleration arises because of a gravitational force or because their reference frame itself is accelerating. In conclusion, Huygens' centrifugal force played a key role in establishing the set of inertial frames of reference and the significance of fictitious forces, even in the development of general relativity.

Now, why is it that we do not experience, while sitting in our chair, the centrifugal force arising from the spin of the Earth? We return to the formula of Christiaan Huygens (Eq. 9.11). The Earth's radius is about 6371 km, so its circumference 40,075 km and the linear velocity v at the surface is $40,075\ \text{km/day} = 463\ \text{m/s}$. Plugging this in Huygens' formula results in a value for the acceleration of $a = 0.034\ \text{m/s}^2$. This means that the centrifugal force experienced by an object on the surface of the Earth is quite small, only about 0.3% of the force of gravity, which is too small for us to notice. With that said, there are many practical issues which arise from the non-inertial frame of reference which is the world around us. Most obviously, there is the weather, wherein centrifugal and Coriolis forces affect the motion of air masses.

Huygens' Governor

In 1659, Huygens derived the formula for centrifugal force in his work *De vi centrifuga*, published in 1703. The formula played a central role not only in the development of classical mechanics, but also in the onset of the Industrial Revolution. This Revolution refers to the rapid development of industry that took place first in Britain and then in continental Europe and America during the late eighteenth century and the nineteenth centuries. It was brought about by the introduction of

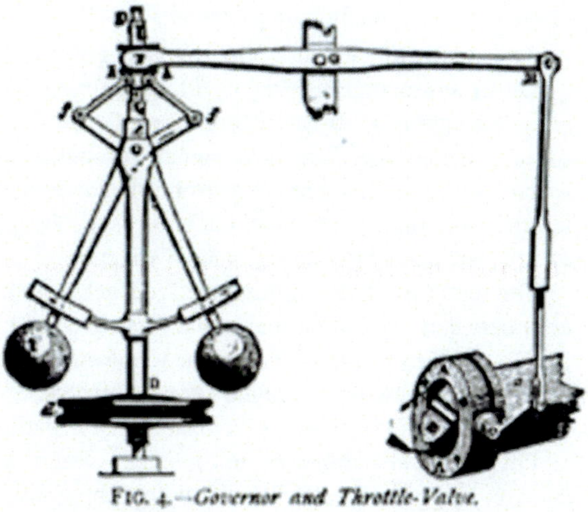

Fig. 9.6 Centrifugal "fly-ball" governor. The balls swing out as speed increases, which closes the valve, until a balance is achieved between demand and the proportional gain of the linkage and valve

machinery. It was characterized by the use of steam power, the growth of factories, and the mass production of manufactured goods.

A centrifugal governor is a device with a feedback system that regulates machinery, for example to regulate the fuel flow to an engine to control engine speed and keep it close to constant. It applies the principle of proportionality. Christiaan Huygens invented centrifugal governors. In the seventeenth century, they were used to control the distance and pressure between millstones in windmills. James Watt modified the Huygens governor in 1788 so that it could manage his steam engine. The governor controlled how much steam was allowed to enter the engine's cylinders. Watt is sometimes referred to as the inventor of this governor, because this advancement was so significant. Steam engines saw the greatest use of centrifugal governors during the Steam Age in the nineteenth century.

Ancient Greek and Roman water clocks had used float valves to control water flow; flush toilets today use float valves of a similar design to control tank water level. Pumping water required the use of early steam engines with purely reciprocating motion. They could tolerate variations in the working speed, so governors were not needed. The Scottish engineer James Watt developed the rotative steam engine to operate factory machinery. Watt's fly-ball governor was able to maintain a constant operating speed. It was a centrifugal feedback valve used to regulate the engine's speed. It measures how quickly a shaft is spinning using weights mounted on spring-loaded arms, and then utilizes proportional control to regulate the engine's speed (see Fig. 9.6).

The theoretical basis for the operation of governors was described by James Clerk Maxwell in 1868 in his seminal paper "On Governors." The term governor was used for any device that automatically regulates the supply of fuel, steam, or water to a machine, ensuring uniform motion or limiting speed. In the twentieth century governors were applied in all kinds of uses in communication and control. The term governor was gradually replaced by the word "servo-mechanism" or simply "servo". A mathematical treatment of the Huygens governor was given by Poor (1925).

Let us summarize this theory of the operation of governors. The desired state is for the engine to run at a "target speed". The engineer communicates the target speed to the engine. The "error signal" represents the deviation between the actual speed and the target speed. This error is then interpreted by the governor to adjust the throttle, resulting in a change that reduces the error in speed. A basic loop consists of a receptor, a control center, and an effector.

The terms positive and negative feedback may be defined as the altering of the error between target value and actual value of a parameter. The parameter in this case is speed. If the error is widening due to the effector, it is called positive feedback. If the error is narrowing it is called negative feedback. Since the steam engine governor narrows the error, it serves as an example of negative feedback.

Physiological parameters must remain within a narrow range for a person to survive. Negative feedback loops within the body keep physiological parameters, such as heart rate, within a target range. For example, the average resting heart rate should remain in the target range from 60 to 100 beats per minute. A negative feedback loop works by adjusting an output, such as heart rate, in response to a change in input, such as blood pressure. In this case, the receptor measures blood pressure, the control center is the brain, and the effector is the heart muscle.

If a person is at rest and blood pressure increases, pressure receptors in the carotid arteries detect this change in input and send nerve impulses to the medulla of the brain. This action tells the brain to reduce nerve impulses that stimulate the heart muscle to contract. The heart contracts more slowly and the heart rate decreases, causing blood pressure to decrease to within target levels.

When exercising, there is an increase in heart rate and blood pressure. The body increases blood flow to muscle tissue in response to the increased demand for oxygen. The homeostatic set points of heart rate and blood pressure are therefore "reset" higher. Vigorous exercise can make heart rate increase to as much as 180 beats per minute or more. Negative feedback loops act to maintain heart rate and blood pressure within these new higher target ranges. After exercise the muscle tissues no longer demand as much oxygen, so homeostatic set points are reset back to the original target.

Chapter 10
Huygens and Curvature

Theorematis quaedam retineri possent item. Reliqua vulcano tradenda. (Certain theorems can be retained. The rest can be given to the volcano)
— *Christiaan Huygens*

Three of the most important works on the science of physics in the seventeenth century are Galileo's *Discourses and Mathematical Demonstrations Relating to Two New Sciences* (1638), Christiaan Huygens' *Horologium Oscillatorium* (1673) and Isaac Newton's *Philosophiæ Naturalis Principia Mathematica* (1687). Huygens dedicated *Horologium Oscillatorium* to his patron King Louis XIV of France. The dedication was an eclogue (a poem in which shepherds converse) of some 280 hexameters, composed by the Dutch poet Vallius (latinized form of Adrianus van der Walle). In honoring Huygens, Vallius tried to emulate Virgil, the Roman poet who honored Julius Caesar by comparing him with Daphnis. Daphnis was a figure from Greek mythology, a shepherd who was said to be the inventor of pastoral poetry. What is striking in this poem is that Huygens is placed on the side of Phoebus Apollo (the Greek god of light, spring, purity, moderation and arts) as opposed to Pan (the Greek god of the forest, nature and boundless sensuality): that is to say that he is glorified as an intellectual hero seeking pure science in the first place.

Nymphs of Scheveningen, dwelling in pure waves,
You who safeguard these adjoining lands,
Delights of all those who frequent your shores,
Beloved by Phoebus, enchanting and wise spirits,
Please come to my aid and inspire my songs.
May neither Pan nor his hideous Fauns
Ever approach these revered places,
And soil the whiteness of your foaming sea.
You, whom Glory elevates even above the heavens,
Deign to lend an ear, Huygens, to this poem
Written in your honor, in the honor of your father,
Your brothers also, and your entire lineage
Of which you are the flower. Your sublime thoughts

T. J. Moser and E. A. Robinson, *Walking with Christiaan Huygens*, History of Physics,
https://doi.org/10.1007/978-3-031-46158-3_10

In a mythic setting will be placed here.
May thus the rich fantasy of the ancients
Adorn noble astronomy with its splendors.

Part 1 of *Horologium Oscillatorium* contains the descriptions of clock designs. The rest of the book is devoted to the analysis of pendulum motion and a theory of curves.

Part 2 gives essentially the law of inertia and the law of composition of "motion". Huygens uses these rules to re-derive Galileo's original study of falling bodies. He then describes constrained fall, obtaining the solution to the tautochrone problem as given by a cycloid curve and not a circle as Galileo had conceived. We have already met the tautochrone problem in Chap. 6.

Part 3 outlines a theory of evolutes and the rectification of curves. The term rectification as used here refers to the determination of the length of a curve. In other words, rectification is the finding of a straight segment equal to a given arclength. In modern terms, we refer to this as line integration. The evolute of a curve is the locus of all its centers of curvature; in this chapter, we will come back to the theory of evolutes in some detail.

Part 4 is concerned with the study of the center of oscillation. The derivations are based on a single assumption: that the center of gravity of heavy objects cannot lift itself, which Huygens used as a virtual work principle. In the process, Huygens obtained solutions to dynamical problems such as the period of an oscillating pendulum as well as a compound pendulum, the center of oscillation and its interchangeability with the pivot point, or suspension point, and the concept of moment of inertia.

The last part of the book gives propositions regarding bodies in uniform circular motion, without proof, and states the laws of centrifugal force for uniform circular motion.

Cycloids

From ancient times onwards, it was understood that material objects are in a state of continual motion. Things move from one place to another. They undergo transformations. They change to other things and move to other places. Nothing remains constant forever. Recall Heraclitus' quote: "One cannot step into the same river twice". There are many kinds of change. Changing position is called motion. However, motion is always relative to something. A tree rooted to Earth is not moving relative to Earth. But, the Earth is in orbit about the Sun, and in turn the Sun is in orbit in the Milky Way galaxy. Thus, relative to an observer in space, the tree is moving. When describing motion, some system of reference must first be specified.

The first great law of motion formulated by Isaac Newton (1642–1727) is called the law of inertia: "Everybody continues in its state of rest, or in uniform motion in a straight line, unless it is compelled to change state by forces impressed on it."

Changing position can take place in many ways. The motion may be rectilinear (that is, along a straight line), it may be curvilinear, or it may be oscillatory, and so forth. Further, motion may occur with constant velocity, or with changing velocity, in which case it is said to be accelerated motion. Motion is most often and easily conceptualized by setting up a coordinate system, for example, a Cartesian representation. This technique, very familiar in the construction of graphs, was invented by René Descartes (1596–1650). The Cartesian axes constitute the frame of reference which gives meaning to statements about motion of objects in a system. In describing the motion of an object, we are concerned with two related changes: (1) in position and (2) in time. The distance travelled is usually described as a "function" of elapsed time, but another equally valid point of view would be that time is a function of distance.

"Function" is a relationship basic to mathematics and science. It is the name of any mathematical expression that describes a relation between variables, that is, between linked values of things which can change. In accordance with a definition that Leonhard Euler (1707–1783) gave in 1749, a function is often explained as a variable quantity that is dependent on another variable quantity. The mathematical conception regarding the nature of functions—expressions that fix the relationship between variable quantities—has had a long evolution. Serious investigation of functions probably did not effectively begin until the seventeenth century when two great advances were made in quick succession—the introduction of analytic geometry by Descartes and the development of the basic underpinnings of calculus by Isaac Newton and Gottfried Wilhelm Leibniz (1646–1716). These new, extremely powerful methods of analysis made more exact knowledge of functions and their properties supremely important.

Leibniz was not only a scientist but a noted philosopher. According to English critic Thomas De Quincey (1785–1859), Leibniz was not like other great thinkers, whom he compared to planets revolving in their orbits. Leibniz, said De Quincey, was like a comet that joined different systems. Leibniz, by age 26, had designed a calculating machine which served as the basis for all mechanical calculating machines for nearly two hundred years; he had devised a program of legal reform for the Holy Roman Empire; and he had developed a plan to divert the French from their attacks on the Rhineland, which would include convincing the French king, Louis XIV, to build a Suez canal. He was well known in England. He had visited London in 1673 on a diplomatic mission, met Henry Oldenburg (1618–1677, German diplomat and natural philosopher), and had been elected to the (British) Royal Society in London (which had been founded by Oldenburg). Most important, Leibniz's European travels put him in touch with Christiaan Huygens. They carried on an extensive correspondence. Huygens, by his study of lines tangent to a curve, was only a small step away from the invention of calculus.

A type of motion that is very frequently observed in nature is vibratory motion; for example, the back-and-forth movement of a pendulum, the up-and-down motion of a bobbing spring, the oscillations of the prongs of a tuning fork, and, much more generally, the behavior of all instruments for emitting sound. Vibratory motion involves the molecules themselves as sound waves are propagated through the air. In fact, vibratory motion is typical of all waves, whether water waves, visible light, radio waves, x-rays, or gamma-radiation; it is also typical of the electrons in the wires of almost all circuits in use.

Historically the Netherlands has always been a great seafaring nation, and Huygens' first scientific and technological contributions involved improvements to the two most important navigational tools of the seventeenth century, the telescope and the clock. Huygens, helped by his brother Constantijn, designed a telescope that was far superior to contemporary devices, and in March 1655 discovered Saturn's moon Titan. In that period, the central problem of navigation was determining longitude. Longitude can, in effect, be measured by time because if the difference in local time at two points is known, the longitudinal distance between them can be computed. However, during the first part of the seventeenth century, this was not a practical option because the existing mechanical clocks were not sufficiently accurate.

A conventional pendulum is a weight suspended from a pivot so that it can swing freely. When a pendulum is displaced sideways from its resting (equilibrium) position, it is subject to a restoring force due to gravity that will accelerate it back toward the equilibrium position. When released, the restoring force acting on the pendulum's mass causes it to oscillate about the equilibrium position, swinging back and forth. The time for one complete cycle, a left swing and a right swing, is called the period. The period depends on the length of the pendulum and also to a slight degree on its amplitude (that is, the width of the pendulum's swing). Huygens is the true inventor of the pendulum clock in which the escapement mechanism counts the swings and a driving weight provides the push, to overcome friction in the pivot and resistance from the air. In effect, the escapement is a feedback regulator that controls the speed of this type of mechanical clock. Huygens, with the help of Salomon Coster, produced his first pendulum clock in December 1656, and it was much more accurate than contemporary clocks. It is not an overstatement to contend that this was one of the great technological breakthroughs in history. Pendulum clocks were the most accurate clocks in the world for the next 300 years. Huygens' invention of the first accurate clock can be considered one of the major discoveries that shaped and determined the modern world, which is based on science and technology. This invention enabled much more sophisticated experiments and detailed measurements (details are given in Chap. 6).

Galileo believed that a pendulum is isochronous; in other words, that the period of a pendulum does not depend on the amplitude of its swing. Huygens, via mathematics, found that a pendulum swinging through the arc of the circle is not isochronous. It only appears isochronous when the length of the arc is quite short relative to the length of the pendulum. This property gives a clock with a long pendulum an advantage over a clock with a short pendulum. However, the pendulums of the early clocks were kept short and light to minimize the amount of energy needed to keep them in motion. As a result, the early pendulum clocks had very wide pendulum swings, which decreased their accuracy.

The word derived from Greek for *same time* is *tautochrone* and, in mathematics, the tautochrone problem consists of finding the curve along which a bead placed anywhere on the curve will fall to the bottom of the curve in the same amount of time. The solution is a cycloid, a fact first discovered by Huygens. The cycloid is intimately related to the circle; a cycloid is the locus of a point on the rim of a circle rolling along a straight line. In *Horologium Oscillatorium* (1673), Huygens gives a complete mathematical description of an improved pendulum clock and calls such a device a cycloidal clock because its pendulum is forced to swing in an arc of a cycloid. Huygens did this by suspending the pendulum (made up of a bob on a wire string) at the cusp of the evolute of the cycloid. The cycloidal clock was extremely accurate, but unfortunately the movement caused an excessive amount of friction.

Johann Bernoulli (1667–1748), also known as Jean or John, was a Swiss mathematician and a prominent member of the Bernoulli family, which consisted of many renowned mathematicians. Johann's elder brother Jakob introduced the term "integral" in the context of integration and discovered the number $e = 2.71828\ldots$, now known as "Euler's number". In June 1696 Johann Bernoulli became the first to solve the brachistochrone problem, which can be stated as:

> Given two points A and B in a vertical plane, what is the curve traced out by a point acted on only by gravity, which starts at A and reaches B in the shortest time.

The statement and solution of the brachistochrone problem can be regarded as the origin of *variational calculus,* a field of mathematical analysis that uses variations to find maxima and minima of functionals, which are "functions of functions" or mappings from a set of functions to the real numbers.

Let us give the solution obtained by Johann Bernoulli. A heavy point object (a bead) starts from rest at the point A and slides without friction along a curve to a lower point B. There are infinitely many such curves. The problem is to find the curve down which a bead sliding from rest and accelerated by gravity will slip (without friction) from A to B in the least amount of time. This curve is called the brachistochrone curve. The term derives from the Greek *brachistos* (shortest or minimum) and the Greek *chronos* (time or delay). The word brachistochrone means shortest time or alternatively minimum delay.

Johann Bernoulli considered an arbitrary curve from A down to B in the vertical x, z plane. The x-axis is horizontal and the z-axis is vertically downward. The bead is subject to the acceleration g of gravity. The numerical value for g is about 9.8m/s^2(ignoring variations over the surface of the Earth in this context). In sliding down the curve the bead accelerates, so its velocity keeps increasing.

Johann Bernoulli's solution divides the vertical plane into layers, with interfaces as shown in Fig. 10.1. He assumes that the bead follows a straight line in each layer. The path is then piecewise linear. Let $v_1, v_2, v_3, \cdots$ be the velocity downward in the successive layers, and let the ray crossing the interfaces successively have incident angles $\theta_1, \theta_2, \theta_3, \cdots$. Application of Snell's law at each interface gives

$$\frac{\sin\theta_1}{v_1} = \frac{\sin\theta_2}{v_2} = \frac{\sin\theta_3}{v_3} = \cdots . \tag{10.1}$$

In the limit, as the layers become infinitely thin, the line segments tend to a curve. If v is the velocity at (x, z) and θ is the angle of incidence, then the curve satisfies

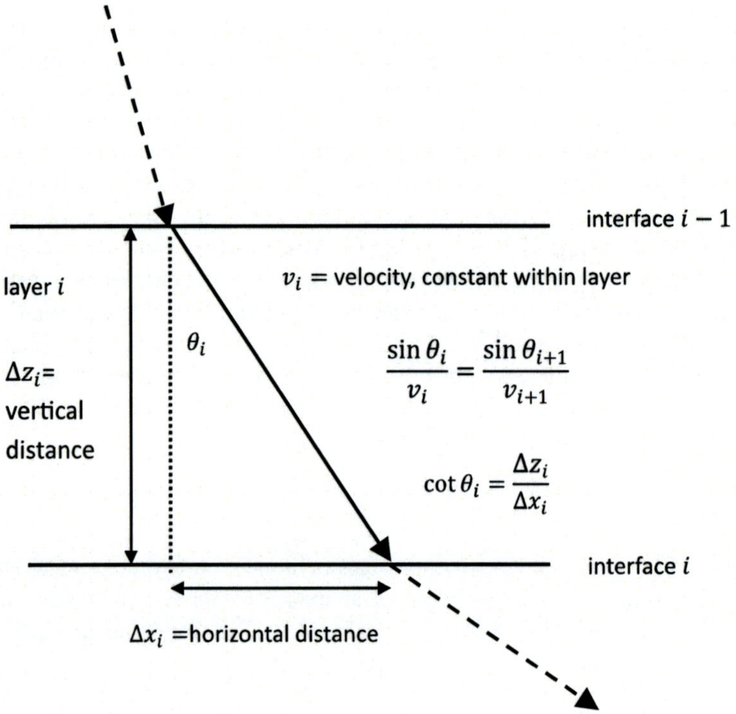

Fig. 10.1 Bead path in layer i

$$\frac{\sin\theta}{v} = p, \tag{10.2}$$

where p is a constant. Now we come to an important point. We need the velocity function, which gives velocity v versus depth z. Fortunately, Galileo had already derived the velocity function. If we combine Eqs. 9.4 and 9.5 from Chap. 9 and substitute $x = z$ and $a = g$, where g is the gravitational acceleration, we obtain

$$v = \sqrt{2gz}\,. \tag{10.3}$$

Substituting Eq. (10.3) into (10.2), we obtain the equation of the curve as

$$z = k^2\sin^2\theta \text{ where } k^2 = \frac{1}{2gp^2}\,. \tag{10.4}$$

We have, from Fig. 10.1, the derivative of z:

$$z' = \frac{dz}{dx} = \cot\theta \text{ and } \sin^2\theta = \frac{\sin^2\theta}{\sin^2\theta + \cos^2\theta} = \frac{1}{1 + \cot^2\theta} = \frac{1}{1 + (z')^2} . \quad (10.5)$$

Thus, Eq. (10.4) for the curve becomes

$$z = k^2 \sin^2\theta = \frac{k^2}{1 + (z')^2} , \quad (10.6)$$

which gives

$$z\big(1 + (z')^2\big) = 2r \text{ where } 2r = k^2 . \quad (10.7)$$

As can be seen by substitution, Eq. (10.7) is satisfied by the cycloid

$$x(t) = r(t - \sin t), z(t) = r(1 - \cos t) , \quad (10.8)$$

where the parameter t plays the roles of both the travel time for motion along the cycloid and the angle of the rolling circle.

Bernoulli makes the analogy to a light ray. Instead we will make the analogy to a seismic ray traveling through the Earth, hence the downward pointing z-axis in Fig. 10.2. The seismic ray traversing a medium is a brachistochrone: that is a minimum-delay curve (it has minimum travel time among all curves connecting the same end points). In the case that the velocity function is given by Eq. 10.3, the brachistochrone is a cycloid. As mentioned already in Chap. 6 in the context of Huygens' pendulum clock, a cycloid is described by a point in the circumference of a wheel that rolls upon a straight line. The variable t measures the angle through which the wheel of radius r has rotated. In our case the straight line is the x-axis and the rolling proceeds upside down (so that the circle rolls under the x-axis). Johann Bernoulli ended his discussion of the brachistochrone problem with these words: "Before I end, I must voice once more the admiration I feel for the unexpected identity of Huygens' tautochrone and my brachistochrone. Nature always tends to act in the simplest way, and so it here lets one curve serve two different functions." (Bernoulli, 1697). Indeed, the cycloid emerges each time, much like a jack from a box.

Bernoulli's method applies to other cases as well, as is shown in Fig. 10.3. In seismology, another well-known velocity function that is often used to model a *diving wave* is given by

$$v(x, z) = v_0 + kz \text{ with } v_0 \text{ and } k \text{ both constant.} \quad (10.9)$$

The x-axis depicts the surface of the ground. We see that the velocity increases linearly with depth but does not vary with horizontal coordinate. At the shot point (the origin of coordinates, where the ray originates), the ray makes the angle θ_0 with

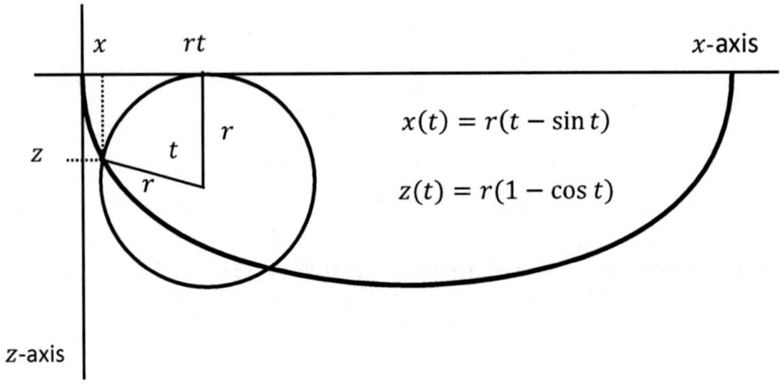

Fig. 10.2 Cycloid (upside down)

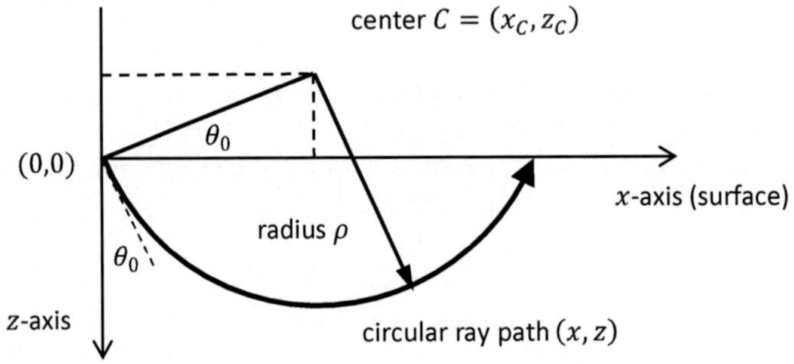

Fig. 10.3 Diving wave

the vertical. This angle plays the role of an initial condition which determines where the ray re-emerges at the surface. The ray travels along the ray path (x, z). In this case, the method of Johann Bernoulli shows that the ray path is an arc of a circle of radius

$$\rho = \frac{v_0}{k \sin \theta_0} = \text{constant.} \tag{10.10}$$

The center of this circle is point C with coordinates

$$(x_C, z_C) = \left(\frac{v_0}{k \tan \theta_0}, -\frac{v_0}{k} \right). \tag{10.11}$$

The tangent to the circle at the shot point makes an angle θ_0 with the z-axis. Note that the vertical component of the center point, z_C, does not depend upon the angle. As a result, the entire family of rays with the given velocity function, and different takeoff angles, is made up of circular paths that pass through the origin (which is the shot point). All of the circular paths have their centers on the line $z = -v_0/k$.

The circle has a slight advantage to the cycloid, namely that points on a circle can be constructed with a compass and straightedge in a finite number of steps. In other words its points can be computed using only four basic arithmetic operations (addition, subtraction, multiplication and division) and square root extraction. The circle is an algebraic curve, unlike the cycloid which is transcendental. For this reason linear Earth models with a velocity function $v = v_0 + kz$ are popular, where v_0 is the velocity at $z = 0$ and k represents the velocity gradient with depth. More complicated Earth models are often built by composite patchwork of linear functions, where each geological layer has its own linear function and its own (v_0, k) pair.

Evolute, Involute and Caustic

Huygens originated what is now known as differential geometry and the idea of curvature, a value that shows how much a curve deviates from a straight line. Consider the curve *MPQ* in Fig. 10.4. At P, PT is the tangent line and PN is the normal line (that is, the line perpendicular to the tangent). The figure also shows four circles (A, B, C, D), each tangent to the curve at P. Such circles are called tangent circles and each has its center on the normal. There is a unique tangent circle that fits curve MPQ at P better than any other. This optimum-fitting circle (D, in this case) may be described as the circle that "kisses" the curve. It is known as the osculating (a term coined by Leibniz and derived from the Latin for kissing) circle. The curvature κ of MPQ at P is defined as $\kappa = 1/r$, where r is the radius of the osculating circle (D). The sharper the curve MPQ at P, the larger its curvature and the shorter its radius of curvature. The flatter the curve at P, the smaller its curvature and the longer the radius of curvature—with the limit being a curve with no curvature, namely a straight line that would coincide with the tangent. In summary, the osculating circle is the circle that touches a curve (on the concave side) at a point and whose radius equals the curve's curvature at the same point. Just as the tangent is the rectilinear line that best approximates a curve at a point, the osculating circle is the circle that best approximates the curve at the point.

Huygens then developed the theory of evolutes. The evolute E of a curve C is defined as the locus of the centers of the osculating circles of C, or equivalently, the locus of centers of curvature of C. That is to say that when the center of curvature of each point on a curve is drawn, the resultant shape will be the evolute of that curve. The evolute E is a companion curve of the original curve C. Some properties are as follows. When C is flat, the radius of curvature becomes infinite and the evolute E disappears into infinity. When the curvature of C shrinks to zero, the radius of curvature becomes zero and E coincides with C. When the curvature of the original curve C has a local maximum or minimum, its evolute E has a singularity which looks like a cusp. The evolute E can also be defined as the envelope of the normals to the original curve C.

In working out the theory of evolutes, Huygens would have asked questions like: "What is the evolute of a circle?" and "What is the evolute of a straight line?" He

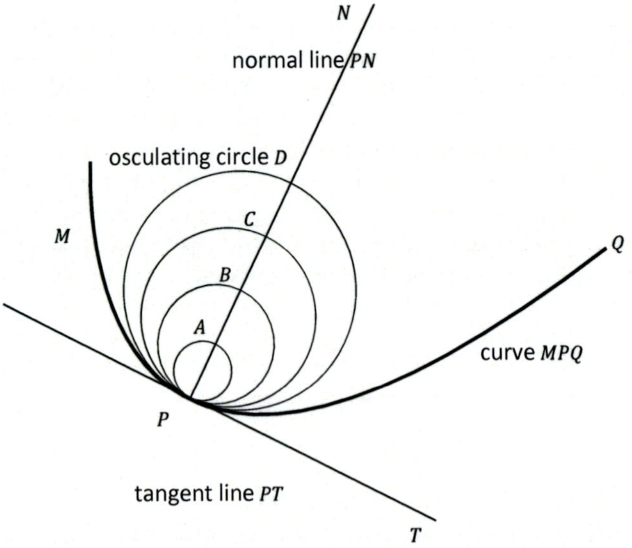

Fig. 10.4 The osculating circle is circle D, fitting the curve MPQ at P better than any other circle

would have come up with something like the diagram in Fig. 10.5, which shows that the evolute of a circle is the center of the circle. In other words, the osculating circle for circle C is circle C itself and, similarly, the evolute of a straight line is a point at infinity. Moreover, Huygens also would have asked "What is the evolute of a cycloid?"—and ingenuously gave the answer: again a cycloid, with the same shape, arclength and altitude, but shifted down and right (see Fig. 10.6 and also Chap. 6).

Of course we can turn such questions around. An involute is the inverse of an evolute. An involute of a given curve is the curve described by the end of a thread which is unwound from the given curve. See Fig. 10.7. On paper it can be constructed as follows: a flat body, one side of which has the shape of the initial curve AB, is placed on a piece of the paper. A thin thread is wound over the curve and stretched taut. A pen is attached to the outer end of the thread, the tip of which rests on the paper. Then the thread is slowly unwound from the curve, always keeping it taut. The curve that emerges on paper is an involute. The tangent at an arbitrary point A or B of the evolute is normal to the involute. As a result, an involute is an orthogonal trajectory of the tangents to the evolute. It is important to note that the evolute of a curve is uniquely determined, but the involute of a curve depends on where you start unwinding and in which direction. In other words, there is a one-parameter family of them. Figure 10.7 shows two of them for the curve AB: BD and AC. The points on the curve AB again are centers of curvature of points on the curves BD and AC—and so on.

The notions of the evolute and involute were introduced by Huygens in his work *Horologium Oscillatorium* (1673). Huygens formulated his theory of evolutes sometime around 1659 to help solve the problem of finding the tautochrone curve, which

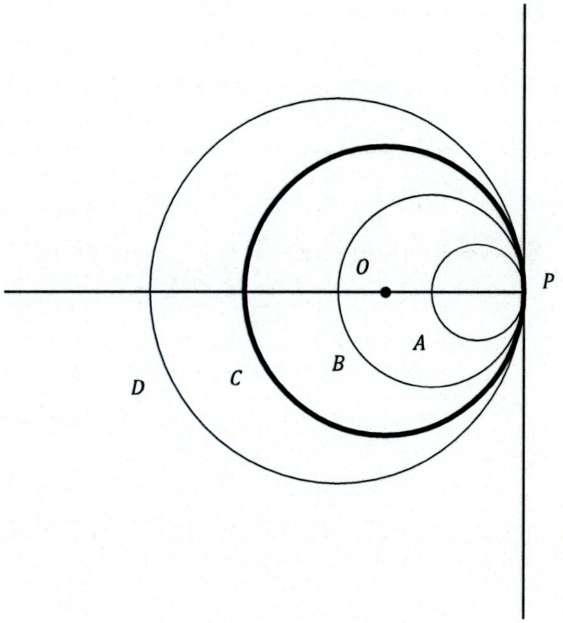

Fig. 10.5 The evolute of a circle C is its midpoint O, its osculating circle is the circle C itself

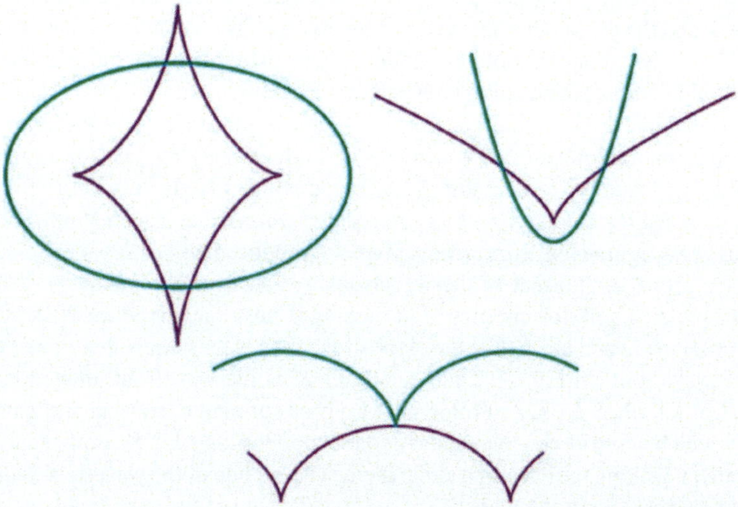

Fig. 10.6 Top left/right, and bottom: evolutes (in purple) of the ellipse, parabola and cycloid (in green). The evolute has a cusp when the original curve has minimum or maximum curvature. As shown by Huygens, the evolute of a cycloid is a cycloid again (bottom)

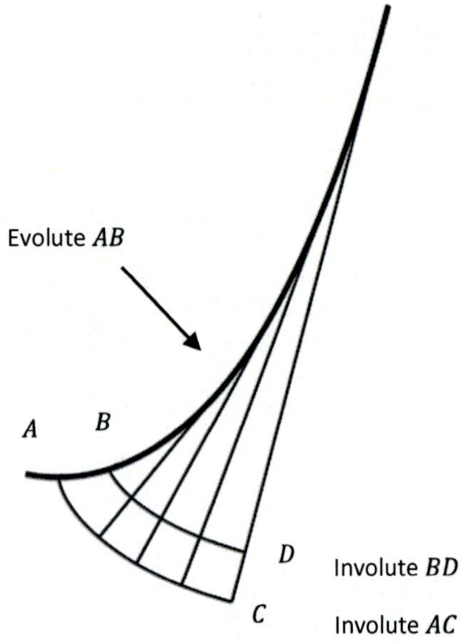

Fig. 10.7 Evolute and involute

in turn helped him construct an isochronous pendulum. The tautochrone curve is a cycloid, and the cycloid has the unique property that its evolute is also a cycloid (see Fig. 10.7). The theory of evolutes, in fact, allowed Huygens to achieve many results that would later be found using conventional calculus.

The involute and evolute have an important implication for wave propagation and the formation of caustics, which was first noticed by Huygens and explained in detail in his *Traité de la Lumière.* A modern treatment is given in *Unrollling Time, Christiaan Huygens and the mathematization of nature* (J.G. Yoder, 1988). We return to Huygens' figure (Fig. 10.8) from Chap. 4 reproduced here.

Rays propagate normal to the wavefronts (in an isotropic medium). When the wavefront is curved, for instance after passing a lens or a curved interface, the rays may cross over each other. The envelope of the rays is then equivalent to the evolute of the wavefront. In Fig. 10.8 the curve EVK is the wavefront after passing the spherical interface $EFGA$. Following Huygens' principle, subsequent wavefronts are formed by considering elementary source points along EVK, propagating them forward and taking the envelope along them. The evolute of the wavefront is given by the curve $NbehC$. It is the "caustic" of the rays. The initial and subsequent wavefronts (EVK, abc, def, ghk and so on) are involutes of the caustic. Regarding Huygens' principle applied to the formation of caustics, in modern terms the caustic is defined as the envelope of refracted rays, but in Huygens' terms, it is formed by the evolute of the wavefront (locus of centers of curvature of the wavefront). So there the evolute appears again.

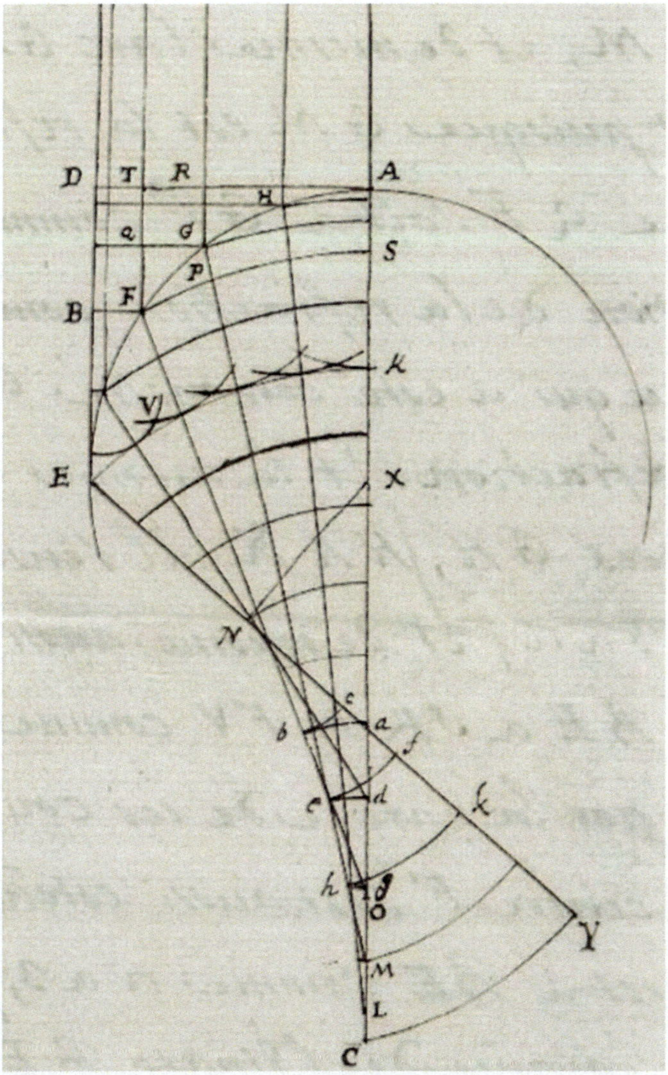

Fig. 10.8 Caustic, given by the curve $NbehC$, as evolute of the wavefront EVK. In turn, the subsequent wavefronts EVK, abc, def, and ghk are involutes of the caustic [with permission from Leiden University Library, HUG 31, ff. 255r.-312v.]

Sherlock Holmes and Huygens

More on curvature. In Arthur Conan Doyle's story "The Adventure of the Priory School", Sherlock Holmes claims to be able to tell from the traces left in the sand by a bicycle in which direction the bicyclist had been riding.

> Sherlock Holmes: "This track, as you perceive, was made by a rider who was going from the direction of the school."
>
> Watson: "Or towards it?"
>
> Sherlock Holmes: "No, no, my dear Watson. The more deeply sunk impression is, of course, the hind wheel, upon which the weight rests. You perceive several places where it has passed across and obliterated the more shallow mark of the front one. It was undoubtedly heading away from the school."

Although Sherlock Holmes was a bit premature in his conclusions, the question has an interesting mathematical aspect which was covered by Huygens. The question directly points to a geometrical curve called the tractrix (or pulling curve, from Latin "trahere" to pull). The rear wheel of the bicycle is the tractrix of the front wheel. The front wheel precedes and pulls the rear wheel behind it at a fixed distance, given by the length of the bicycle. The rear wheel is always pointed towards the front wheel, whereas the front wheel has variable orientation. Hence, at each point of the track of the rear wheel, we can construct the tangent, and at a fixed distance forward, the tangent will cross the track of the front wheel.

To answer Watson's question: given a pair of bicycle tracks it is in general possible to determine the direction of the bicycle. From the two curves only one can qualify as the rear wheel track, since its tangent must always cross the other curve. As soon as we have established this, the direction of movement follows from the fact that the tangent must cross the other curve at a fixed distance. Moreover, as long as the front wheel track remains either convex or concave (and therefore its curvature does not change sign), the rear wheel track will approach it in the direction of movement. In other words, it is indeed possible to deduce the direction of the bicycle from its wheel tracks. See Fig. 10.9. Incidentally, it is worth noting that the system is decidedly *not* invariant with respect to time-reversal.

Of course, curvature plays an essential role here. If the suspect had taken care to ride his bicycle along a straight line and avoid any oscillations, the front and rear wheel tracks would coincide and Holmes would be unable to follow him.

Coincidence or not, the tractrix was studied by none other than Huygens. The tractrix was introduced by Claude Perrault (1613–1688), a French physician, who was also interested in physics and, as an amateur architect, designed the Paris Observatory and participated in the design of the east façade of the Louvre. His brother was Charles Perrault, the famous story-teller of stories like Cinderella. Actually, Newton and Huygens both worked on the tractrix. Huygens found its equation in 1693 (in parametric form and Cartesian coordinates, see Fig. 10.10)

$$x = t - \tanh t, \quad y = 1/\cosh t \; . \tag{10.12}$$

In 1693, Huygens found another amazing property related to the tractrix. When it is revolved around its x-axis in three dimensions, we obtain the surface of revolution of the tractrix, which is also called tractroid or pseudo-sphere. Even although this body extends to infinity, Huygens discovered that its volume and surface area are finite. For a given radius R (corresponding to the line segment AB in Huygens'

Fig. 10.9 Bicycle tracks, front wheel track (blue), rear wheel track (red). In the picture, the bicycle was moving to the right

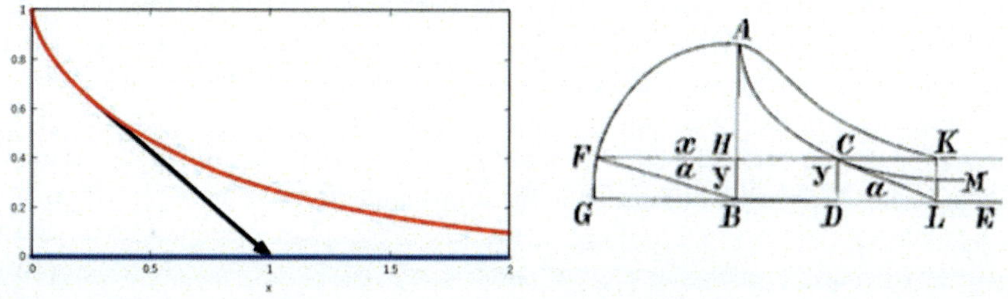

Fig. 10.10 Tractrix. As the point moves along the x-axis (blue curve) it pulls the point on the tractrix (red curve) behind it. Right: diagram from Huygens

diagram, Fig. 10.10 right) the surface area is $4\pi R^2$—that is, equal to the surface area of a sphere with radius R—and its volume $\frac{2}{3}\pi R^3$—that is, equal to half the volume of a sphere with radius R.

The theory of curvature was generalized to three-dimensional surfaces and completed in the nineteenth century by Carl Friedrich Gauss (1777–1855). Surfaces have a system (2×2 matrix) of curvatures at each point, corresponding to curvatures in perpendicular coordinate directions. The pseudo-sphere has a constant negative Gaussian curvature, resulting from the product of its two main curvatures: one along the x-axis, which is negative, and the other perpendicular to it, which is positive. As such, the pseudo-sphere has applications in hyperbolic or non-Euclidian geometry and in the Minkowski description of space–time in the special theory of relativity (see Chap. 8).

Proto-Calculus

Calculus is the study of continuous change in mathematics, just as geometry is the study of shape and algebra is the study of generalizations of arithmetic operations. Calculus was first known as infinitesimal calculus or "the calculus of infinitesimals." Infinitesimal calculus was developed independently in the late seventeenth century by Isaac Newton and Gottfried Wilhelm Leibniz. The term "proto-calculus" refers to methods of infinitesimal analysis appearing before the calculus of Leibniz-Newton. In *La Géométrie* (1637), René Descartes details a "geometrical calculus" (calcul géométrique) that rests on a distinctive approach to the relationship between algebra and geometry. Specifically, Descartes offers a new way of understanding the connection between a curve's construction and its algebraic equation. His introduction of coordinates, later named Cartesian coordinates after him, is the cornerstone of the algebraization of mathematics.

Descartes, known for introducing skepticism as an essential component of natural sciences, invented analytic geometry. For example, from ancient times a circle has been recognized *geometrically* as a picture of a round curve in a plane consisting of all points that are a given fixed distance from a given fixed point (the center). The distance between any point of the circle and the center is called the radius. Descartes changed the world of mathematics. He showed how a picture could be replaced by an equation. He recognized a circle as the curve defined *algebraically* by the Cartesian coordinates (x, y) and the constant r for radius in the form of the algebraic equation

$$x^2 + y^2 = r^2. \tag{10.13}$$

Descartes' analytic geometry was a major conceptual advance, because it related the previously separate fields of geometry and algebra to each other. Descartes showed that he was able to solve previously unsolvable problems in geometry by converting them into simpler problems in algebra. In a sense, solving problems became more dependent on algebraic endurance than on geometrical dexterity. He represented the horizontal direction as x and the vertical direction as y. This concept is now indispensable in mathematics and other sciences.

Analytic geometry made possible the next great advance in mathematics: namely, infinitesimal analysis (commonly called calculus). In effect, Descartes' analytic geometry gave Huygens and others the tools needed for the development of proto-calculus, which served as a precursor to calculus. Huygens' teacher Frans van Schooten (1615–1660), who was recommended to him by Descartes (a friend of his father Constantijn Huygens), played an important role in the advent of analytic geometry. The correspondence between Huygens and Van Schooten on geometrical problems and solutions makes for interesting reading. Subsequently, Leibniz and

Newton further refined and developed the ideas of proto-calculus, ultimately giving rise to the calculus we know today.

The product rule and chain rule, the notions of first and second derivatives and of analytic functions were used by Isaac Newton in an idiosyncratic notation which he applied to solve problems of mathematical physics. Since Newton was in possession of his own version of calculus (using what he called fluents and fluxions) well before 1687, we might expect to find something like his calculus in his masterpiece *Principia.* In fact Newton composed the *Principia* entirely without the use of his method of fluents and fluxions. In other words, the *Principia* contains no explicit use of the calculus of fluents and fluxions. This is usually attributed to Newton's preference for the geometrical methods of the ancients, such as Archimedes, Euclid, and so on. The ancients made use of limiting processes, such as those used in the determination of the area of the circle, by inscribing the circle inside a regular polygon and increasing the number of its sides.

These ancient applications were quite simple as compared to those used in *Principia.* In turn, Newton's calculus of fluents and fluxions was quite primitive as compared to modern calculus with its rigorous proofs, formalized algorithms and systematic procedures for dealing with continuous functions. It took nearly 100 years for such mathematicians as the Bernoullis, d'Alembert, and Euler to establish the modern methods of calculus that are used today to solve the problems in *Principia.* Newton had available Huygens' solution of the centrifugal force problem for which Huygens used intricate limiting forms of the geometrical methods of the ancients. Explicitly, Huygens' forms include ways to create second derivatives geometrically. These forms were exactly what Newton needed for the solutions of the problems in *Principia.* In fact, Lemma 11, Book One, Sect. 1 of *Principia* is essentially a reworking of Huygens' way to create second derivatives geometrically. It is said that *Principia* is based on the thought processes of calculus, merely disguised. However, the same can be said of Huygens' work. Newton was absolutely right in choosing the geometric methods of Huygens to use in *Principia.* He had no other choice; the bulky notation of fluents and fluxions was unsuitable for problems that required methods of vector calculus. Vector calculus, or vector analysis, is concerned with differentiation and integration of vector fields, primarily in three-dimensional Euclidean space, and is of great use in three-dimensional gravitational problems. Vector analysis would be fully developed in the nineteenth century by Gauss, Green and Stokes, and play an essential role in the completion of Huygens' wave principle by Fresnel and Kirchhoff (see Chap. 7), as well as in the formulation of the laws of electrodynamics by Maxwell (see Chap. 3).

Generally, we can say that although both Huygens and Newton had developed notions of infinitesimal calculus ("proto-calculus"), they were still embedded in the ancient tradition of geometrical reasoning ("*more geometrico*"). With them, the geometrical reasoning achieved its final stage of development, but at the same time it carried the seeds of the calculus revolution.

Taylor and the Vibrating String

The English mathematician Brook Taylor (1685–1731) was educated at St. John's College, Cambridge, and was among the most enthusiastic of Newton's admirers. He began publishing papers in the *Philosophical Transactions of the Royal Society (London)* in 1712, and these papers covered a variety of topics, including the motion of projectiles, the center of oscillation, and the shapes that liquids take when they are raised by capillarity. He stepped down from his position as Royal Society secretary in 1719 and abandoned the study of mathematics. His *Methodus Incrementorum Directa et Inversa*, which was published in London in 1715, was his first publication and the one for which he is best known. In 1719, Taylor published a treatise on perspective which marked the first comprehensive statement on the principle of vanishing points (although the concept of vanishing points for horizontal and parallel lines in a picture placed in a vertical plane had previously been proposed by Guidobaldi Del Monte in his work *Perspectivae Libri VI* (Pesaro, 1600) and by Simon Stevin in *Sciagraphia* (Latin translation from Dutch by Willebrord Snellius, Leiden, 1608).

At the time of Newton, it was generally assumed that all functions were continuous, because contemporary observations of natural events seemed to indicate a continuous relationship between physical variables. This view was reinforced by the frequent use of the new tool calculus to formulate natural laws in terms of differential equations. For example, Newton's second law of motion, which says that force is the product of mass and acceleration, is usually expressed in the simple equation $F = ma$. However, this is really a second order differential equation because acceleration is the second derivative of the function which relates distance and time. If x is distance and t is time, then the familiar $F = ma$ can be written as the differential equation

$$F = m\frac{\mathrm{d}^2x}{\mathrm{d}t^2} . \tag{10.14}$$

As a result, it became commonplace to assume that any function describing a relationship between physical variables would be differentiable. The notion that such a function could change in a capricious or random manner, and therefore might not be differentiable at all points, was not part of mathematical thinking at that time. It is now understood that functions are not restricted, and some may exhibit unexpected behavior. However, the simplistic idea of universal continuity and unlimited differentiability most likely prompted Taylor to investigate the possibility that a function could be expressed in terms of its derivatives. Taylor found this was true and in 1715 introduced analytic functions into mathematics.

The core of his discovery was the Taylor series, which allows a wide class of functions (called analytic functions) to be expressed as the summation of other elementary functions (monomials, that is power functions of the form x^n). More precisely, the Taylor series expands the function $f(x)$ to the point $x + h$. It is assumed that the function is differentiable to all orders in the neighborhood of the point x. The infinite series has coefficients given by the function $f(x)$ and its successive derivatives $f'(x)$, $f''(x)$,... at the given point x. The Taylor series can be expressed as

$$f(x+h) = f(x) + hf'(x) + \frac{h^2}{2} f''(x) + \cdots . \tag{10.15}$$

This is an astonishing result because the definition of the derivative of any order at the point x requires nothing more than knowledge of the function in an arbitrarily small neighborhood of this point. The Taylor series, therefore, establishes that the shape of the function at any finite distance h from the point x is uniquely determined by the function's behavior in the infinitesimal vicinity of the point. This property of the Taylor series implies that an analytic function has a strong interconnected structure, making it possible to predict precisely what will happen at any point a finite distance from a certain point by merely studying the function's behavior in a small vicinity of the point. This property is extremely valuable in mathematical analysis because, although it is not common to all functions, it is shared by many of the most useful and regularly encountered, such as polynomials, the sine and cosine functions, and the rational functions (away from their poles). As a consequence, the Taylor series was instrumental in both interpolation—the estimation of values of a function between known data points—and extrapolation—estimation of the values of a function beyond known data points, or into the future if the x-axis plays the role of time.

In his book *Methodus Incrementorum Directa et Inversa* (1715), Taylor introduced the Taylor series, which largely remained unrecognized until Lagrange (1772) realized its usefulness and termed it "the main foundation of differential calculus". We could say that the Taylor series is built on first-, second-, and higher derivatives, and therefore reaps the fruits of Huygens' work. Taylor's book contains a proof of the Taylor series expansion. In addition, the book also includes several theorems on interpolation. Taylor was the earliest writer to deal with theorems on the change of the independent variable. Taylor was perhaps the first to realize the possibility of an operational calculus (which turned calculus problems into algebraic problems), and he is usually recognized as the creator of the theory of finite differences. The *Methodus Incrementorum* also contains the earliest determination of the differential equation of the path of a ray of light when traversing a heterogeneous medium; and, assuming that the density of the air (and hence the velocity of light in air) depends only on its distance from the Earth's surface, Taylor obtained by means of integration the approximate form of the curve. The form of the catenary and the determination of the centers of oscillation and percussion are also discussed: the catenary is the curve which occurs when a chain is hanging under its own weight between two end points; the center of oscillation of a system is the point where its entire mass can be concentrated for analyzing its oscillatory motion; the center of percussion is the point in a body where an applied force causes only rotation without any translation. The applications of the calculus to such and other questions given in the *Methodus Incrementorum* have hardly received the attention they deserve. If they did, Taylor would undoubtedly be established as one of the great mathematicians of all time. As it now stands, Taylor is in effect hidden in Newton's shadow.

Among the problems that Taylor discusses in the second part of *Methodus Incrementorum* are two that deal with the vibrating string. They had already been discussed in Taylor's paper "De motu nervi tensi" ("On the motion of a tense sinew"), in *Philosophical Transactions* in 1713. The first problem is to determine the motion of a tense string; the second is, given the length and weight of the string, as well as the stretching weight, to find the period of vibration. Taylor concludes that at any point of the arc, the normal acceleration $\partial^2 u/\partial t^2$ is proportional to the curvature $\partial^2 u/\partial x^2$. This means that, for small vibrations, Taylor has in principle discovered the wave equation now written as

$$\rho \frac{\partial^2 u}{\partial t^2} = F \frac{\partial^2 u}{\partial x^2}. \tag{10.16}$$

The quantity ρ is the mass per unit length of the string and F is the tension as determined by the weight which stretches the string. However, there is no evidence that Taylor had any notion of partial derivatives (as displayed in Eq. 10.16). Nonetheless, he did find, that the motion of an arbitrary point is like that of a simple pendulum. More specifically, Taylor showed that the period is equal to

$$T = 2\sqrt{\frac{mL}{F}}, \tag{10.17}$$

where L is the length of the string and m its mass. Because $m = \rho L$ we have

$$T = 2\sqrt{\frac{\rho L^2}{F}} = 2L\sqrt{\frac{\rho}{F}}. \tag{10.18}$$

Let us now define the constant c as

$$c = \sqrt{\frac{F}{\rho}}. \tag{10.19}$$

Then the period becomes

$$T = \frac{2L}{c}. \tag{10.20}$$

This equation shows that the period T is proportional to the length L of the string. Taylor took the form of the curve to be sinusoidal. In conclusion, Taylor made two great discoveries in reference to the vibrating string, namely that the motion satisfies

(10.16), which we know today as the wave equation, and that sinusoidal curves with period given by (10.17) are solutions of the wave equation.

The quantity F/ρ has the dimension of the velocity squared. We let the constant $c = \sqrt{F/\rho}$ be this velocity. Then (10.16) becomes

$$\frac{1}{c^2}\frac{\partial^2 u}{\partial t^2} = \frac{\partial^2 u}{\partial x^2}. \tag{10.21}$$

This equation for free vibrations of a string is the wave equation as initiated by Taylor. Taylor determined that the wave equation has sinusoidal solutions. A sinusoidal solution is also known as simple harmonic motion. The legendary Greek mathematician and philosopher Pythagoras (ca 600 BC) was the first to consider a purely physical problem in which harmonic analysis made its appearance. Pythagoras studied the laws of musical harmony by generating pure sine vibrations on a vibrating string, fixed at its two endpoints. This problem excited scientists for almost 2500 years before the mathematical turning point arrived when Taylor recognized that the normal displacement $u = (x, t)$ of the vibrating string satisfies the wave equation.

The constant c is a physical quantity characteristic of the material of the string. Later d'Alembert (1717–1783) showed that c represents the velocity of the traveling waves on the string, not of the displacement u. D'Alembert is generally credited with deriving the one-dimensional wave equation from combining Hooke's law and Newton's acceleration law (1746). Euler is credited with the three-dimensional wave equation. Taylor did very useful preparatory work on the physics of the wave equation, but indeed there is no evidence that he had a notion of partial derivatives. D'Alembert's observation that c is both a material constant and the wave velocity would cast its shadow forward to Maxwell's observation for electromagnetic waves that their propagation velocity is directly related to the electric and magnetic constants of the vacuum (see Chapter 3, Eq. 3.1 and discussion around it).

After the work of Taylor, the problem of constructing the solution of the wave equation was attacked by some of the greatest mathematicians of all time, and in so doing, they paved the way for the theory of spectral analysis.

D'Alembert's Solution

Jean-le-Rond d'Alembert was born in Paris on November 16, 1717, and died there on October 29, 1783 (Rouse Ball, 1908). He was the illegitimate child of the chevalier Destouches. His real mother abandoned him on the steps of the little church of St. Jean-le-Rond, which was close to the cathedral of Notre-Dame. He was taken to the parish commissary, where he received the Christian name of Jean-le-Rond. He was boarded out by the parish to the wife of a glazier who lived near the cathedral, and it was in this humble environment that he found a true home. His father appears to have covered some of his education costs. Otherwise, d'Alembert was almost entirely autodidactic.

Essays he wrote in 1738 and 1740 attracted attention to d'Alembert. In 1740 he was elected a member of the French Academy, probably due to the influence of his father. Despite his father's career assistance, he absolutely refused to leave his adopted mother; he continued to live with her until her death in 1757. It cannot be said, however, that she sympathized with his success, for at the height of his fame she remonstrated with him for wasting his talents on such work: "You will never be anything but a philosopher," she said, "And what is a philosopher? He is a madman who torments himself during his life so that people will speak of him after he's gone."

Nearly all his major mathematical works were produced during the years 1743–1754. The first of these was his *Traité de Dynamique*, published in 1743, in which he enunciates the principle known by his name, namely, that the "internal forces of inertia" (that is, the forces which resist acceleration) must be equal and opposite to the forces which produce the acceleration. This equality may be inferred from Newton's third law of motion, but the full consequences had not been realized previously. D'Alembert's principle enabled mathematicians to obtain the differential equations of motion of any rigid system.

When d'Alembert applied his principle to fluids in his 1744 publication *Traité de l'Équilibre et du Mouvement des Fluides*, it produced partial differential equations that he was unable to solve at the time. In his *Théorie Générale des Vents*, published in 1745, he focused on the study of air motion, which once more brought him to partial differential equations. In 1746, a second edition was dedicated to King Frederick the Great of Prussia, and it resulted in an invitation to Berlin and an offer of a pension; he declined the former but later pocketed his pride and the latter after some pressure. In 1747, he applied the differential calculus to the problem of a vibrating string, and arrived at the *wave equation* (with $c = 1$):

$$\frac{\partial^2 u(x,t)}{\partial x^2} = \frac{\partial^2 u(x,t)}{\partial t^2} . \tag{10.22}$$

D'Alembert succeeded in showing that it was satisfied by

$$u(x,t) = f(x+t) + g(x-t) , \tag{10.23}$$

where f and g are arbitrary (twice) differentiable functions. This solution was published in the transactions of the Berlin Academy for 1747. The proof begins by saying that, if $\partial u/\partial x$ be denoted by p and $\partial u/\partial t$ by q, then we have the exact differential

$$\mathrm{d}u = p\mathrm{d}x + q\mathrm{d}t . \tag{10.24}$$

In these quantities p and q the wave (10.23) becomes

$$\frac{\partial q}{\partial t} = \frac{\partial p}{\partial x}. \tag{10.25}$$

Therefore $p\mathrm{d}t + q\mathrm{d}x$ is also an exact differential; denote it by $\mathrm{d}v$. Therefore

$$\mathrm{d}v = p\mathrm{d}t + q\mathrm{d}x \,. \tag{10.26}$$

Hence

$$\mathrm{d}u + \mathrm{d}v = p\mathrm{d}x + q\mathrm{d}t + p\mathrm{d}t + q\mathrm{d}x = (p+q)(\mathrm{d}x + \mathrm{d}t) \tag{10.27}$$

and

$$\mathrm{d}u - \mathrm{d}v = p\mathrm{d}x + q\mathrm{d}t - p\mathrm{d}t - q\mathrm{d}x = (p-q)(\mathrm{d}x - \mathrm{d}t). \tag{10.28}$$

Thus $u + v$ must be a function of $x + t$, and $u - v$ must be a function of $x - t$. We may therefore put

$$u + v = 2f(x+t) \tag{10.29}$$

and

$$u - v = 2g(x-t) \,. \tag{10.30}$$

By adding, we can eliminate the v terms, leaving

$$u = f(x+t) + g(x-t) \,. \tag{10.31}$$

which is the d'Alembert solution (10.23). Substitution of (10.23) in (10.23) offers a direct verification.

The two subsequent areas in which d'Alembert made the most important contributions to mathematics were in physical astronomy, particularly the precession of the equinoxes and changes in ecliptic obliquity. These were collected in his *Système du Monde*, published in three volumes in 1754. Together with Denis Diderot, he spent the latter years of his life working on the great French encyclopedia (full title *Encyclopédie, ou dictionnaire raisonné des sciences, des arts et des métiers* or Encyclopedia, or Systematic Dictionary of the Sciences, Arts, and Crafts, published in France between 1751 and 1772). D'Alembert also wrote numerous philosophical and mathematical articles for the *Encyclopédie*, the best of which are those on geometry and probabilities. Euler then took up the matter and showed that for the wave equation (10.21) the general solution is

$$u = f(x+ct) + g(x-ct), \tag{10.32}$$

where f and g are arbitrary functions. Euler also formulated the wave equation in three dimensions (see Chap. 7).

Leibniz

We turn back to the seventeenth century. Infinitesimals are quantities that are closer to zero than any standard real number, but are not zero. They do not exist in the standard real number system, but are used in many applications. The use of infinitesimals often provides insight. Entities can retain certain specific properties, such as angle, even though the defining entities are infinitely small. Infinitesimals have been used in mathematics since the time of the ancient Greeks. Archimedes used infinitesimals in his work *The Method of Mechanical Theorems* to find areas of regions and volumes of solids. In the seventeenth century, infinitesimals were used in the development of calculus, where the derivative was regarded as a ratio of two infinitesimal quantities. Infinitely many infinitesimals are summed to produce an integral. However, the infinitesimal viewpoint in calculus was soon dropped in favor of limits, which are performed using the standard real numbers.

In 1672, Leibniz met Huygens who convinced him to dedicate significant time to the study of mathematics. By 1673 Leibniz had finished reading Pascal's *Traité des sinus du quart de cercle*, or *Treatise on the sine of the quadrant*, and it was during this research that Leibniz famously said "a light turned on". Leibniz, like Newton, saw the tangent as a ratio but defined it as the ratio between ordinates and abscissas. He expanded on this logic to claim that the integral was, in fact, the sum of the ordinates for infinitesimal intervals in the abscissa; in other words, the sum of an infinite number of rectangles. To emphasize the summation property Leibniz introduced the integral sign $\int$, as an elongated letter "S" and abbreviation for the Latin word "summa", meaning sum. From these notions the inverse relationship of differentials and integrals, or antiderivatives, became clear and Leibniz quickly realized the potential to form a whole new system of mathematics. Whereas Newton used several approaches over the course of his career, Leibniz made differentials and integrals the foundation of his notation and calculus. For a vivid description of their priority dispute, see Sonar (2018): *The History of the Priority Di∫pute between Newton and Leibniz*—note the particular typography of this book title.

Huygens, the senior to both Leibniz and Newton, took a reserved attitude towards calculus. In his view, all that calculus contributed could be found equally well by traditional (geometrical) methods. This is best expressed in Huygens' letter to Leibniz of 22 October 1679:

> I have taken a close look at your new characteristic, but to be quite honest, I do not understand, by what you are trying to do, why you expect so much of it. For your examples are truths that are well known to us, and you give a rather obscure proof of the proposition that the intersection of a sphere with a plane is a circle. Finally, I do not see how you could apply your characteristic to such different things as the squares, finding curves from their tangent lines, the roots from negative numbers, the problems of Diophantus, and the shortest and

> most beautiful constructions in geometric problems. And what seems most strange to me, is your application to the invention and explanation of machines. I tell you plainly that these are beautiful wishes, and I would like to see evidence before I believe your claims. However, I am careful to say that you were mistaken, knowing in fact the subtleness and depth of your mind.

Note that Huygens, despite his reservations, still keeps a door open for Leibniz' new ideas. He agrees that there might be value in Leibniz' proposals. He continues by urging Leibniz to keep busy with ongoing problems. Huygens continues to report on his own research, which covered a range of problems, from theoretical to very practical, for instance on improvements of Descartes' proposals for lens construction. Nevertheless, the statement in his letter is a very important one, in which Huygens agrees that there might be value in Leibniz' proposals. Clearly, Huygens was still rooted in the traditional geometrical approach—but he was the champion at it. Remember the skilled way Huygens differentiated and integrated the cycloid (Chap. 6).

> I am only asking you to keep busy with the things you have already given us, such as the arithmetic quadrature and what you have discovered for the roots of equations beyond the third degree, if you are satisfied with that yourself. For the one, which you propose, which is $x^x - x = 24$, it is determined by integers, but otherwise by its nature it does not seem to be because there are exponents that are fractions, as we can understand by logarithms, and so your number can also be a fraction or irrational number that satisfies the said equation as well as 3. All summer I have been working on my refractions, especially those in the Icelandic crystal, in which phenomena are so strange that I have not yet understood them all (especially the effects of polarization). But what I have understood greatly confirms my theory of light and ordinary refractions.

> Among other things, I have found the solution to Descartes' problem: if the shape of one lens surface is known, what is the shape of the other surface so that together they make a given point a parallel beam? I have solved it even more generally than he did, because he only assumes a sphere or a cone. If my health permits, I will try to have this Traité printed in the coming winter. I would like to take your advice to make my reflections known in abbreviated form, without proof, but I cannot get around to it, because in these matters people will not take my word for it.

> Furthermore, I have nothing more to say than my invention of a convenient spirit level that straightens and checks the image in one go, so that people always know that they are not making mistakes. This is not the case with any of the common spirit levels, at least with the ones that use a viewer, like I am doing. I will publish about it in the Journal (des scavans) and send you the article at the earliest opportunity.

Let us recall that, as evidenced by this letter, most scientific discoveries were communicated by bilateral correspondence, rather than by broadly distributed journals. The Journal des savants (*Journal of the Learned, Journal des sçavans from 1665 to 1790, Journal des Scavans from 1791 to 1830*) was one of the earliest scientific journals, and the oldest still surviving today. As we noted in Chap. 3, this journal anonymously published Ole Roemer's results on the finite speed of light.

As stated, calculus (also known as infinitesimal calculus or "the calculus of infinitesimals") is the mathematical study of continuous change, similar to how geometry studies shape and how algebra studies generalizations of arithmetic operations. Before Newton and Leibniz the word "calculus" referred to any body of mathematics, but in the following years, "calculus" became a popular term for a field of mathematics based upon the work of Newton and Leibniz. Newton completed no definitive publication formalizing his fluents and fluxions. Instead, many of his mathematical discoveries were transmitted through correspondence, smaller papers, or as embedded material in his other sources, such as *Principia* and *Opticks*. Leibniz began his mathematics with a comprehensive background. He invented the step reckoner which was a digital mechanical calculator. It was the first calculator that could perform all four arithmetic operations. In effect Leibniz, like Pascal, was a computer scientist. His reasoning in the development of his calculus can now be brought to mind. Alfred Lord Tennyson wrote "Ours not to reason why, ours but to do and die."

The job of a computer is not to reason why. In the same way, the job of the calculus of Leibniz is not to reason why. To this end, Leibniz removed the mystery of infinitesimals from infinitesimal calculus and replaced the material with a set of computational rules that do not involve infinitesimals at all. No longer was it the calculus of infinitesimals but it was now simply calculus. No other mathematician had this insight. Calculus, now a set of differentiation rules, belonged to Leibniz.

The Basic Differentiation Rules

- Constant rule: the derivative of a constant function is zero. If $f = c$ where c is a constant, then $\mathrm{d}f/\mathrm{d}x = 0$.
- Power rule: the derivative of a power function $f = x^n$ is obtained by reducing the power by 1 and putting it as a factor in front of x: $\mathrm{d}x^n/\mathrm{d}x = nx^{n-1}$ (for instance $\mathrm{d}x^5/\mathrm{d}x = 5x^4$).
- Sum rule: the derivative of a sum is the sum of the derivatives:: $\mathrm{d}(f+g)/\mathrm{d}x = \mathrm{d}f/\mathrm{d}x + \mathrm{d}g/\mathrm{d}x$.
- Difference rule: the derivative of a difference is the difference of the derivatives: $\mathrm{d}(f-g)/\mathrm{d}x = \mathrm{d}f/\mathrm{d}x - \mathrm{d}g/\mathrm{d}x$.
- The constant, sum and difference rules are summarized in saying that differentiation is a *linear* operation: $\mathrm{d}(af(x) + bg(x))/\mathrm{d}x = a\mathrm{d}f/\mathrm{d}x + b\mathrm{d}g/\mathrm{d}x$.

Augustin Louis Cauchy

Augustin Louis Cauchy (1789–1857) brought precision and rigor to mathematics. He is credited for 16 fundamental concepts and theorems in mathematics and mathematical physics, probably more than any other mathematician. Cauchy set the foundation of mathematical analysis. His collected works comprise 27 volumes. Cauchy was born in Paris in 1789, only a month after the storming of the Bastille. His father, who was a royalist, quickly moved his family to Arcueil, a small town in the southern suburbs of Paris. The family lived in poverty and life was hard. Cauchy suffered from the harsh conditions and was in a state of ill-health for the rest of his life. During his eleven years stay in Arcueil, Augustin was educated by his parents. Cauchy wrote: "I shall never flaunt the little learning that I have acquired through the care and help my father has given me. If I have learned anything, it is only because he took care to teach me. Had he not taken upon himself the trouble of instructing me, I would be as ignorant as many other children."

In 1813, Cauchy returned to Paris. He was persuaded to devote himself entirely to mathematics by Joseph-Louis Lagrange and Pierre-Simon Laplace. In 1816 Cauchy published the memoir on definite integrals that became the basis of the theory of complex functions. In the same year he won the grand prize from the French Académie des Sciences for a 300-page paper on waves at the surface of a liquid, now accepted as a classic in hydromechanics. In 1822 he laid the foundations of the mathematical theory of elasticity. His lectures and researches in analysis during the 1820s represent the first phase of modern rigor in mathematics. He clarified and solidified the principles of calculus by developing them with the help of limits and continuity, concepts now considered essential to analysis.

These contributions are embodied predominantly in his three great treatises: *Cours d'analyse* (Cours in analysis, 1821); *Résumé des leçons sur le calcul infinitésimal* (Summary of lessons on infinitesimal calculus, 1823); and *Leçons sur les applications du calcul infinitésimal à la géométrie* (Lessons on the applications of infinitesimal calculus on geometry, 1826–28). To the same period belongs his development of the theory of functions of a complex variable (a variable involving a multiple of the square root of minus one, $\mathrm{i} = \sqrt{-1}$), today indispensable in applied mathematics, ranging from physics to aeronautics.

In calculus, the (ε, δ) definition of limit is a formalization of the notion of limit. It provides rigor to the following informal notion: the dependent expression $f(x)$ approaches the value L as the variable x approaches the value c if $f(x)$ can be made as close as desired to L by taking x sufficiently close to c, in notation $\lim_{x \to c} f(x) = L$. The concept is due to Cauchy, who occasionally used ε, δ arguments in proofs in his *Cours d'Analyse* (using ε as abbreviation for error, French "erreur"). The first formal definition of limit was given by Bernard Bolzano in 1817, and the definitive modern statement was ultimately provided by Karl Weierstrass. To get a flavor of these arguments: the limit of $f(x)$ as x approaches c is defined as the number L, if for every $\varepsilon > 0$ we can find a $\delta > 0$ such that for all x with $|x - c| < \delta$ we have $|f(x) - L| < \varepsilon$. In other words, f gets arbitrarily close to L if x only gets close enough to c. Weierstrass was the one who completed the development of calculus to full rigor.

Cauchy made significant contributions to wave theory, particularly in the study of dispersion. Dispersion during wave propagation occurs when the wave's velocity depends on its frequency spectrum, which means that the wave changes shape as it propagates. Cauchy derived the first known dispersion formula, which established a simple empirical dispersion law. The "Cauchy Transparent" dispersion of light works best when the material has no optical absorption in the visible spectral range, and thus undergoing what is called "normal dispersion", meaning that the refractive index decreases monotonically with increasing wavelength. A second formulation of the Cauchy model is the "Cauchy Absorbing" dispersion. It is more suitable to describe the optical properties of weakly absorbing materials (Cauchy, 1830).

Cauchy used infinitesimals both in his research and teaching. In *Leçons sur les applications du calcul infinitésimal à la géométrie* (1826–28), Cauchy used infinitesimals neither as variable quantities nor as sequences but rather as numbers. He also applied infinitesimals in an 1832 article on integral geometry. Using infinitesimals, he defined the center of curvature as the intersection point of two infinitely close normal lines to the curve. The locus of centers of curvature for each point on the curve is the evolute of the curve. This term is generally used in physics for the study of lenses and mirrors.

Cauchy wrote, "As for methods I have sought to give them all the rigor that one requires in geometry, so as never to have recourse to the reasons drawn from the generality of algebra." Huygens could have used the same words to introduce his masterpiece *Horologium Oscillatorium* published in 1673. Up to then proto-calculus had been used only to solve problems involving first derivatives. Huygens developed a proto-calculus based upon infinitesimals by which he solved problems involving second derivatives. We mentioned above that Cauchy showed that the center of curvature is the intersection point of two infinitely close normal lines to the curve. A century and a half previously, Christiaan Huygens solved the same problem in PROPOSITION XI in Part III, of *Horologium Oscillatorium.* We now want to show how Huygens used geometry to create a second derivative in his derivation of the radius of curvature. However, first we give the derivation of the radius of curvature in modern calculus.

Modern Curvature

See Fig. 10.11 and consider a function $y(x)$ and a circle with radius r centered at (x_m, y_m):

$$(x - x_m)^2 + (y - y_m)^2 = r^2. \tag{10.33}$$

We want to match the circle to second order to the curve, in other words to make them *osculate.* The first derivative $y' = \mathrm{d}x/\mathrm{d}y$ of the circle is given by

$$(x - x_m) + (y - y_m)y' = 0, \tag{10.34}$$

and its second derivative $y'' = \mathrm{d}^2x/\mathrm{d}y^2$ is given by

$$1 + (y')^2 + (y - y_m)y'' = 0. \tag{10.35}$$

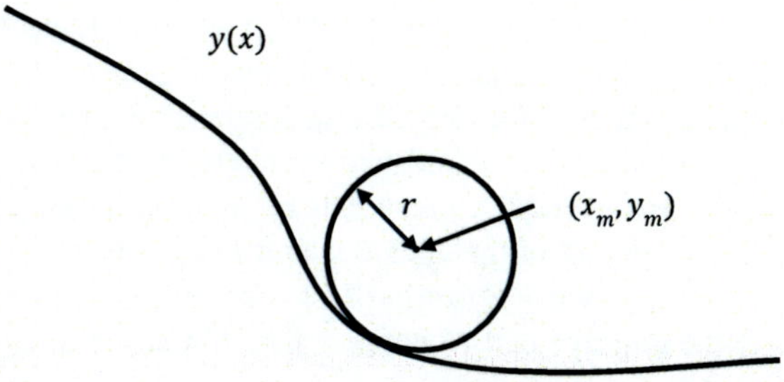

Fig. 10.11 Derivation of curvature

Now we want to fit the circle and its first and second derivatives to the function $y(x)$. That means Eqs. (10.33), (10.34) and (10.35) form a system of three equations with three unknowns: x_m, y_m and r, and we take y' and y'' from the function $y(x)$. We eliminate x_m, y_m from (10.34) and (10.35):

$$(x - x_m) = -(y - y_m)y'. \quad (10.36)$$

and

$$(y - y_m) = -(1 + (y')^2)/y''. \quad (10.37)$$

Substituting (10.36) and (10.37) in (10.33) leads to

$$r = \frac{(1 + (y')^2)^{3/2}}{y''}, \quad \kappa = 1/r = \frac{y''}{(1 + (y')^2)^{3/2}}. \quad (10.38)$$

Here r represents the radius of curvature at a point $(x, y(x))$ of the curve, in terms of its first and second derivatives y' and y''. The curvature κ is the inverse of r: $\kappa = 1/r$. κ can be given a ± 1 sign depending on which side of the curve the center of curvature is located, in other words depending on whether the curve is concave or convex. Note that the curvature is generally not equal to the second derivative ($\kappa \neq y''$); this is only the case for flat parts of the curve ($y' = 0$). Now let us see how Huygens used his proto-calculus to find the curvature.

Two Infinitely Close Normal Lines

Let us now consider PROPOSITION XI. First Huygens describes the geometry of two infinitely close normal lines. See Fig. 10.12 and Fig. 10.13. Let line HN be the x-axis (with positive direction to the left) and FL be the y-axis (with positive direction up). As noted already, the notations "x" and "y"-axis came from Descartes. Let ABF be a part of some curve, turning in one direction. It is required to find another curve DE, which is the evolute of curve ABF, that is, the collection of all of its centers of curvature. All tangents to the curve DE by necessity cross the curve ABF at right angles. For example the tangents BD and FE touch the evolute CDE in turn. Moreover, the points B and F are understood to be arbitrarily close together. In other words, BD and FE are two infinitely close normal lines. The ordering of points of the curve ABF corresponds to the ordering of points on the evolute CDE.The intersection G of the lines BD and FE lies beyond the point D on the line BD.

For BD and FE to cross each other it is necessary that they remain at right angles to the concave part of curve ABF. Moreover, when the point F is in the vicinity

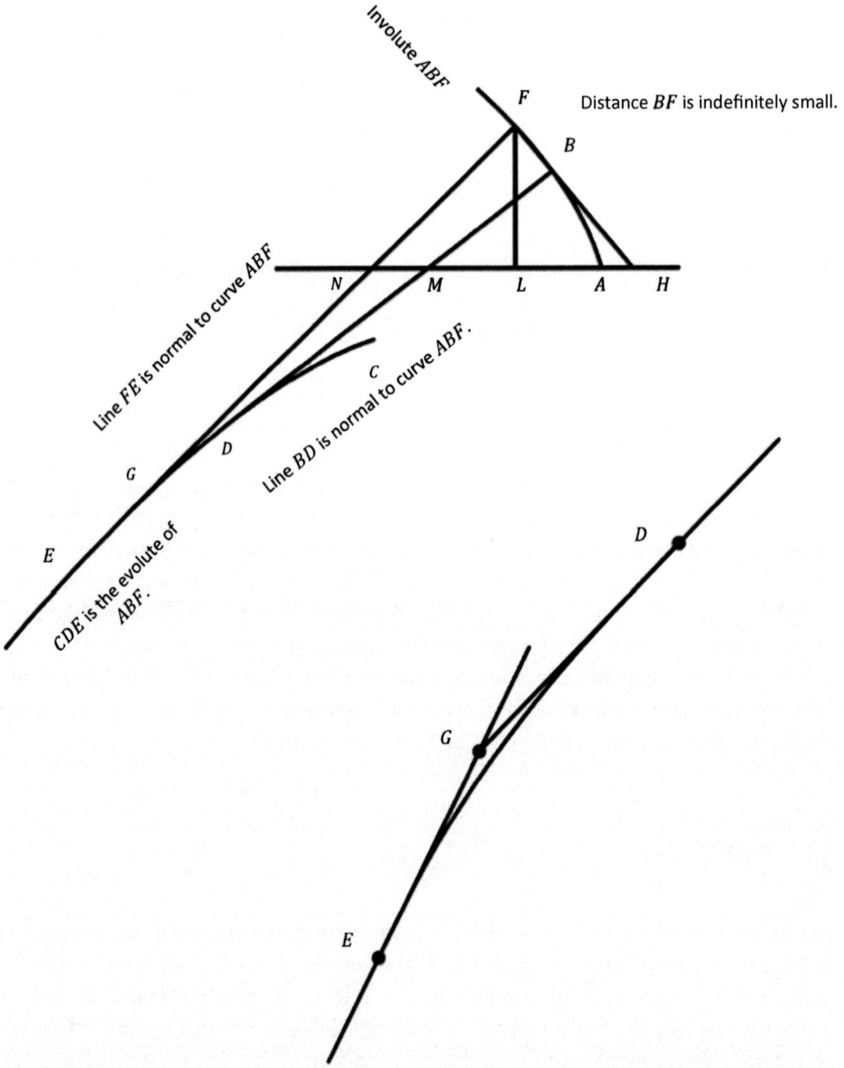

Fig. 10.12 Left: Two infinitely close normal lines. Right: A magnification of the curve DE

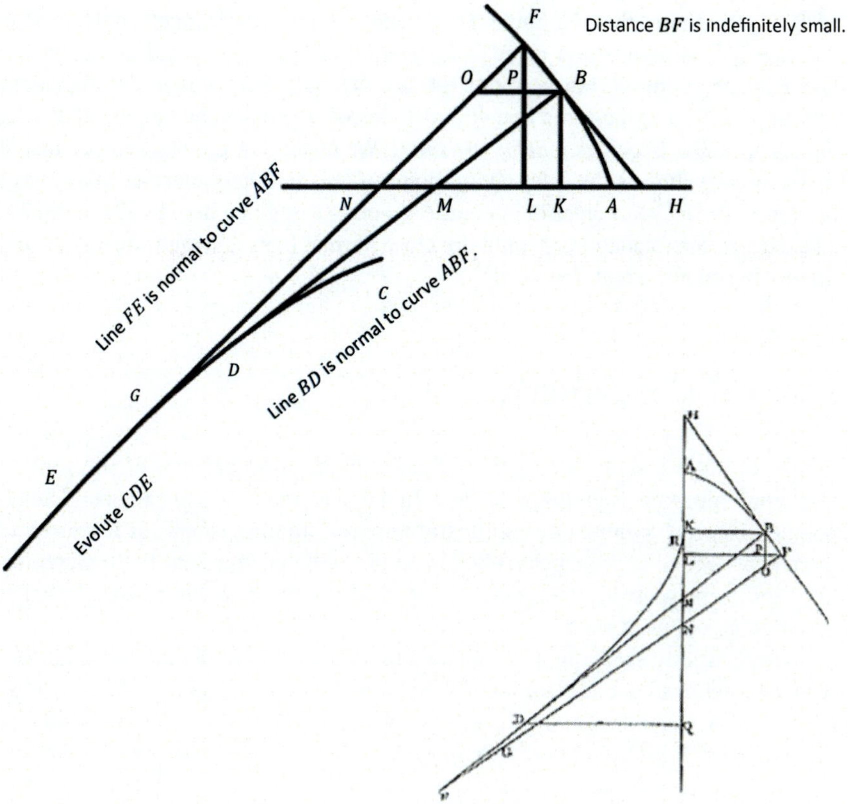

[Fig. 71.]

Fig. 10.13 Top left: Same as Fig. 10.12 but with additional points and lines. Bottom right: Huygens' original drawing (transposed)

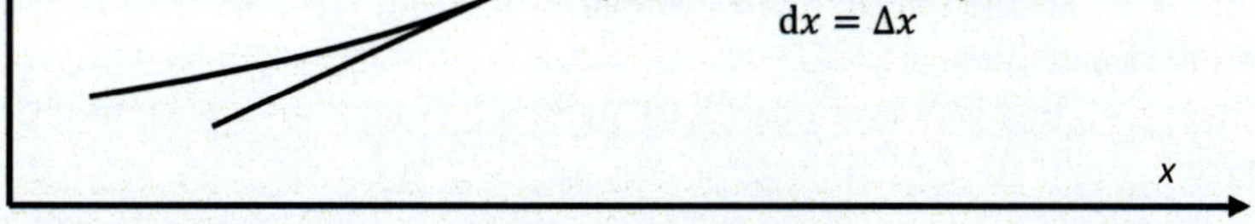

Fig. 10.14 Leibniz's notation in the modern version of calculus

of B, it is apparent that the points D, G, and E come closer together. Thus, if the distance BF is understood to be *indefinitely small*, the three said points become one and same point G. Furthermore, the line BH is drawn, which is the tangent to the curve at B, and the same line will also be considered as the tangent at F when the distance BF becomes *indefinitely small*. We recall that Cauchy showed that the center of curvature is the intersection point of two infinitely close normal lines to the curve. In fact, the center of curvature is point G as given here by Huygens. This ends Huygens explanation of infinitely close normal lines. Huygens now goes on to discuss his own method.

Proto-Calculus Curvature

In this section we want not only to continue with the proto-calculus of Huygens, but also to introduce the calculus of Leibniz. In the modern version of calculus, Leibniz' notation uses the symbols $\mathrm{d}x$ and $\mathrm{d}y$ to represent infinitely small (or infinitesimal) increments of x and y, respectively, just as Δx and Δy represent finite increments of x and y, respectively (see Fig. 10.14). In the proto-calculus of Huygens these symbols are not used at all.

Indeed, strictly speaking, to avoid the precise meaning of the infinitesimal 'd', it would be better to use 'Δ' for a finite difference and then take the limit $\Delta \to 0$ at the end. However, historically it is more correct to use Leibniz' notation 'd' because the concept of limit had not yet been introduced in Huygens' time. As we have noted already Huygens cleverly formulates his statements in such a way that they imply a limiting process, without actually carrying it out explicitly. Leibniz himself was aware that a precise definition of the infinitesimal 'd' was still missing and that this would create much criticism. Finite differences were introduced by Taylor in 1715.

In terms of notation, the differential d, used in differential quotients and integrals, is sometimes written in italic and sometimes upright. Historically, d was always written in italic, d. The use of the upright d seems to have been originating from the late twentieth century, based on the argument that d is a constant. In this book, for consistency we try to adhere to the upright convention (the same applies to the imaginary unit i discussed earlier).

Let us now turn to Huygens' derivation. Let K be the projection of B on the x axis, BO be parallel to KL, and the perpendiculars BK and FL intersect line HN. The line FL cuts the line BO at P. The lines BD and FE cross line NH at points M and N respectively.

Since triangle BOG is similar to triangle MNG (as can be seen in Fig. 10.15), we have

$$\frac{BG}{MG} = \frac{BO}{MN}. \tag{10.39}$$

The right-hand side of Eq. (10.39) may be written as

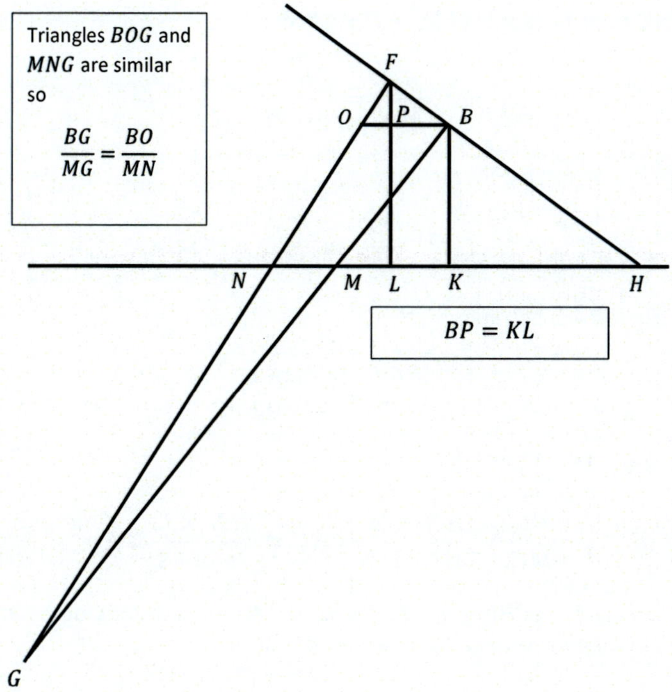

Fig. 10.15 Triangle BOG and triangle MNG are similar

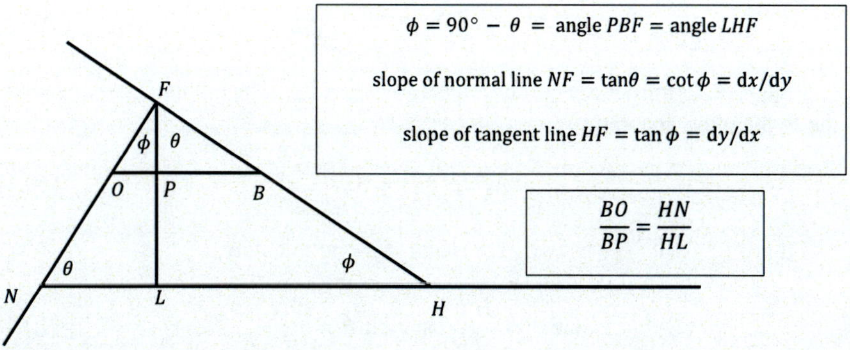

Fig. 10.16 Triangle OFB and triangle NFH are similar, with further definitions

$$\frac{BO}{MN} = \frac{BO}{BP}\frac{BP}{MN}. \tag{10.40}$$

From Fig. 10.15 we also see

$$BP = KL\,. \tag{10.41}$$

Thus the right-hand side of Eq. (10.39) becomes

$$\frac{BO}{MN} = \frac{BO}{BP}\frac{KL}{MN}. \tag{10.42}$$

From Fig. 10.16 we see

$$\frac{BO}{BP} = \frac{HN}{HL}. \tag{10.43}$$

Thus Eq. (10.42) becomes

$$\frac{BO}{MN} = \frac{HN}{HL}\frac{KL}{MN}. \tag{10.44}$$

We now use Eq. (10.39) to obtain

$$\frac{BG}{MG} = \frac{HN}{HL}\frac{KL}{MN} \text{ or } BG = \frac{HN}{HL}\frac{KL}{MN}MG. \tag{10.45}$$

Equation (10.45) is Huygens' fundamental formula for deriving point G on the evolute. We want to evaluate the two components:

$$\text{first component} = \frac{HN}{HL}$$
$$\text{second component} = \frac{KL}{MN}. \tag{10.46}$$

Figure 10.17 introduces the arc-length s along ABF. The infinitesimal $\mathrm{d}s$ satisfies the Pythagoras' theorem:

$$\mathrm{d}s^2 = \mathrm{d}x^2 + \mathrm{d}y^2. \tag{10.47}$$

Figure 10.17 also gives two expressions for sinθ; namely,

$$\sin\theta = \frac{HF}{HN} \text{ and } \sin\theta = \frac{HL}{HF}. \tag{10.48}$$

Thus the first component (10.46) is

$$\frac{HN}{HL} = \frac{HN}{HF}\frac{HF}{HL} = \frac{1}{\sin\theta}\frac{1}{\sin\theta} = \frac{\mathrm{d}s}{\mathrm{d}x}\frac{\mathrm{d}s}{\mathrm{d}x} = \left(\frac{\mathrm{d}s}{\mathrm{d}x}\right)^2. \tag{10.49}$$

Figure 10.18 further illustrates the following. As given by the second line of (10.46), the second component is KL/MN. Its reciprocal is

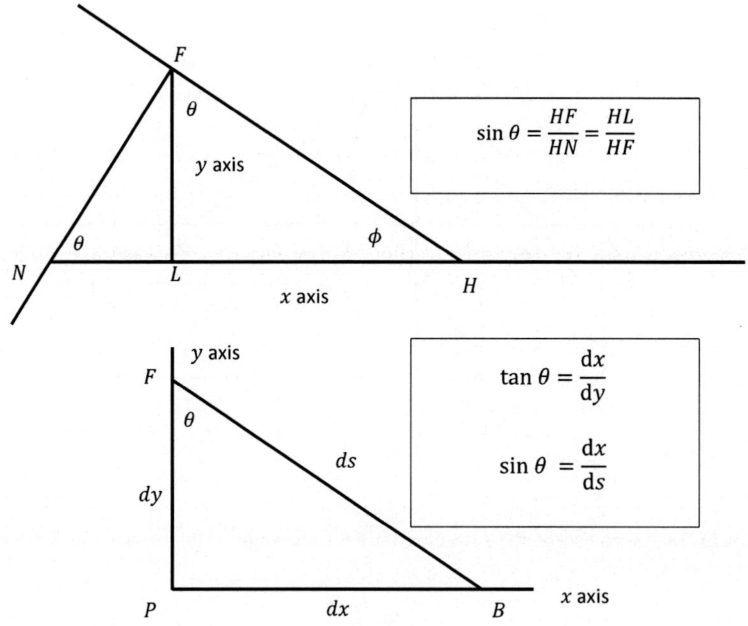

Fig. 10.17 Triangle FNL, triangle HNF, and the infinitesimal triangle FPB are similar. Because θ is the angle of the normal line NF , it follows that $\tan\theta = BP/PF = \mathrm{d}x/\mathrm{d}y$ the arclength. BF is given by the infinitesimal $\mathrm{d}s$ where $\mathrm{d}s^2 = \mathrm{d}x^2 + \mathrm{d}y^2$

$$\frac{MN}{KL} = \frac{KL + LN - KM}{KL} = 1 + \frac{LN - KM}{KL}. \tag{10.50}$$

The second term on the right is

$$\frac{LN - KM}{KL}. \tag{10.51}$$

Because θ is the angle of the normal line NF, it follows that

$$\tan\theta = \frac{FL}{LN} = \frac{\mathrm{d}x}{\mathrm{d}y}, \tag{10.52}$$

and we have

$$LN = \frac{FL}{\tan\theta} = \frac{y}{\left(\frac{\mathrm{d}x}{\mathrm{d}y}\right)} = y\frac{\mathrm{d}y}{\mathrm{d}x}. \tag{10.53}$$

The line KM is an infinitesimally small distance away from line LN. Thus we have the differential

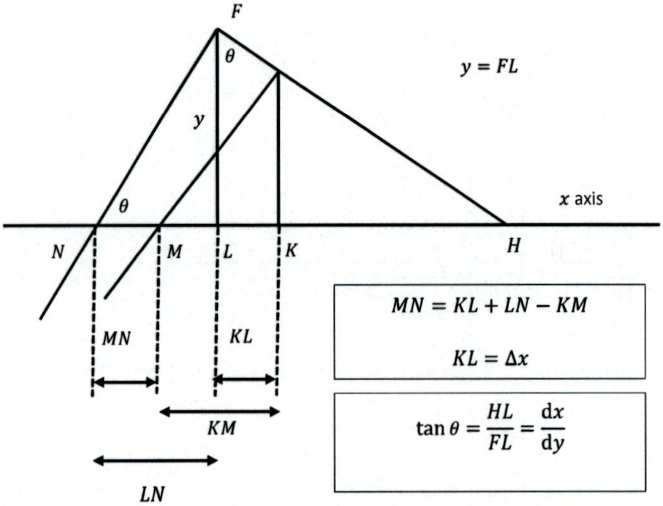

Fig. 10.18 Line segments on the x axis. Because θ is the angle of the normal line NF, it follows that $\tan\theta = \mathrm{d}x/\mathrm{d}y$ (see Fig. 10.16)

$$LN - KM = \mathrm{d}\left(y\frac{\mathrm{d}y}{\mathrm{d}x}\right) . \tag{10.54}$$

We also have the differential

$$KL = \mathrm{d}x . \tag{10.55}$$

Thus Eq. (10.49) becomes

$$\frac{MN}{KL} = 1 + \frac{LN - KM}{KL} = 1 + \frac{\mathrm{d}\left(y\frac{\mathrm{d}y}{\mathrm{d}x}\right)}{\mathrm{d}x} . \tag{10.56}$$

If we use Leibniz' product rule for derivatives, we obtain

$$\frac{\mathrm{d}\left(y\frac{\mathrm{d}y}{\mathrm{d}x}\right)}{\mathrm{d}x} = \left(\frac{\mathrm{d}y}{\mathrm{d}x}\right)^2 + y\frac{\mathrm{d}\left(\frac{\mathrm{d}y}{\mathrm{d}x}\right)}{\mathrm{d}x} = \left(\frac{\mathrm{d}y}{\mathrm{d}x}\right)^2 + y\frac{\mathrm{d}^2y}{\mathrm{d}x^2}. \tag{10.57}$$

Equation (10.56) becomes

$$\frac{MN}{KL} = 1 + \left(\frac{\mathrm{d}y}{\mathrm{d}x}\right)^2 + y\frac{\mathrm{d}^2y}{\mathrm{d}x^2}. \tag{10.58}$$

Because

$$\mathrm{d}x^2 + \mathrm{d}y^2 = \mathrm{d}s^2 \tag{10.59}$$

Equation (10.58) becomes

$$\frac{MN}{KL} = \left(\frac{\mathrm{d}x}{\mathrm{d}x}\right)^2 + \left(\frac{\mathrm{d}y}{\mathrm{d}x}\right)^2 + y\frac{\mathrm{d}^2y}{\mathrm{d}x^2} = \left(\frac{\mathrm{d}s}{\mathrm{d}x}\right)^2 + y\frac{\mathrm{d}^2y}{\mathrm{d}x^2}. \tag{10.60}$$

The inverse of (10.60) is the second component given by the second line of Eq. (10.46); that is,

$$\text{second component} = \frac{KL}{MN} = \frac{1}{\left(\frac{\mathrm{d}s}{\mathrm{d}x}\right)^2 + y\frac{\mathrm{d}^2y}{\mathrm{d}x^2}}. \tag{10.61}$$

The first component is given by first line of Eq. (10.46); that is,

$$\text{first component} = \frac{HN}{HL} = \left(\frac{\mathrm{d}s}{\mathrm{d}x}\right)^2. \tag{10.62}$$

Combining the two components gives

$$\frac{MG}{BG} = \frac{HL}{HN}\frac{MN}{KL} = \frac{\left(\frac{\mathrm{d}s}{\mathrm{d}x}\right)^2 + y\frac{\mathrm{d}^2y}{\mathrm{d}x^2}}{\left(\frac{\mathrm{d}s}{\mathrm{d}x}\right)^2} = 1 + \frac{y\frac{\mathrm{d}^2y}{\mathrm{d}x^2}}{\left(\frac{\mathrm{d}s}{\mathrm{d}x}\right)^2}. \tag{10.63}$$

Note that

$$BM = y\frac{\mathrm{d}s}{\mathrm{d}x}. \tag{10.64}$$

An alternative expression for MG/BG is

$$\frac{MG}{BG} = \frac{BG - BM}{BG} = 1 - \frac{y\mathrm{d}s/\mathrm{d}x}{BG}. \tag{10.65}$$

Setting Eq. (10.63) equal to Eq. (10.65) we obtain

$$1 + \frac{y\frac{\mathrm{d}^2y}{\mathrm{d}x^2}}{\left(\frac{\mathrm{d}s}{\mathrm{d}x}\right)^2} = 1 - \frac{y\mathrm{d}s/\mathrm{d}x}{BG}. \tag{10.66}$$

Thus

$$\frac{\frac{\mathrm{d}^2y}{\mathrm{d}x^2}}{\left(\frac{\mathrm{d}s}{\mathrm{d}x}\right)^2} = -\frac{\mathrm{d}s/\mathrm{d}x}{BG}. \tag{10.67}$$

Solving for BG we obtain

$$BG = -\frac{\left(\frac{ds}{dx}\right)^3}{\frac{d^2y}{dx^2}}, \tag{10.68}$$

which is

$$BG = -\frac{\left(\left(\frac{ds}{dx}\right)^2\right)^{3/2}}{\frac{d^2y}{dx^2}} = -\frac{\left(\left(\frac{dx}{dx}\right)^2 + \left(\frac{dy}{dx}\right)^2\right)^{3/2}}{\frac{d^2y}{dx^2}}$$

$$= -\frac{\left(1 + \left(\frac{dy}{dx}\right)^2\right)^{3/2}}{\frac{d^2y}{dx^2}} = -\frac{\left(1 + (y')^2\right)^{3/2}}{y''}. \tag{10.69}$$

(Here $y' = dy/dx$ and $y'' = d^2y/dx^2$). The minus sign, which is a matter of convention, may be omitted. Equation (10.69) is the modern formula for the radius of curvature of a twice continuously differentiable curve (compare with Eq. (10.38)). Huygens' geometric equivalent is, to repeat,

$$BG = \frac{HN}{HL}\frac{KL}{MN}MG. \tag{10.70}$$

By Eq. (10.70) one can determine the location of G on the normal BMG drawn to the curve at point B. Thus, the position of each point G on the evolute is defined by the characteristics of the original curve (the involute) infinitely close to the point B, which corresponds to the end of the unwinding cord in Huygens' physical representation of the process. Cauchy wrote in his *Sept Leçons de physique générale* (published in 1868): "We must first study the facts, multiply the observations, then seek to link them to each other by formulas, and thus recognize the particular laws which govern a certain class of phenomena. It is only after having established these particular laws that we can generally hope to note and discover the more general laws which complete the theories, by relating to the same principle a multitude of apparently very diverse phenomena."

To recall, the radius of curvature may be thought of as the measure of the flatness or sharpness of a curve at a point; the smaller the radius of curvature, the sharper the curve. The curvature of a curve at a point is the rate at which the inclination of the curve is changing with respect to the length of arc. At a given point on a curve, there is, in general, but one tangent and one normal. An infinite number of circles, all having their centers lying on the normal, can be drawn through this point. Of these circles, the one whose radius equals the radius of curvature for the curve at that point is called the *circle of curvature for the point.* Obviously, each point on the curve has a different circle of curvature (except in special cases such as when the curve itself is a circle). It can be shown that the circle of curvature at any point "fits" the curve more closely, near that point, than any other circle. The circle of curvature is also known as the *osculating circle.* The center of the circle of curvature is known as the *center of curvature.*

In *Horologium Oscillatorium*, Huygens was prompted by the pendulum clock to develop a mathematical theory of evolutes and involutes. A clock was the physical object that provided the motivation. Huygens wrote in a letter to Leibniz on 1 September 1691:

> I have often considered that the curved lines which nature presents to our view, and which she describes, as it were, with precision, have truly remarkable properties. Such are the circles encountered everywhere, the parabola drawn by jets of water, the ellipse and the hyperbola, which the shadow of the end of a stile follows and which one also finds elsewhere. The cycloid described by a nail in the circumference of a wheel. And finally our catenary, which has been observed without investigation for so many centuries. In my opinion, such curved lines deserve to be presented as exercises, but not as exercises newly created just to use geometric calculations

Huygens could have added the tractrix, discussed earlier, to his list of natural curves. As an aside, the catenary is the line that is described by a flexible uniform chain hanging freely between two points under the influence of gravity. Galileo and others thought it to be a parabola. Huygens, in a letter to Mersenne in 1646, proved that this is not the case. In 1691 Leibniz, Johann Bernoulli and Huygens, responding to a challenge launched by Jakob Bernoulli, demonstrated almost simultaneously that the exact form of the catenary is a transcendent curve by determining its equation. The catenary arch is frequently used in architecture and engineering, because it evenly distributes loads and forces along the curve, and appears as the cross section of the soap film formed by two circular rings. End of aside.

In effect, *Horologium Oscillatorium* was a treatise on calculus; not the calculus of Leibniz and Newton, but the proto-calculus formulated by Huygens. Huygens' formulation had the distinct advantage of being able to solve problems involving second derivatives, as we have shown above. This achievement was a significant accomplishment in the history of mathematics. Also, we see here Huygens as the applied physicist, studying curves with physical applications, rather than studying them for their own sake (*l'art pour l'art*). Huygens would say: "*Reliqua vulcano tradenda*"—the rest can be given to the volcano.

In formulating his physical theories in the *Principia*, Newton gave many of his proofs in a geometric form of infinitesimal calculus, based on limits of ratios of vanishing small geometric quantities. In fact, Newton used the same mathematical methods as Huygens. This is pointedly illustrated by the recollection of one of Newton's collaborators, Henry Pemberton (1694–1771): "But Sir Isaac Newton has several times particularly recommended to me Huygens' style and manner. He thought him the most elegant of any mathematical writer of modern times, and the most just imitator of the ancients." (Pemberton, 1728). Newton would usually refer to Huygens as '*Summus Hugenius*', the eminent Huygens.

Conversely Huygens, fourteen years older than Newton, had high esteem for many (though not all) of Newton's contributions. He highly praised Newton's parabolic mirror telescope which solved the problems of chromatic and spherical aberration. In his letter to Fatio de Duillier of July 11 1687 Huygens wrote, referring to action at distance: "I do not mind at all that Newton is not a Cartesian provided he does not offer us suppositions like that of attraction." A well known quote by Huygens on Newton's gravitation theory comes from his *Discours de la Cause de la Pesanteur* (Discourse on the Cause of Weight, published in 1690 together with Traité de la Lumière): "I had not thought of this regular decrease of gravity, namely that it is as the inverse square of the distance; this is a new and highly remarkable property of gravity, for which we have to find the reason." Nevertheless, a bit more critical later in this text: "There is only this difficulty, that Mr. Newton wants the celestial spaces to contain only a very rare material, in order to that the Planets and the Comets meet all the less obstacle in their course." To his brother Constantijn he wrote (December 30 1688), referring to the *Principia*: "I would like to be in Oxford, if only to meet Mr. Newton, of whom I admire the wonderful inventions I found in the book he sent me." One year later they would indeed meet.

As we have seen in this chapter, Huygens gave a geometrical way of finding the second derivative. To Newton, this procedure was essential because the force on a moving body is proportional to the second derivative of position with respect to time. The methods of Newton's calculus were conspicuously absent from *Principia.* Yet, for about one hundred year England demanded that its mathematicians used Newton's calculus. With the exception of the work of Brook Taylor and a few other English mathematicians, no significant results in this time period were produced. The European continent used the calculus of Leibniz and flourished.

As to why Newton did not use his calculus in *Principia* is controversial. The question explicitly presumes that calculus is missing from the *Principia* published in 1678. It is true that Leibniz-Newton calculus is missing, but not the proto-calculus of Huygens. In 1678, both Leibniz and Newton calculus were in their original cumbersome forms. With its inadequate notation, Newton's form of calculus would not have introduced greater clarity or shorter proofs. Instead it would have compounded the difficulty of understanding. In fact, the work of the Bernoulli family in the development of Leibniz's form of calculus could effectively be used in physics. Calculus is not missing from the *Principia.* It is the calculus of Huygens, not Newton.

Chapter 11
What Huygens Could Have Written on Diffraction

> Not go back is somewhat to advance, And men must walk, at least, before they dance.
> —*Alexander Pope (1688–1744), An Essay on Man*

This is a highly speculative chapter. As far as we know Huygens never said or wrote anything on diffraction. Yet Huygens' principle is perfectly fit for the modeling of diffraction phenomena and it had a very strong imprint on the historical development of diffraction theory. Diffraction as a wave propagation effect had been identified by Huygens' contemporaries Grimaldi, Newton and Leibniz, but it was ignored by Huygens. Intentionally? We do not know.

Let us first try to define diffraction and then give a historical account. Diffraction is the disruption of a wave propagating around and beyond an obstacle. It plays a role in many branches of science, for instance electromagnetic, optical, acoustical and water wave propagation, and has been a topic of scientific investigation for many centuries. Diffraction of light and sound can be observed anytime in daily life.

Prominent scientists have exhibited significant ambiguity regarding the definition of diffraction. Here are a few well-known quotes:

- Sommerfeld—Diffraction is any deviation of light rays from rectilinear paths which cannot be interpreted as reflection or refraction.
- Hecht—Diffraction occurs whenever a portion of a wavefront, be it sound, matter or light, is obstructed in some way.
- Landau and Lifshitz—Diffracted waves are phenomena which are the consequence of deviations from geometrical optics.
- Feynman—No-one has ever been able to define the difference between interference and diffraction satisfactorily.

All these statements hold elements of truth and encompass certain aspects of the essence of diffraction. What they share, is that diffraction involves a distortion of the regular wave propagation. We will illustrate this using some simple examples. In the photograph shown in Fig. 11.1, sunlight that strikes the edge of the window is diffracted around it, rather than creating a perfectly dark shadow zone. Likewise, the fact that one can hear sound around a corner from a neighboring room without

T. J. Moser and E. A. Robinson, *Walking with Christiaan Huygens*, History of Physics,
https://doi.org/10.1007/978-3-031-46158-3_11

Fig. 11.1 Diffraction at a window edge. Light is broken and scattered into the shadow

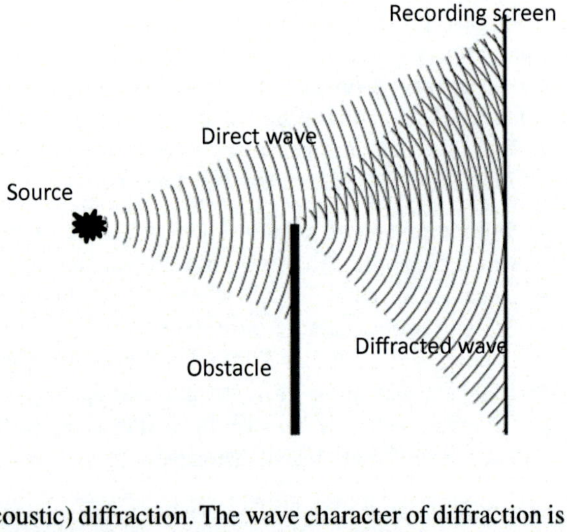

Fig. 11.2 Diffraction patterns beyond an opaque half screen, demonstrating the wave character of diffraction

seeing its source relates to (acoustic) diffraction. The wave character of diffraction is illustrated in Fig. 11.2: on the viewing screen behind an opaque half screen interference patterns between the direct and diffracted waves are seen with increasing wave length into the shadow zone. *Geometrical optics* deals with the direct wave only and is incomplete because it predicts a perfectly dark shadow and a sharp discontinuity at the shadow boundary which are both physically infeasible. Diffraction is the physical correction to geometrical optics.

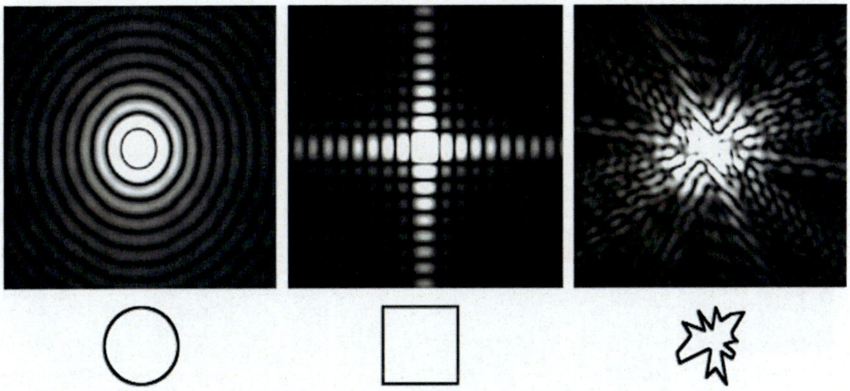

Fig. 11.3 Diffraction patterns as function of aperture: circular, rectangular and irregular aperture

Fig. 11.4 Diffraction patterns for rectangular, triangular and curvilinear triangle (from Sommerfeld, Optics, 1949). Waves diffracted by an aperture occur in pairs; as a result odd-cornered stars used in ancient mythology and some well-known modern flags are physically impossible

Diffraction patterns through an aperture strongly depend on its shape (Fig. 11.3). They always come in pairs—an interesting consequence of this is that the five-cornered star often used in heraldry is physically impossible (Fig. 11.4).

Fig. 11.5 Camera Obscura: diffraction through a slit in a curtain, building across the street and images of it on the wall of a hotel room (clockwise from left)

Diffraction through a slit can be used in very simple imaging, referred to as Camera Obscura (Fig. 11.5).

Diffraction is strictly speaking not the same as scattering. They are both phenomena related to the interaction of the waves with obstacles and inhomogeneities in a medium, but they differ in the underlying principles. Diffraction is the bending of waves as they hit a sharp discontinuity, the diffractor, and geometrical optics no longer applies (see Fig. 11.6, left). In diffraction theory, as we will see in this chapter, the diffracted wave is modeled by an additional wave. The diffractor is discontinuous and can have an arbitrary contrast with the surrounding medium. The diffraction modeling is typically accomplished using another set of diffracted rays.These rays also adhere to the principles of geometrical optics again, but their amplitudes and polarities account for the discontinuity. They serve as a physical correction to the geometrical optics in the original medium.

Scattering assumes that the medium is decomposed into a background model and a scatterer on top of it, and it describes the perturbation of the wavefield resulting from the presence of the scatterer. In scattering theory, the scattered wave is modeled by a perturbation series, such as the Born series. This involves an integral, the scattering integral, which is carried out over the support of the scatterer (the part where it differs from zero). The scatterer may, or may not be smooth (see Fig. 11.6, right). Usually the assumption is that the scatterer has a small amplitude compared to the background medium; this is referred to as the weak-scattering approximation.

In this chapter, we only deal with diffraction. Diffraction, unlike scattering, lends itself in a very straightforward way to Huygens' principle.

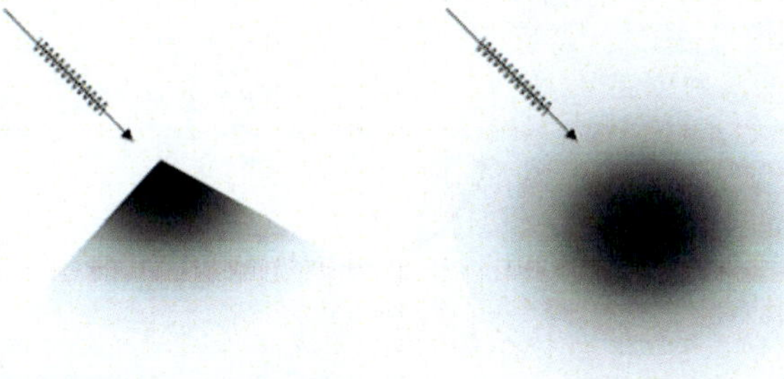

Fig. 11.6 Diffraction is not the same as scattering. Incident wave on a diffractor (left) and a scatterer (right)

Before Huygens

The history of diffraction before Huygens can be broadly characterized by the following sequence of episodes. Initially, the phenomenon of light refraction attracted much attention, before diffraction came into the forefront.

In antiquity, Euclid, Heron of Alexandria and Ptolemy studied light rays. They found the law of reflection and they observed refraction. The first simple camera obscura was invented by the Chinese philosopher Mo-Tzu: rays from an object traveling through a small pinhole cast an image of the object at a screen behind the pinhole (see Fig. 11.5).

During the Islamic Golden Age (about 700–1200), there was an active interest in optics. Around 984 Abu Sa'd al-'Ala Ibn Sahl, alias Ibn Sahl (ca. 940–1000) published "On Burning Mirrors and Lenses" on the design of lenses that focus incoming light at a given distance without geometric aberration. He studied refraction and presented a geometrical relation between incident and refracted rays, which can be written in modern trigonometric notation as the law of refraction.

Around the year 1000 Abu Ali al Hassan ibn al Hasan ibn al Haitham, alias Alhazen (ca. 965–1039, see Fig. 11.9) published "Book of Optics" (Kitab al-Manazir) establishing a theory of vision as well as a theory of optics, which expanded on Ptolemy's theory of refraction. Alhazen was the first to decompose reflected and refracted light rays into normal and tangential components. He also studied the problem of reflection on a circular mirror, now called Alhazen's problem: given two points (a source and receiver point) to find the reflection point on the mirror, such that angle of incidence equals the angle of reflection (thereby satisfying what later would be known as Snell's law). Alhazen's problem is also called the circular billiard problem: given a circular billiard table, find the direction to hit a first ball such that it hits a given second ball with exactly one bounce on the table edge.

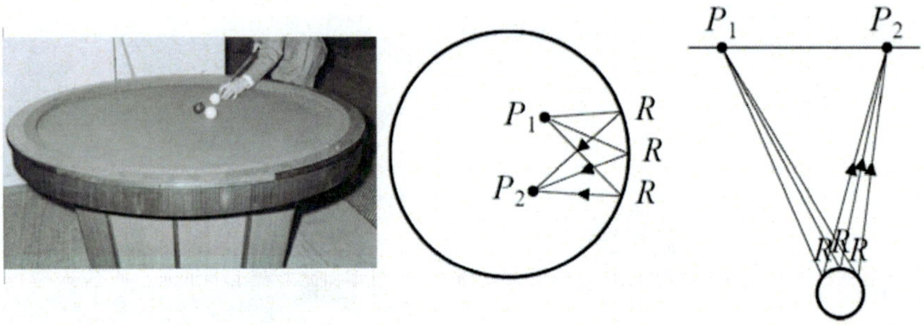

Fig. 11.7 Alhazen's problem or the circular billiard problem: to find the reflection point R on a circular reflector between two given endpoints P_1 and P_2. For high reflector curvature reflection turns into diffraction [with permission by the Nationaal Archief]

Reflection on a spherical reflector becomes relevant for diffraction theory when the radius of the sphere approaches zero, since its response becomes a diffracted rather than a reflected wave. Alhazen's problem leads to an algebraic equation of fourth-order which was fully solved only in recent times (Neumann, 1998) (see Fig. 11.7). Huygens was highly interested in Alhazen's problem, as his notebooks demonstrate (see Fig. 11.8).

The Renaissance first of all witnessed the solution of the refraction problem and a priority dispute around its discovery. In 1602, Thomas Harriot (1560–1621) discovered, or better, rediscovered Snell's law, but did not publish it (incidentally, Thomas Harriot introduced the potato in Europe). In 1621, Willebrord Snel van Royen (1580–1626, Latin Snellius, English Snell) experimentally found the law of angles of incidence, reflection and refraction of light between two media. In 1637, René Descartes (1596–1650) published "La Dioptrique" in which he explained the theory of the rainbow, based on a combination of light reflection and refraction in raindrops. Finally, in 1644 Descartes derived Snell's law of refraction, which in the francophone world bears the name law of Snell-Descartes.

In 1657, Pierre de Fermat (1601–1665) derived the law of refraction from the principle of least time, assuming the propagation of light inside optically dense media to be slower than outside. The principle of least time, or Fermat's principle, states that the path taken between two points by a ray of light is the path along which the travel time is minimal, or more generally, stationary among all paths connecting the two points. As we shall see, Fermat's principle plays a key role in Keller's extension of geometrical optics to the Geometrical Theory of Diffraction (1962).

In 1665, Robert Hooke (1635–1703) formulated a wave theory of light and published "Micrographia", explaining the appearance of color in light by wavefront refraction. At that time, wave interference was still unknown and Hooke was not able to explain the reflection of light on both sides of a thin film.

Some early references to diffraction phenomena can be traced in work of Leonardo da Vinci (1452–1519). The first milestone in the history of diffraction, however, is the work of Francesco Grimaldi (1618–1663, see Fig. 11.9). He performed a set of experiments which clearly proved that geometrical optics is not able to explain

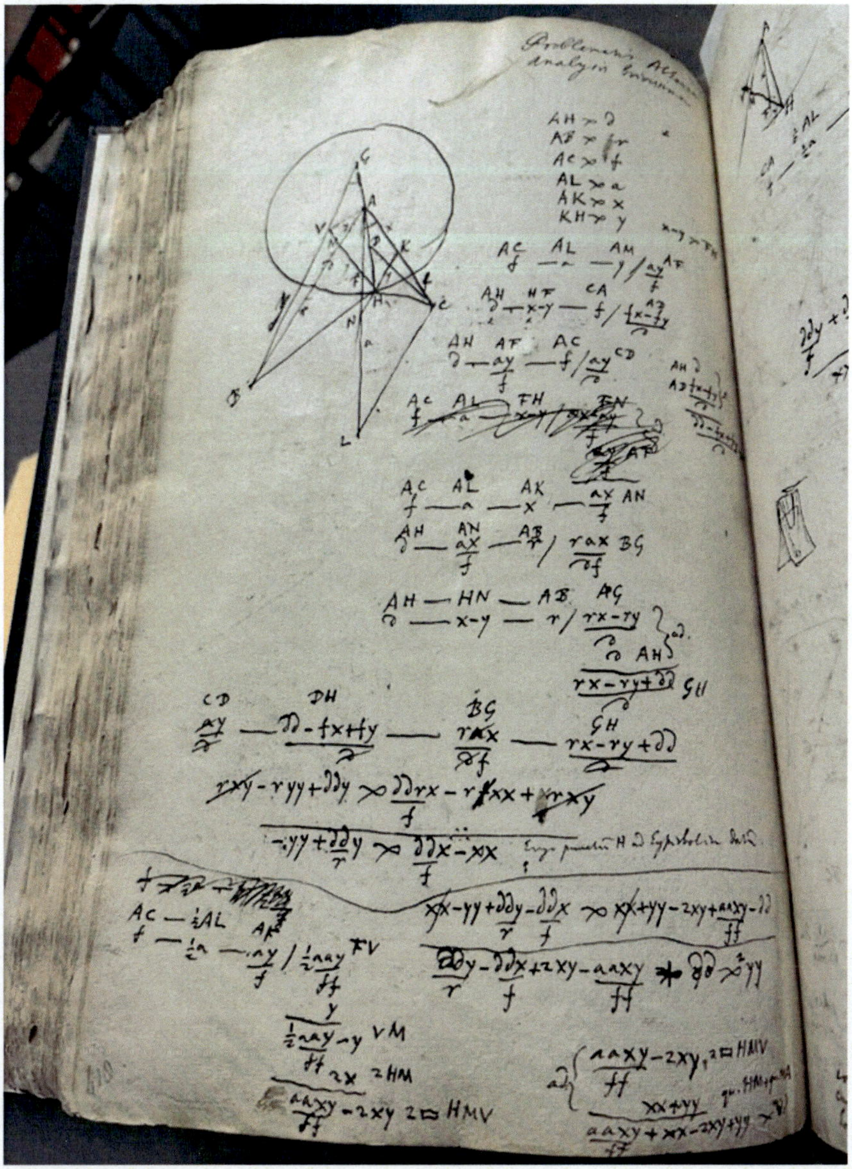

Fig. 11.8 Notes on Alhazen's problem in Huygens' notebooks [with permission from Leiden University Library, HUG 2, fol. 208 v.]

Fig. 11.9 Alhazen and Francesco Grimaldi

a number of observations. Most importantly, he demonstrated that light penetrates into the shadow zone, where geometrical optics incorrectly predicts full darkness. Geometrical optics describes light propagation in terms of rays. Rays are straight lines in a homogeneous medium and energy propagates along the rays, not across them.

Grimaldi's Experimentum Primum (Fig. 11.10, top) studies the shadow of a small opaque (impenetrable) body (EF) placed in a bundle of sunlight passing through a small aperture (AB). Geometrical optics predicts that the rays AFG and BEH cannot bend so that the line GH is in a perfectly dark shadow. However, the observation is that the line MN is at least partially illuminated all through, implying that rays must indeed bend around the obstacle. Grimaldi also observed diffraction fringes in the regions CM and ND (he called them *seriae lucidae*), which were later confirmed by Newton. The existence of diffraction fringes is a clear manifestation of *interference* and the wave character of light, and Grimaldi's observation appears to be the first evidence of interference recorded in history.

Grimaldi's Experimentum Secundum (Fig. 11.10, bottom) shows how a cone of light passing through two small apertures (CD and GH) and falling onto a white screen produces an illuminated zone (IK) which is much larger than predicted by geometrical optics (ON). The penetration of diffracted energy into the geometrical shadow zone is a very fundamental property.

The second contribution of Grimaldi was the introduction of the word "diffraction" itself. "Diffraction" derives from Latin diffringere, which means to break into different directions ("dis" means "apart", "frangere" means "to break"). Grimaldi's work was published posthumously in "Physico-mathesis de lumine, coloribus, et iride, aliisque annexis" in 1665. Proposition I in the book reads: Light propagates or scatters not only directly, by reflection or refraction, but also in a fourth mode, *diffraction.* Hooke and Newton possibly read Grimaldi's work, but referred to diffraction

PHYSICO-MATHESIS
DE LVMINE.
COLORIBVS. ET IRIDE.
ALIISQVE ADNEXIS
LIBRI DVO.
AVCTORE
P. FRANCISCO MARIA GRIMALDO
SOCIETATIS IESV.
OPVS POSTHVMVM.
BONONIÆ. M.DC.LXV.

PROPOSITIO I.

Lumen propagatur seu diffunditur non solùm Directè, Refractè, ac Reflexè, sed etiam alio quodam Quarto modo, DIFFRACTE'.

HActenus nonnisi tribus modis exerceri luminis diffusionem Optici communiter agnouerunt, Directè scilicet, Refractè, ac Reflexè. Certissimis namq; experimentis obseruatum est, lumen per lineam quidem rectam propagari, seu diffundi, donec medium, per quod diffunditur, maneat idem

rit densitatem ab ea, quam habet corpus item diaphanum, per quod radius AB. descendit. Quin immò si transitus fiat à medio rariore ad densius, radius refringetur versùs perpendicularem, hoc est versùs lineam rectam, quæ concipiatur educi per punctum ingressûs B, ita vt faciat cum superficie CD. angulos omnes æquales. At

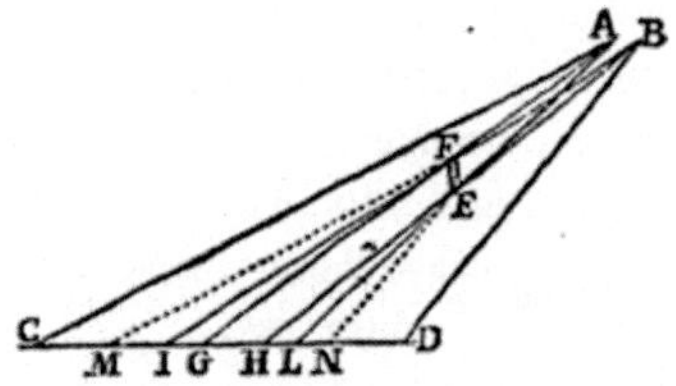

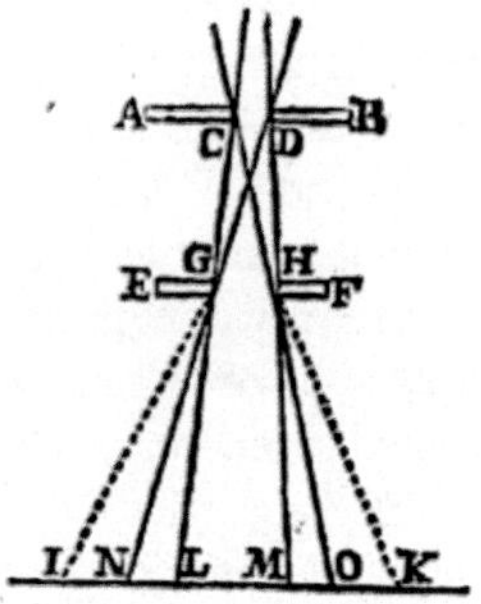

Fig. 11.10 Grimaldi's book with the first mention of the word diffraction (left), and his Experimentum Primum and Secundum (right top/bottom), demonstrating diffraction energy penetration into the shadow and interference

instead as *inflexion*. It took until the work of Fresnel in the early nineteenth century for the term diffraction to be used as the standard.

The full title of Grimaldi's book in English reads "A physico-mathematical treatise on light, colors and the rainbow, and other related topics in two books, in the first new experiments are described and arguments are deduced from them in favor of the substantiality of light. In the second, however, the arguments introduced in the first are refuted and it is shown that the peripatetic theory of the accidentality of the light might be true". Here, the terms substantiality and accidentality have a scholastic origin, meaning "having material substance" and "belonging to a subject without affecting its essence", respectively ("peripatetic" refers to Aristotle's school). In Grimaldi's language, the substantiality of light points to its corpuscular character and accidentality to its wave character, and so here we have a first mention in modern times of the *particle-wave debate*: whether light consists of particles (corpuscles) or waves (Thomas Young in 1802 pointed out that the wave theory in some form goes back to Aristotle, the corpuscular theory to Empedocles (Greek philosopher, 494–434 BC)).

With the appearance of Isaac Newton (1643–1727) the particle-wave debate became acute. Descartes had suggested that light consists of small particles that propagate in a straight line at a finite speed (Ole Roemer (1644–1710) had experimentally found that the speed of light is finite in 1676). On the other hand, Hooke had suggested that light has a wave character and is a rapid vibration movement of a medium. Newton was more undecided. In his book "Opticks" (1704) he described light propagation both in terms of wave theory and corpuscular theory. Newton later rejected wave theory because he could not explain the rectilinear propagation of light with waves (although he accepted that sound waves obviously have wave characteristics). In this respect he did not see that geometrical optics is consistent with wave theory in the limit of a small wavelength. In certain respects we can consider Newton as one of the founders of ray theory.

Newton had a keen interest in diffraction phenomena. An often-quoted statement is the reference to Grimaldi in his book "Opticks": "Grimaldo has inform'd us, that a beam of the Sun's light be let into a dark room through a very small hole, the shadows of things in this light will be larger than they ought to be if the rays went on by the bodies in straight lines." He conducted many experiments on diffraction, such as the knife edge and wedge diffraction experiment, which demonstrate the existence of waves propagating into the geometrical shadow and the formation of fringes on an observation screen perpendicular to the direction of the incoming light. As noted in relation to Grimaldi's work, the occurrence of diffraction fringes points to *interference*, which was fully explained by Fresnel in the early nineteenth century.

In addition to his work on ray theory, Newton can be regarded as an early pioneer of the concept of the *diffracted ray*, which was picked up again only later in the twentieth century. Still he remained skeptical about the ability of light to bend around obstacles and had to use artificial arguments to explain diffraction effects.

Huygens and Diffraction?

What had Huygens to do with diffraction? Everything and nothing. Huygens' main oeuvre on light is "*Traité de la Lumière*" ("*Treatise on Light*", written in 1677, published in 1690). In *Traité de la Lumière*, he proposed what is now called *Huygens' principle* for wave propagation, which was so fruitful that it can explain most propagation effects relevant even today: reflection, refraction, anisotropy, caustics, and, ultimately, diffraction.

Let us now shift our focus to diffraction. It is easy to demonstrate that Huygens' principle can effectively explain not only reflection and refraction, but also the classical diffraction phenomenon: the diffraction of a plane wave passing through an aperture in an opaque screen (as depicted in Fig. 11.11). When the plane wave has reached the screen and aperture (Fig. 11.11, 1), wavefronts at later instants (Fig. 11.11, 2–4) can be constructed by considering elementary secondary point sources at the aperture and drawing the envelope around them. In front of the aperture this results in an undisturbed plane wave (denoted by 'P' in Fig. 11.11, 2–4), at the two edges of the aperture this results in additional diffracted waves bending around the edges (denoted

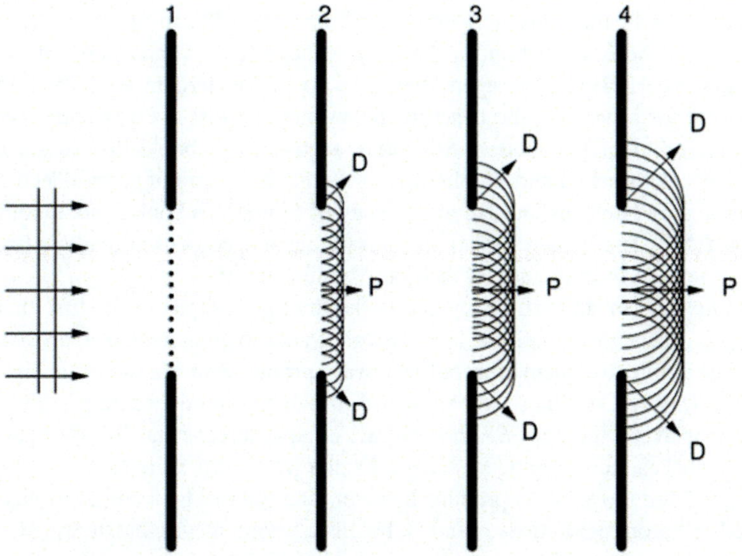

Fig. 11.11 Huygens' principle: diffraction

by 'D' in Fig. 11.11, 2–4). Again, the resulting enveloping wavefront (direct as well as diffracted) is at any point tangent to one of the secondary waves.

The behavior of diffraction at an aperture is thus correctly predicted by Huygens' principle, especially when interference is taken into account. Yet it remains one of the major mysteries of the history of science, why he never studied diffraction or even mentioned it. There is indeed no explicit reference to diffraction, neither in his correspondence, nor in his published work or written manuscripts (see his Complete Works, Huygens 1888–1950). Also there is no mention of the work of Grimaldi, nor of his experiments, which had demonstrated the existence of diffraction in the early seventeenth century. This omission had already occurred to Huygens' contemporary and correspondent Gottfried Wilhelm Leibniz (1646–1716). In a draft letter to Huygens, which he probably never sent, Leibniz wrote: "*I would have liked you to have given us at least your conjectures on the colors, and I would like to know also what is your thought of the attraction that Mr. Newton recognized after Father Grimaldi in the light at p. 231 of his Principles*" (Huygens, correspondence, 1888–1950). Nevertheless, Huygens must have known that Newton speaks of Grimaldi in his "*Principia*" of 1687.

We also know with certainty that in 1695, Huygens was in possession of the work of Grimaldi. This is evidenced by the catalog of the sale of his books, which took place a few months after his death. A copy of this catalog is held by the Royal Library of The Hague. It mentions Grimaldi's work (editors of Huygens' Complete Works, 1888–1950; see Fig. 11.12).

No doubt, a recovery of Huygens' copy of Grimaldi's book could shed light on Huygens' view on diffraction. Did he make notes in the margin (marginalia)?

Or did he leave the book unopened? Leibniz sent a Dutch agent to the auction of Huygens' book collection, in the hope to acquire them and find written notes by Huygens. However, the agent was held up in Amsterdam and sent a student, who went for a walk on the beach of Scheveningen, lost his way and missed (an essential?) part of the auction. Still, they acquired 121 books. These are now in the Niedersächsische Landesbibliothek in Hannover; eleven of them indeed contain marginalia by Huygens, but Huygens' copy of Grimaldi's book is not among them (Hess, 1980). The Leiden University Library has one copy of Grimaldi's book, but it was acquired from the library of Isaac Vossius in 1690.

A faint hint of diffraction appears in the final paragraphs of the first chapter of Huygens' *Traité de la Lumière.* Here Huygens goes to great lengths to prove that his principle is able to explain the rectilinear wave propagation. He considers his famous diagram (Fig. 4.1 in Chap. 4 on Huygens' principle, reproduced here in Fig. 11.13) with an aperture: the lines BH and GI are opaque screens and BG is the aperture. Now he makes the sweeping statement that the wave parts which scatter outside the area ACE are too weak to produce light. In other words, he recognized that waves could bend around the edges B and G, but his assessment was that they are too weak to be observed (but still non-zero!).

In 1679, experiments had been presented by Mariotte and de la Hire at the Royal Academy of Sciences In Paris, showing the diffraction of sunlight through an aperture in a dark screen. The Academy's report read: "On Saturday 6th May 1679 Messrs.

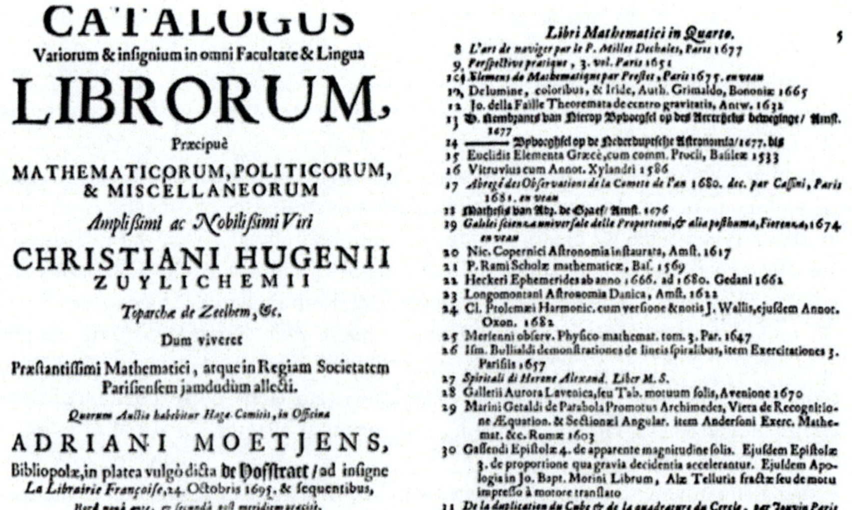
CATALOGUS
Variorum & insignium in omni Facultate & Lingua
LIBRORUM,
Præcipuè
MATHEMATICORUM, POLITICORUM,
& MISCELLANEORUM
Amplissimi ac Nobilissimi Viri
CHRISTIANI HUGENII
ZUYLICHEMII
Toparchæ de Zeelhem, &c.
Dum viveret
Præstantissimi Mathematici, atque in Regiam Societatem
Parisiensem jamdudum allecti.
Quorum Auctio habebitur Hagæ-Comitis, in Officina
ADRIANI MOETJENS,
Bibliopolæ, in platea vulgò dicta de Hofstraet / ad insigne
La Librairie Françoise, 24. Octobris 1695. & sequentibus,
Horâ nonâ ante, & secundâ post meridiem præcisè.
HAGÆ-COMITUM,
ADRIANUM MOETJENS, 1695.

Libri Mathematici in Quarto. 5
8 *L'art de naviger par le P. Milles Dechales, Paris 1677*
9. *Perspective pratique, 3. vol. Paris 1651*
10 *Elemens de Mathematique par Prestet, Paris 1675. en veau*
11 De lumine, coloribus, & Iride, Auth. Grimaldo, Bononiæ 1665
12 Jo. della Faille Theoremata de centro gravitatis, Antw. 1632
13 D. Rembrants van Nierop Wysbegeert op des Aerdtrijcks bewegingé / Amst. 1677
14 ——— Wysbegeert op de Nederduytsche Astronomia / 1677. bis
15 Euclidis Elementa Græcè, cum comm. Procli, Basileæ 1533
16 Vitruvius cum Annot. Xylandri 1586
17 *Abregé des Observations de la Comete de l'an 1680. dec. par Cassini, Paris 1681. en veau*
18 Mathesis van Abr. de Graef / Amst. 1676
19 *Galilei scienza universale delle Proportioni, & alia posthuma, Fierenza, 1674 en veau*
20 Nic. Copernici Astronomia instaurata, Amst. 1617
21 P. Rami Scholæ mathematicæ, Bas. 1569
22 Heckeri Ephemerides ab anno 1666. ad 1680. Gedani 1662
23 Longomontani Astronomia Danica, Amst. 1622
24 Cl. Ptolemæi Harmonic. cum versione & notis J. Wallis, ejusdem Annot. Oxon. 1682
25 Mersenni observ. Physico mathemat. tom. 3. Par. 1647
26 Ism. Bullialdi demonstrationes de lineis spiralibus, item Exercitationes 3. Parisiis 1657
27 *Spiritali di Herone Alexand.* Liber M. S.
28 Galletii Aurora Lavenica, seu Tab. motuum solis, Avenione 1670
29 Marini Getaldi de Parabola Promotus Archimedes, Vieta de Recognitione Æquation. & Sectionæ Angular. item Andersoni Exerc. Mathemat. &c. Romæ 1603
30 Gassendi Epistolæ 4. de apparente magnitudine solis. Ejusdem Epistolæ 3. de proportione qua gravia decidentia accelerantur. Ejusdem Apologia in Jo. Bapt. Morini Librum, Alæ Telluris fractæ seu de motu impresso à motore translato
31 *De la duplication du Cube & de la quadrature du Cercle, par Jouvin Paris 1658*
32 *Opere di Galileo Galilei 2. part. Bologna 1656*
33 Aristotelis Loca Mathematica explicata à Joseph. Blancanot accessiv: de natura Mathem. scientiarum, & clarorum Mathematicorum Chronologia Bon. 1615
A 3 34 Pla-

Fig. 11.12 Catalog of auction of books from Huygens' legacy. Item 11 ("11 De lumine, coloribus, and Iride, Auth. Grimaldo, Bononiae 1665") is his copy of Grimaldi's book

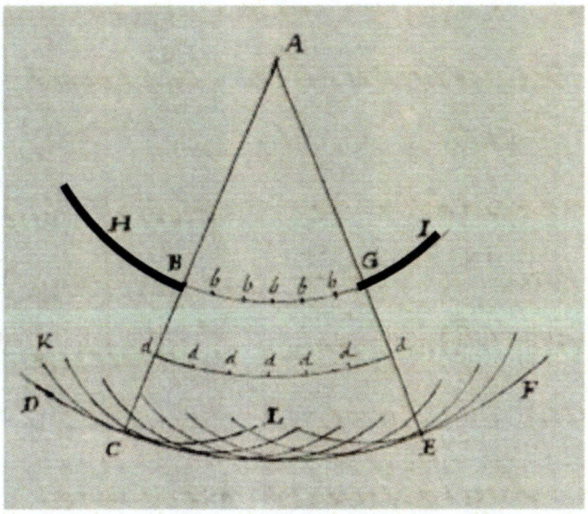

Fig. 11.13 Huygens' principle once again. Suppose BG is the aperture in an otherwise opaque screen $HBGI$ (indicated in black). Are waves diffracted by the edge points B and G observable at D and F? (from manuscript of *Traité de la Lumière*) [with permission from Leiden University Library, HUG 31, ff. 255r.-312v.]

Mariotte and de la Hire made their report of their experience of the sunlight passing through a hole in a dark place, and they found that the images of the Sun were proportionate to the distances without appearing any diffraction, and causing the light to pass through a prism placed close to the hole, the horizontal width of the light which is colored was proportionate to that which was done without a prism. But at a distance from the vertical diameter, it was three or four times larger than when there was no prism." These experiments were still inconclusive, but as a prominent member of the Academy Huygens must have been present. What was his view?

So Huygens essentially ignored diffraction. Still, he used the principle of the Camera Obscura for his invention of the Laterna Magica (1659) see the movie of the self-beheading skeleton (Fig. 11.14; possibly inspired by designs of the German printmaker Hans Holbein the Younger (1497–1543)). A well-known story tells that Huygens was somewhat embarrassed when his brother wanted to show this to the French royal court, as he thought this was too frivolous. Nevertheless, we are tempted to refer to these images as the first diffraction images: they constitute the remote detection and imaging of the bones of the skeleton by their diffractive response. See Fig. 12.9 of Chap. 12, where the same is done for the structural skeleton of the subsurface geology of the Earth.

Young and Fresnel

As we have already related in Chap. 7, the time following Newton and Huygens is characterized by slow progress in wave theory and the wave-particle debate for propagation of light. The same applies to diffraction theory in particular. Newton studied diffraction, but his corpuscular theory could not adequately explain it. Huygens' principle could, although it still had limitations, such as lack of interference. Also Huygens did not study diffraction as far as we know, and Newton's prestige helped

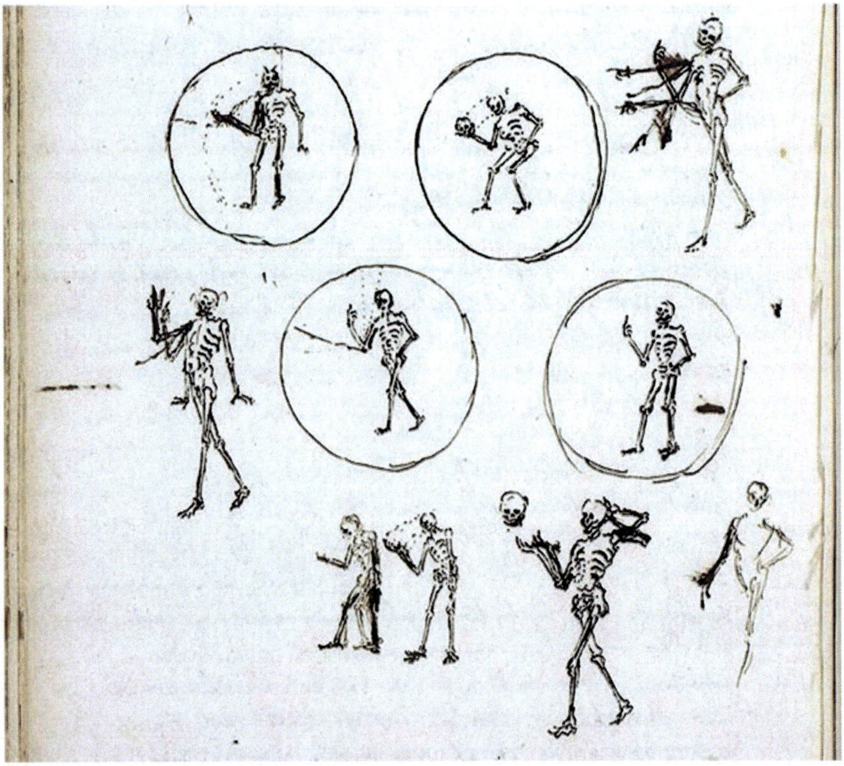

Fig. 11.14 Huygens Laterna Magica

to support the particle theory. These viewpoints may have impeded progress for most part of the eighteenth century.

The turn of the nineteenth century is characterized, as regards diffraction theory, by three developments:

1. the introduction of the concept of interference,
2. the dominance of the wave theory for propagation of light and
3. the study of the canonical problem of diffraction through an aperture.

A historical milestone in wave and diffraction theory was set by the work of Thomas Young (1773–1804). As already mentioned in Chap. 7 in his lectures to the Royal Society of London (1800–1804) he presented two major contributions to wave theory: *interference* and *diffraction*. His main innovation was the *boundary diffraction wave*: to consider the diffracted wave as a separate wave from the main wavefield and investigate its different properties in relation to the main wavefield (Fig. 11.15). This was based on the observation, already made by Grimaldi, that the edge of an aperture appears as a luminous line and source of diffracted light when looking from the shadow zone. The idea of the boundary diffracted wave was a precursor to modern diffraction theory. The prevailing treatment of diffraction up to the nineteenth century was to consider it as an integral part of the total wavefield, starting with Young's French contemporary Augustin Jean Fresnel (1788–1827, see Chap. 7).

Young's ideas were still mainly qualitative. Fresnel tried to put them in mathematical form but, after several attempts, failed. He then turned to Huygens' principle and set up his integral formalism to investigate diffraction by putting an obstacle in the free space. For reference, we copy the Huygens-Fresnel integral Eq. (7.2) from Chap. 7:

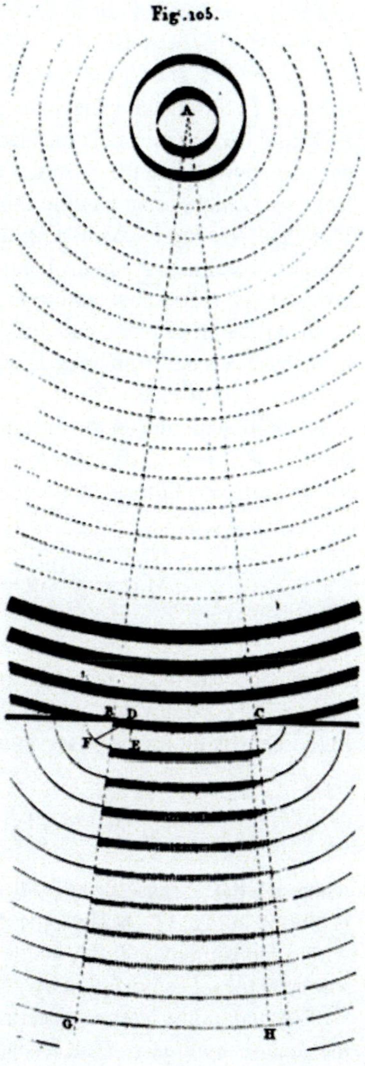

Fig. 11.15 Young's lectures, showing diffraction at the edge of the aperture

$$U(P) = \frac{e^{ikr_0}}{r_0} \int_S \frac{e^{ikr_1}}{r_1} K(\chi) \mathrm{d}S. \tag{11.1}$$

In his "Mémoire sur la diffraction de la lumière" presented at the French Academy of Sciences in 1819 Fresnel considered diffraction at an infinite half-screen (see Fig. 11.16). Here, a point source C generates a spherical wave incident on the half-edge *A–G*, where G is thought to be at infinity. The problem is to find an expression for the wavefield at any point P on the observation plane *D–B*. The wavefield is a superposition of the direct wave from C and the diffracted wave from A. In order to find the total wavefield at P, Fresnel applied Huygens' principle using the wavefront *A–M–I* right at the edge AG as primary wavefront and proceeded to carry out the integral (11.1) over the circular arc *A–M–I* to find the resultant of the secondary waves at P. Still lacking the mathematical apparatus which would be later developed by Kirchhoff, Fresnel met considerable difficulties finding manageable integrals. He began by postulating the existence of elementary waves at each point along the wavefront arc, which, upon passing through the diffractor, undergo mutual interference. The challenge lay in calculating the resulting vibration generated by the collective wavelets reaching any point behind the diffractor. The mathematical complexities proved to be substantial, and took an exceptional time to solve.

In particular, he found that the straightforward choice of a fixed coordinate system at the edge A leads to an integral which is impossible to evaluate analytically. Instead he chose a coordinate system at the pole M at the wavefront between P and C, and as integration parameter the distance z from M along the wavefront *A–M–I*. For each location P, there is a different pole M and a different coordinate system. In this way, Fresnel arrived at the integrals of the form

$$S(x) = \int_0^x \sin \frac{\pi}{2}\xi^2 \mathrm{d}\xi, \quad C(x) = \int_0^x \cos \frac{\pi}{2}\xi^2 \mathrm{d}\xi, \tag{11.2}$$

which since then are referred to as *Fresnel integrals* (see for more details on their history Buchwald, 1989). The Fresnel integrals can be used to express the light intensity along the observation plane *D–B* (Born and Wolf, 1999, Sect. 8.7):

$$I(x) = \frac{1}{2}(\frac{1}{2} + C(x))^2 + \frac{1}{2}(\frac{1}{2} + S(x))^2, \tag{11.3}$$

where x is the distance along *D–B* and $x = 0$ at the shadow boundary B. The intensity is plotted in Fig. 11.16e. Into the illuminated zone (left of B) the intensity oscillates with diminishing amplitude as the distance from the edge increases and approaches a constant level as predicted by geometrical optics.

The maximum intensity occurs at some distance into the illuminated zone. At the shadow boundary (B) it has a (normalized) value of 1/4. Into the shadow zone (right of B) the intensity decreases monotonically towards zero. All these effects were experimentally confirmed by Fresnel. We refer to Feynman's lectures (1989,

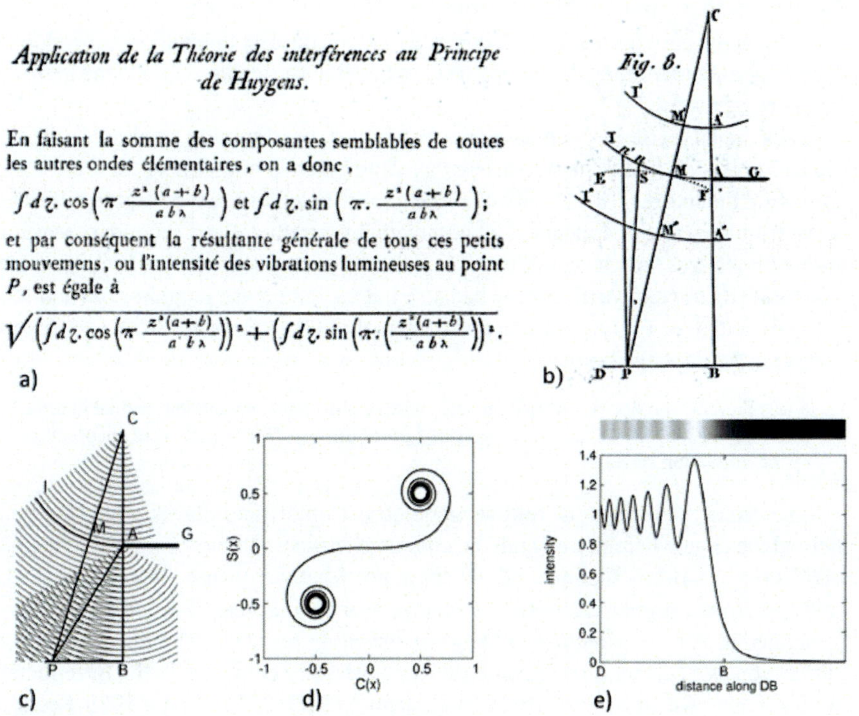

Fig. 11.16 Fresnel diffraction at a straight edge: "Application of the theory of interference to Huygens' principle". **a** Text portion from Fresnel's price winning Memoir for the French Academy of Sciences (1819), **b** Diagram of the set-up considered by Fresnel for diffraction on a straight edge. Here, C is a luminous point. AG is an opaque body, so large that no light comes around the edge G (quasi infinity). DB is the observation screen. A–M–I is the primary wavefront at the moment it reaches the edge AG. The problem is to find light intensity at a point P on DB, by superposition of secondary waves from A–M–I. This is solved by carrying out the Huygens-Fresnel integral (11.1) over A–M–I. The argument of the integral (11.1) depends on the distance nS as n moves along A–M–I. Fresnel derives for the distance nS the expression $z^2(a+b)/2ab$, where z is the distance from M along A–M–I, $a = |CA|$, $b = |AB|$, and arrives at the integrals mentioned in **a** now referred to as *Fresnel integrals*. **c** Fresnel considers the wavefield at P as a superposition of the initial wave from C and diffracted wave from A. **d** Cornu's spiral obtained by plotting the two Fresnel integrals $C(x)$ and $S(x)$ against each other. **e** Intensity along the observation screen DB as evaluated by the Fresnel integrals. Note the fringes in the illuminated zone (left of B) and the exponential damping into the shadow zone (right of B) [with permission from the Académie des Sciences]

Vol. 1, Sect. 30–6) for more details on Fresnel diffraction by a screen and Born and Wolf (1999, Sect. 8.8.4) for details on its phase properties.

From now on, Fresnel integrals would recur in almost every self-respecting publication on diffraction theory. In a way, we can say that Fresnel integrals are the hidden messengers of Huygens' principle in diffraction theory. When plotted against each other ($C(x)$ against $S(x)$ as a function of the parameter x) they form the well-known *Cornu spiral* (see Fig. 11.16d), named after Marie Alfred Cornu (1841–1902). The

Cornu spiral has a number of intriguing properties, such as that its curvature depends linearly on the arclength along the curve. Since this implies that the centripetal force increases smoothly along the curve, the Cornu spiral is often used in rail and highway engineering.

Even in spite of his considerable difficulties in obtaining analytical expressions, Fresnel obtained the diffracted wave immediately and almost for free. When executing the integral (11.1) only over the unobstructed wave *A–M–I*, the diffracted wave from the edge *A* appears as an automatic by-product. Each secondary wave on *A–M–I* interferes with its neighbor, except the secondary wave at *A*, which generates a separate diffracted wave, simply because it does not have a neighbor beyond *A* to interfere with (see Fig. 11.16c). In this context the statement from Baker and Copson (1939) is direct to the point:

> It was Fresnel who first discovered the real cause of diffraction, namely the mutual interference of the secondary wave emitted by those parts of the wavefront which were not obscured by the diffraction screen.

The direct wave from *C* follows geometrical optics and leaves a shadow zone behind the edge where there would be complete darkness (the region right to the line *AB* in Fig. 11.16bc). Complete darkness is not what we observe behind the screen and therefore geometrical optics is incorrect or incomplete. The diffracted wave generated at *A* is a *correction to the geometrical optics wave*. The interpretation of diffraction as a correction on geometrical optics is very common. In the Introduction to this chapter we already referred to statements by Richard Feynman (1989, Lectures on Physics, Vol. I Chap. 30) that diffraction and interference are closely related and in fact nearly identical—"when there are only a few sources then the result is usually called interference, when there are many the word diffraction is more often used".

Another note to make is that Fresnel's construction essentially depends on the wavelength of the wave generated at *C*. For wavelengths going to zero, it appears that the diffracted wave goes to zero and in the limit $\lambda \rightarrow 0$ we are back at geometrical optics. As noted in Chap. 7, Thomas Young appears to have been the first to use the word wavelength and Fresnel the first to use the Greek letter λ for it. Returning to Young's double-slit experiment (Chap. 7, Fig. 7.3), Fresnel was able to explain the observed fringes by a superposition of two diffracted waves, taking into account the difference in their phase shift upon arrival at the observation screen (Fig. 11.17). As we noted already in Chap. 7, Young and Fresnel maintained an active collegial correspondence in which they expressed the highest mutual respect. After receiving Fresnel's results, Young appreciated the formalism based on Huygens' principle without reservation.

A famous event in the history of research of light and diffraction was the report that Dominique François Arago (1786–1853) made of Fresnel's Memoir in (1819) for the commission of the French Academy of Sciences. One of its members, Siméon Denis Poisson (1781–1840, see Fig. 11.19), had concluded that Fresnel's wave theory of light and diffraction would imply a bright spot behind a circular opaque disk positioned in a parallel beam of light (see Fig. 11.18). Poisson considered this to be absurd and a refutation of wave theory. Arago conducted the experiment, the bright

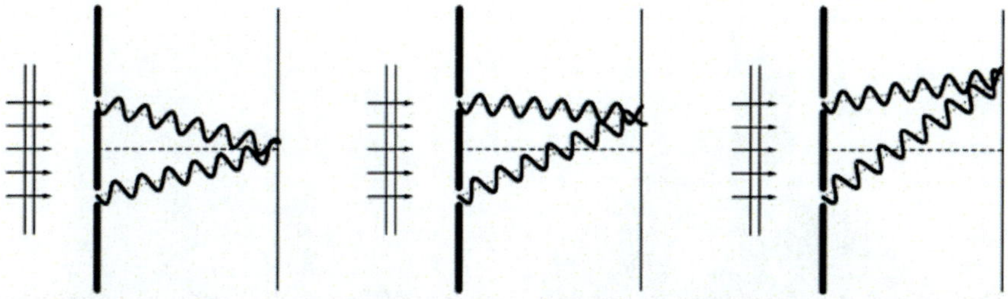

Fig. 11.17 Fresnel interference. Left: The two waves travel the same distance. Therefore, the waves arrive in phase. Constructive interference occurs at this point and a bright fringe is observed. Middle: The lower wave travels one-half of a wavelength farther than the upper wave. Therefore the waves arrive out phase. Destructive interference occurs at this point and a dark fringe is observed. Right: The lower wave travels one wavelength farther. Therefore, the waves arrive in phase. Constructive interference occurs at this point and a bright fringe is observed

spot appeared and Poisson became a supporter of the wave theory of light. Note that a detailed study of history teaches us that Poisson should have known better, because as early as 1723 the same phenomenon had been observed by the French-Italian astronomer and mathematician Giacomo Filippo Maraldi (1665–1729). In any case, Arago's experiment marked a decisive victory of Huygens' wave theory, for at least until the advent of quantum mechanics. The spot is since then called Poisson's spot.

The Fresnel formulation of diffraction has an interesting consequence, which is based on the local nature of the diffraction phenomenon. In 1837, Jacques Babinet (1794–1872, see Fig. 11.19) published *Optical Meteorology*, in which he expanded what is now known as *Babinet's principle* or theorem. The observation is that the diffraction pattern from an aperture in a screen is complementary to the diffraction pattern of a complementary screen, that is, a screen where the openings exactly correspond to the opaque parts of the first screen and the opaque parts to the openings on the first screen. The diffracted wave from the first screen added to the diffracted wave from the complementary screen leads to the wave which would result if there was no screen at all. In short: *complementary screens result in complementary diffraction*

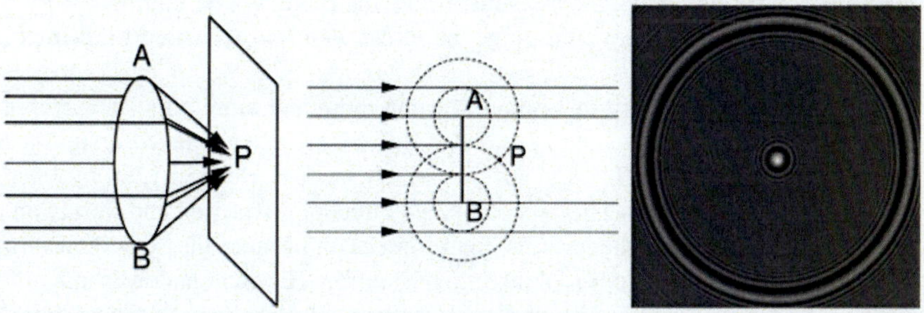

Fig. 11.18 Poisson's spot: a plane wave hitting the circular opaque disc (AB) leaves a shadow behind it in which the diffracted waves from the edges A and B (dashed lines) penetrate interfering constructively and causing a distinctly bright spot at the point P

Fig. 11.19 Siméon Denis Poisson and Jacques Babinet

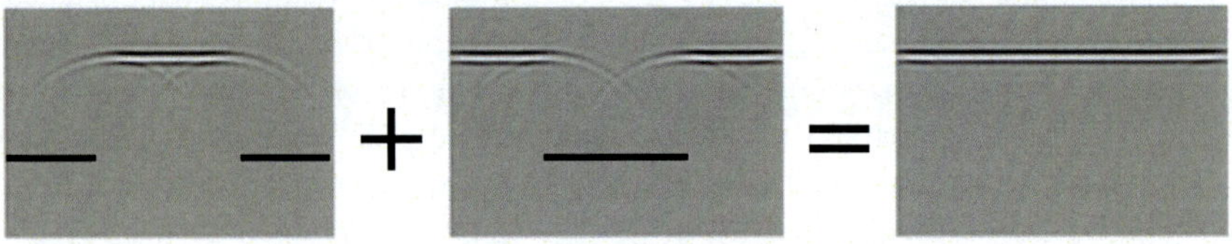

Fig. 11.20 Babinet's principle. Complementary screens lead to complementary diffraction patterns

patterns (see Fig. 11.20). This can be seen from the integral (11.1): if the screen S is subdivided into complementary screens S_1 and S_2, the total integral is the sum of the integrals over S_1 and S_2. Babinet's principle is therefore a direct consequence of the superposition as formulated by the Huygens-Fresnel principle.

Babinet appears to have been inspired by a paper by Necker (1832), which describes illumination effects if the Sun rises behind a mountain covered with trees seen in the Alps. If the observer is located in the shadow of the mountain looking to the ridge at a certain angle, the trees just below the ridge will be illuminated by the Sun even before rising above the ridge, an effect which is attributed to diffraction (Fig. 11.21).

It is probably fair to say that optics became more and more popular during the eighteenth and nineteenth century and that diffraction effects played a significant role. One example is the Brocken specter, which was described in 1780 by Johann Silberschlag (1721–1791). This is a combined effect of diffraction and refraction of visible light, that occurs whenever the light source (for instance the Sun) shines from behind on the observer who is looking into mist or fog. His own shadow is magnified and surrounded by a rainbow-type "glory", often with sudden movements because of the changes in the fog, giving the appearance of a specter or ghost. The phenomenon was first described occurring in the German Harz mountains (Fig. 11.22).

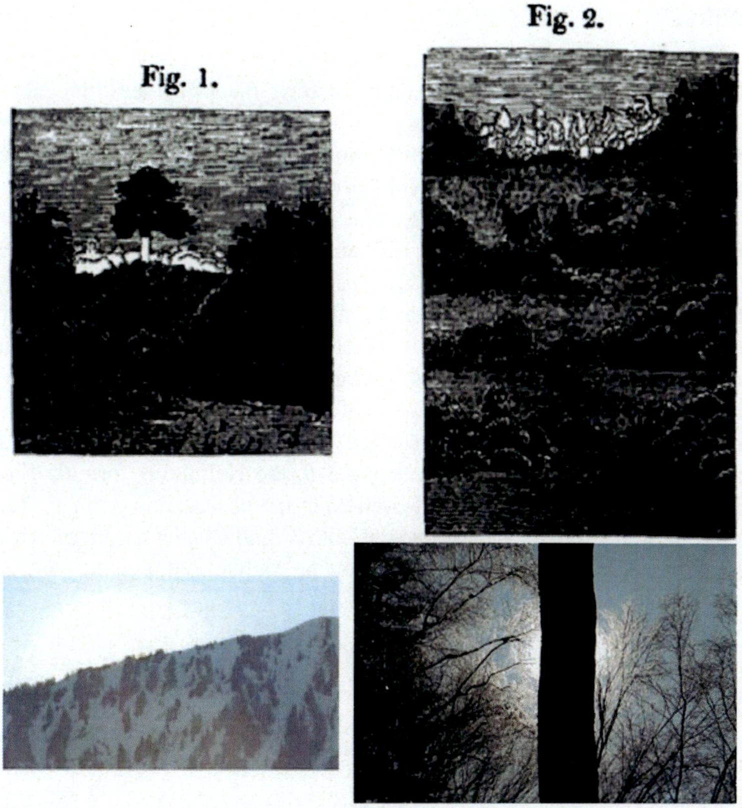

Fig. 11.21 Top: L.A. Necker, The London and Edinburgh Philosophical Magazine and Journal of Science 1832. ".. all the trees bordering the margin are entirely, branches, leaves, stems and all, of a pure and brilliant white, appearing extremely bright and luminous, although projected on a most brilliant and luminous sky..." [with permission from Taylor and Francis Group] Bottom: photo taken in Swiss Alps (2012) right before sunrise and around a tree, both illustrating Babinet's principle [with permission from www.parasalene.de]

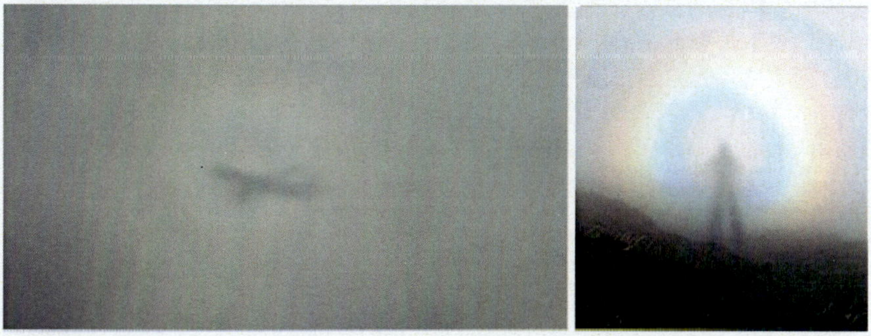

Fig. 11.22 Brocken specter (left: taken from an airplane; right: taken by the person depicted)

Kirchhoff

As stated in Chap. 7, the mathematization of wave theory achieved its culmination with Kirchhoff. For diffraction theory this is no different. The canonical problem for diffraction studied in the nineteenth century was to predict the wave propagation through an aperture in a screen. We consider the set-up of Fig. 11.23.

Here we have a point source at P_0 generating a spherical monochromatic wave, incident on an opaque screen S_1, S_2 with an aperture A. P is the observation point beyond the screen where the wave disturbance is to be determined. Q is a generic point in the aperture and vectors r and s are given by Q–P_0 and P–Q. Kirchhoff approached the problem by setting up the Huygens-Kirchhoff integral (or rather its monochromatic version the Helmholtz integral) over a specially constructed closed surface $S = S_1 + S_2 + A + S_R$ consisting of 1. the two parts of the screen S_1 and S_2, 2. the aperture A, and 3. a spherical surface S_R with radius R centered at the middle of the aperture. Over the composite surface S, the Helmholtz formula (Eq. 7.29, Chap. 7) is evaluated to obtain the wavefield at the observation point P. The great difficulty in doing so is that the values of both V and $\partial V/\partial n$ are unknown over S. Kirchhoff resolved this by making the following, rather bold, assumptions:

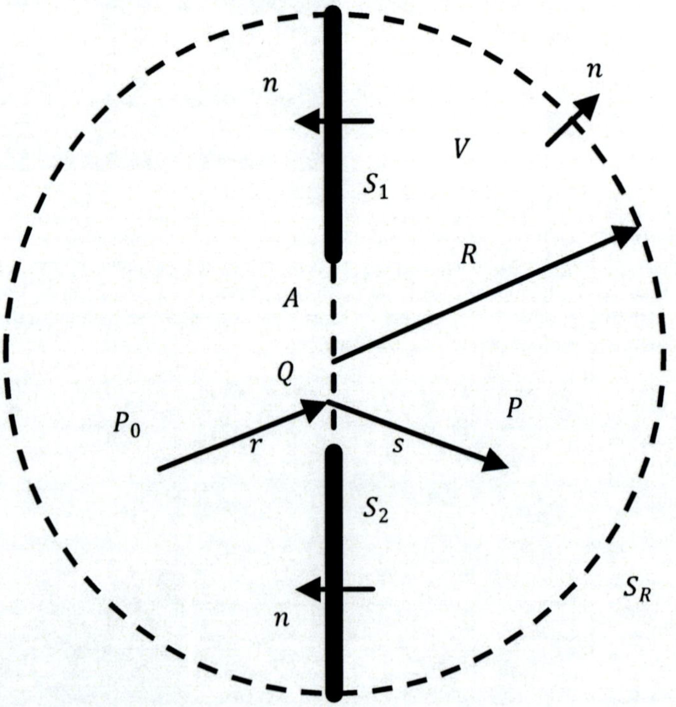

Fig. 11.23 Kirchhoff diffraction formula

$$
\begin{array}{lll}
V = V_{inc}, & \partial V/\partial n = \partial V_{inc}/\partial n, & \text{on } A, \\
V = 0, & \partial V/\partial n = 0, & \text{on } S_1 \text{ and } S_2, \\
R(\partial V/\partial R - ikV) \to 0 & R \to \infty & \text{on } S_R,
\end{array}
\tag{11.4}
$$

where V_{inc} is the incident wavefield on A and $\partial V_{inc}/\partial n$ its derivative normal to A, and R the radius of the sphere S_R. The first two conditions in (11.4) are called *Kirchhoff boundary conditions.* The first assumption in (11.4) implies that the wavefield in the aperture is equivalent to the incident wave and that there is no interaction from the screen S_1—this assumption is also referred to as St. Venant's hypothesis, named after Adhémar Jean Claude Barré de Saint–Venant (1797–1886). The second assumption in (11.4) implies that the wavefield in the shadow part of the screen is zero. In other words, the backside of the screen is assumed to be a totally absorbing, black body. As a result of the Kirchhoff boundary conditions, the integration over the screen and aperture $S_1 + S_2 + A$ is reduced to an integration of the input wavefield over the aperture A only.

The third condition in (11.4) is necessary to ensure that the contribution of the integration over the surface S_R can be safely ignored. This is referred to as *Sommerfeld's radiation condition.* If this condition is met the integration over S_R goes to zero as R grows to infinity. The radiation condition (German: 'Ausstrahlungsbedingung') is historically joined by the finiteness condition (German: 'Endlichkeitsbedingung'), which requires that RV remains bounded as $R \to \infty$; however it can be proven that the first implies the latter (Wilcox, 1956; about Sommerfeld later in this chapter). As a result of the Kirchhoff boundary conditions and the Sommerfeld radiation condition the integration of the input wavefield over the entire closed surface S is reduced to the aperture A only.

Finally, Kirchhoff applied a *high-frequency approximation* (which is very common in diffraction theory): if the source P_0 is located far away from the aperture in terms of the wavelength, we can assume $kr \gg 1$ and the following expressions apply:

$$
\begin{aligned}
V_{inc} &= A\frac{\exp ikr}{r}, \\
\frac{\partial V_{inc}}{\partial n} &= A\frac{\exp ikr}{r}\left[ik - \frac{1}{r}\right]\cos(\mathbf{n},\mathbf{r}) \approx A\frac{\exp ikr}{r}\cos(\mathbf{n},\mathbf{r}),
\end{aligned}
\tag{11.5}
$$

where A is the amplitude of the incident wave, $r = \|\mathbf{r}\|$ the distance from the source P_0, $\mathbf{n}$ the normal to the screen and $\cos(\mathbf{n}, \mathbf{r})$ the cosine of the angle between $\mathbf{n}$ and $\mathbf{r}$. Using this set of assumptions (11.5), Kirchhoff arrived at the following expression for the diffracted wave at P:

$$
V_P = -\frac{ikA}{4\pi}\int_A \frac{\exp ik(r+s)}{rs}[\cos(\mathbf{n},\mathbf{r}) - \cos(\mathbf{n},\mathbf{s})]\,\mathrm{d}S, \tag{11.6}
$$

which is known as the *Fresnel-Kirchhoff diffraction formula* (following notation of Born and Wolf (1999, Sect. 8.3.2).

Let us make a few notes to the Fresnel-Kirchhoff diffraction formula (11.6). First, it goes almost without saying that it bears a strong imprint of Huygens. Second, the diffraction formula (11.6) has a clear physical interpretation in terms of the familiar amplitude decay and phase rotation depending on propagation distance, or, as Larmor has remarked (quoted by Baker and Copson, 1939), (11.6) "puts in evidence the factor of attenuation $(rs)^{-1}$ and the phase depending on the path $r + s$."

Another physical interpretation follows from the observation that the diffraction formula (11.6) has the inclination factor $\cos(\mathbf{n}, \mathbf{r}) - \cos(\mathbf{n}, \mathbf{s})$. This factor is maximal for the forward direction $\mathbf{s} = -\mathbf{r}$, zero for the backward direction $\mathbf{s} = \mathbf{r}$, and between them varies proportionally to $K(\chi) = 1 + \cos(\chi)$ where χ is the angle between $\mathbf{s}$ and $-\mathbf{r}$. Comparing Kirchhoff's diffraction formula with the expression (11.1) originally used by Fresnel, we see that for the first time the inclination factor is based on physical principles, rather than heuristics; for instance the inclination for $\chi = \pi/2$ is not zero, as Fresnel incorrectly had assumed. Let us note that the inclination factor is a general property of the Kirchhoff integral formula, not necessarily limited to diffraction problems, as can be seen by expanding the aperture as necessary. Perhaps it is a historical peculiarity that it was primarily derived in the context of diffraction through an aperture. Another historical peculiarity is that the diffraction formula (11.6) was still derived for monochromatic waves, whereas Kirchhoff had generalized the Helmholtz solution (7.29) to polychromatic waves (Kirchhoff's integral formula (7.34)). It is straightforward to show, however, that the same Fourier integral representation can be used to provide a polychromatic version of the diffraction formula (11.6). A polychromatic representation implies that the waves can have a finite (transient) duration and therefore a wavelet (pulse) character as opposed to a single-frequency wave of infinite duration. Incidentally, if the radiation does not exist at all times, Sommerfeld's radiation condition is satisfied automatically, as contributions from the surface S_R with radius $R \to \infty$ will be too late to contribute to the wave measured at P (Born and Wolf, 1999, Sect. 8.3.2).

In relation to the above diffraction approximations and their validity range in terms of wavelength, aperture and distance from the diffracting object, the approximation of geometrical optics or ray theory is reaching much farther. As noted by Baker and Copson, (1939, Sect. I.2.2), a general theorem formulated by Kirchhoff states that geometrical optics is the limiting case of physical optics, for frequency going to infinity or (equivalently) wavelength going to zero. In the limit of geometrical optics/ray theory, the aperture is so wide and the wavelength so small, that diffracted waves can be neglected and the wavefield is fully described by the direct ray-theoretical wavefield. In this case the diffuse boundary of the shadow cast by the aperture becomes infinitely sharp and no wave energy penetrates into the shadow (perfect darkness).

Kirchhoff's diffraction theory was subject to many criticisms, some of which we will detail in the following section. However, it is a fair statement to say that Kirchhoff diffraction is entirely adequate for a wide range of problems, mainly due to the fact that the wavelength can be assumed to be smaller than the size of diffracting objects. In this sense, it provides the correct asymptotic behavior and as such is being studied continuously until today. Kirchhoff was of course fully aware of the limitations of his approximations. We end this section by returning to the quote from Kirchhoff's paper on the theory of light rays (Chap. 7, Fig. 7.11), which continues as follows: "To develop a fully satisfactory theory of these objects from the hypotheses of wave theory does not seem to be possible even today; yet these conclusions can be given more sharpness."

After Kirchhoff

The history of diffraction following Kirchhoff, from around the end of the late nineteenth century to the middle of the twentieth century, can be broadly characterized by two developments: *perfecting the theory* and the *rediscovering of the boundary diffraction wave.* In fact these developments are closely related since the boundary diffraction wave appeared as a natural by-product of efforts to perfect the theory.

Let us start with the imperfections of diffraction theory as it stood towards the end of the nineteenth century. Kirchhoff's diffraction theory had the following problems. From Kirchhoff's exact integral theorem for wave propagation (7.34), it became clear that the study of diffraction could be formulated as an ordinary boundary value problem of mathematical physics, depending on a proper choice of boundary conditions. A wavefield that satisfies Helmholtz's equation and for which Kirchhoff's boundary conditions, $V = 0$ and $\partial V/\partial n = 0$, hold on any surface must be zero everywhere in space. As a result, Kirchhoff's assumption of darkness at the backside of the screen implies, paradoxically, that the total wavefield is zero everywhere.

In fact, the assumptions formulated in the boundary conditions (11.4) are mutually incompatible, as the French theoretical physicist Jules Henri Poincaré (1854–1912) indicated. A consequence of the Helmholtz integral formula (7.29) with the surface boundary conditions (11.4) would be that the Fresnel-Kirchhoff diffraction integral (11.6) is zero for points outside the surface S, that is, to the left of the screen $S_1 + A + S_2$, which is generally not the case. In addition, the assumption that the total wave inside the aperture A would be equal to the incident wave (no interaction with the aperture edge) leads to a discontinuity in the total wavefield, which is physically impossible (because a solution of the wave equation must be smooth). Finally, experimental evidence showed that the assumption that the total field at the backside of the screen is zero is only approximate, that is, the screen is never completely dark.

Several authors suggested solutions for the shortcomings of Kirchhoff's diffraction theory. Kottler (1923, quoted in Baker and Copson, 1939) considered Kirchhoff's solution to be the rigorous solution to a slightly different problem (called "saltus problem"), where the blackness of the backside of the screen is defined in a different way than Kirchhoff did, the screen is assumed to be infinitesimally thin and the discontinuity across the boundary is prescribed, instead of the boundary values themselves. Although this gives analytically correct expressions for the diffracted wave, it is not possible to give a satisfactory physical definition of a thin black screen

(Baker and Copson, 1939, Sect. II.4; the topic was taken further by Marchand and Wolf, 1966).

Another way to fix the problem of incompatible boundary conditions was found by Lord Rayleigh (1897). He constructed a doublet of mirrored Green's functions for the Helmholtz equation (valid in the half space) that solves boundary conditions, where *either* the field *or* its derivative is zero on the boundary—and not *both*, as Kirchhoff assumed. These are called Rayleigh-Sommerfeld diffraction formulae of the first and second kind (the conditions that the field or its derivative is zero at the boundary are the well-known Dirichlet and Neumann boundary conditions, respectively). The Kirchhoff diffraction field is decomposed as

$$U^{(K)} = \frac{1}{2}(U^{(I)} + U^{(II)}), \tag{11.7}$$

where $U^{(K)}$ is the field predicted by Kirchhoff theory, and $U^{(I)}$ and $U^{(II)}$ the fields by Rayleigh-Sommerfeld integrals of the first and second kind. In the limit of small angles of incidence and diffraction and large distance from the aperture, $U^{(K)} \approx U^{(I)} \approx U^{(II)}$. The Rayleigh-Sommerfeld integrals indeed resolve the problem of the contradictory boundary conditions $U = \partial U/\partial n = 0$, but in fact still both assume that $U = U_{inc}$ inside the aperture; moreover the field behind the screen is still unknown. As pointed out in Born and Wolf (1999): no matter how we write the integrals, the boundary values as the field approaches the aperture are not recovered. They also state that "there is some evidence that the Kirchhoff theory gives results that are in closer agreement with observation" (than the Rayleigh-Sommerfeld theory). Due to the mirror construction the Rayleigh-Sommerfeld solution is limited to diffraction at aperture in a plane screen, unlike Kirchhoff's solution which applies to screens and apertures of any geometrical shape (Sommerfeld, 1949). As we shall see with Sommerfeld's exact solution (below), a mirror construction of Green's functions helps to achieve a fully rigorous solution, but implies a loss of generality.

Several authors hoped that Kirchhoff's solution could be interpreted as a first step in a sequence of successive approximations, which would converge to an accurate and consistent solution. Unfortunately, as noted by Bouwkamp (1954), investigations presented by Franz (1949) and Schelkunoff (1951) showed that this was wishful thinking.

In his lectures of 1887–88 at the Faculté des Sciences in Paris (1889), Poincaré arrived at a similar diffraction formula as Kirchhoff. Although he was not aware of Helmholtz's and Kirchhoff's memoirs, he was indeed aware of the internal contradictions, as noted above. In particular, he demonstrated that the Fresnel-Kirchhoff diffraction integral (11.6) is approximately valid if

1. the width of the aperture A is much larger than the wavelength (*high frequency approximation*) and
2. the observation point P is both not too close to the aperture A(*far field approximation*) and not too deep into the shadow (*boundary layer approximation*).

Similar conclusions were drawn by Sommerfeld (1896), about whom more below. Experiments with wide-angle diffraction performed by Stokes, Fizeau and Gouy showed that the Kirchhoff approximation breaks down when the observation point is too deep in the shadow (Darrigol, 2012). Under the assumptions 1 and 2 formulated by Poincaré, Kirchhoff diffraction theory provides a good agreement with experiments, justifying its use (for a recent overview see Buchwald and Yeang (2016)).

Sommerfeld and the Ultimate Diffraction Formula

A decisive step towards rigorous diffraction theory was finally made by the German physicist Arnold Sommerfeld (1868–1951), who derived an exact expression for diffraction on a perfectly conducting (and therefore perfectly reflecting) half-plane. All earlier theories (Young, Fresnel, Kirchhoff) assumed the diffracting screen to be perfectly black, that is, all energy incident on it would be totally absorbed. As noted for instance by Born and Wolf (1999, Sect. 11.1), the notion of total blackness is incompatible with the electromagnetic theory of light propagation. On the other hand, although the perfectly conducting screen is an idealization, it is compatible with electromagnetic theory. Moreover, since it permits an exact mathematical formula for diffraction, it has attracted particular fame. Most modern diffraction theories rely, implicitly or explicitly, on Sommerfeld's exact solution.

Figure 11.24 gives the set-up for Sommerfeld's construction for the derivation of the exact wave response from a perfectly conducting semi-infinite screen. The screen is assumed to be infinitesimally thin, perfectly reflecting and extending to infinity on the right. We consider a plane incident wave at angle α_0. For the geometrical optics approximation, we have three regions (see Fig. 11.24, top):

1. the region in front of the screen, illuminated by both the incident and reflected waves;
2. the region on the left of the screen edge, illuminated by the incident wave only; and
3. the region behind the screen which is a shadow zone where no energy penetrates.

At the shadow boundary between regions 2 and 3 the incident wavefield has a sharp discontinuity. At the boundary between regions 1 and 2, the reflected wave has a sharp discontinuity, and even although both 1 and 2 are illuminated (by the incident wave), we refer to this boundary as the reflected wave shadow boundary. Along both shadow boundaries, the wave discontinuities as predicted by geometrical optics are clearly unphysical.

Sommerfeld's solution for the total wavefield is achieved by a rather elaborate derivation involving complex integration along carefully chosen contours. Inspired by Rayleigh's double Green's function solution (11.7), it uses a mirror construction in such a way that the solution vanishes for $\theta = 0$ and $\theta = 2\pi$ (the screen), but not for $\theta = \pi$, and has a periodicity of 4π. We refer to Baker and Copson (1939)

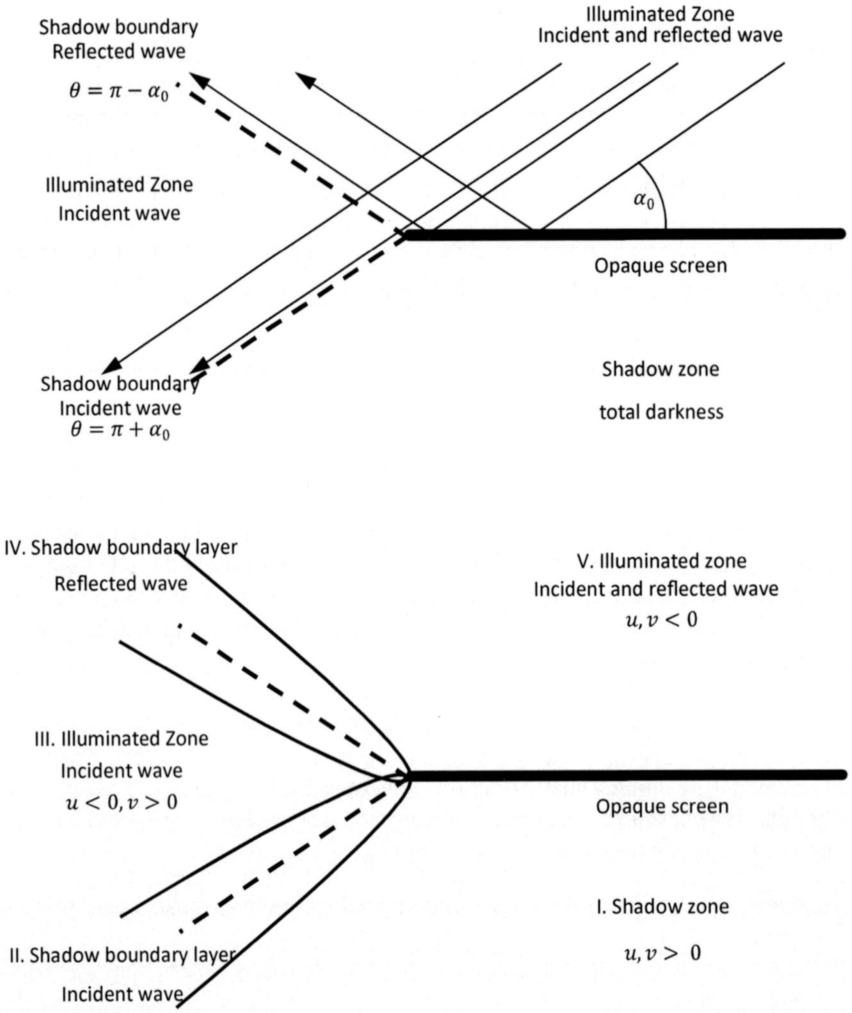

Fig. 11.24 Sommerfeld's diffraction on perfectly conducting half-plane. Top: geometrical optics; Bottom: five regions of Sommerfeld diffraction theory

and Born and Wolf (1999) for details of the derivation and content ourselves with presenting and interpreting the final, surprisingly compact, expression:

$$\begin{gathered} U = \frac{e^{-i\pi/4}}{\sqrt{\pi}}(G(u) - G(v)), \\ \text{where } u = -\sqrt{2kr}\ \cos\tfrac{1}{2}(\theta - \alpha_0) \text{ and } v = -\sqrt{2kr}\cos\tfrac{1}{2}(\theta + \alpha_0), \\ G(a) = e^{-ia^2}\int_a^{\infty} e^{-i\mu^2} d\mu. \end{gathered} \quad (11.8)$$

Here, α_0 is again the angle of the incident plane wave, an observation point in the plane is given by polar coordinates (θ, r), where θ is the angle with the screen and r is the distance from the edge of the screen. The wave number is given by $k = 2\pi/\lambda$. The solution for the total wave U is given by the difference between values of the function G for two arguments u and v, which vanish at the incident- and reflected wave shadow boundaries, respectively (see Fig. 11.24, bottom). The solution (11.8) is only part of the full solution for the electromagnetic case (U being the out-of-plane component of the electric field and the other components for the electromagnetic field following directly). The solution can be shown to satisfy the Helmholtz equation and necessary boundary conditions at the screen. It is smooth everywhere in space, especially at the shadow boundaries. The only discontinuity occurs at the screen, where it accounts for the surface charge density. As a result, the solution is physically compatible. As shown in Born and Wolf (1999) the generalization to oblique incidence on the edge of a half-plane in three dimensions is straightforward and similar (again rigorous) expressions apply.

Let us interpret Sommerfeld's solution (11.8). A first observation is that the Fresnel integrals (11.2) are back again: they enter into the solution (11.8) in disguised but prominent form through the function G, showing that they are indeed fundamental mathematical functions for diffraction problems (and messengers of Huygens' principle!).

An asymptotic analysis for $kr \to \infty$ (far field, high frequency) is most revealing. It leads to five different regions with different wave behavior (labeled I–V in Fig. 11.24, bottom). The behavior of the function $G(a)$ for $kr \to \infty$ depends on whether its argument a is positive or negative:

$$\begin{aligned} G(a) &= \tfrac{\mathrm{i}}{2a} + O(\tfrac{1}{a^3})(a \to +\infty) \\ G(a) &= \sqrt{\pi}\mathrm{e}^{\mathrm{i}\pi/4}\mathrm{e}^{-\mathrm{i}a^2} + \tfrac{\mathrm{i}}{2a} + O(\tfrac{1}{a^3})(a \to -\infty) \end{aligned}. \tag{11.9}$$

This behavior is called the *Stokes phenomenon* and has far-reaching consequences for diffraction theory: the asymptotic behavior of functions can differ in different regions of the complex plane (Stokes, 1864). As a result, in the far-field/high-frequency approximation, the total wave U (solution (11.8)) can be decomposed in a geometrical optics part $U^{(g)}$ and a diffraction part $U^{(d)}$:

$$U = U^{(g)} + U^{(d)}. \tag{11.10}$$

For three separate regions depending on the signs of the arguments u and v (I, III, V in Fig. 11.24, bottom) the geometrical optics part of the solution reads:

$$\begin{aligned} U^{(g)} &= \mathrm{e}^{-ikr\cos(\theta-\alpha_0)} - \mathrm{e}^{-ikr\cos(\theta+\alpha_0)}, && u, v < 0, \quad \text{region V} \\ U^{(g)} &= \mathrm{e}^{-ikr\cos(\theta-\alpha_0)}, && u < 0, v > 0, \text{ region III}, \\ U^{(g)} &= 0, && u, v > 0, \quad \text{region I} \end{aligned} \tag{11.11}$$

which is zero in the shadow zone (I), equal to the incident wave in zone (III) and equal to the sum of incident and reflected wave before the screen (region V). For all three regions (I, III and V) the diffraction part of the solution reads:

$$U^{(d)} = \sqrt{\frac{2}{\pi}} e^{i\pi/4} \frac{\sin\frac{1}{2}\alpha_0 \sin\frac{1}{2}\theta}{\cos\theta + \cos\alpha_0} \frac{e^{ikr}}{\sqrt{kr}} \tag{11.12}$$

At the shadow boundaries ($\theta = \pi \pm \alpha_0$, regions II and IV), the denominator $\cos\theta + \cos\alpha_0$ tends to zero, so that the approximation (11.12) becomes singular, whereas the exact total field U is regular and reads:

$$\begin{aligned} U &= \tfrac{1}{2} e^{ikr} + O(\tfrac{1}{\sqrt{kr}}), \qquad \theta = \pi + \alpha_0, \text{ inside region II} \\ U &= e^{ikr\cos(2\alpha_0)} - \tfrac{1}{2} e^{ikr} + O(\tfrac{1}{\sqrt{kr}}), \ \theta = \pi - \alpha_0, \text{ inside region IV}, \end{aligned} \tag{11.13}$$

which is a decomposition again in the direct, reflected and diffracted waves.

A general statement is that both contributions U^g and U^d to the total field U (11.10) are discontinuous, but that these discontinuities exactly counterbalance each other in U. The total field displays reciprocity again (source and observation can be interchanged). In all five regions, the diffracted waves decay as $1/\sqrt{kr}$, so that they can be interpreted as cylindrical waves originating at the edge. Near the shadow boundaries, the diffracted waves have amplitudes of the same order as the associated geometrical optics wave. In regions where the geometrical optics and diffracted waves are comparable, they give rise to interference fringes.

Further inspection of the diffracted wave (11.12) shows that it changes sign between the regions I, III and V. This points to the *polarity reversal*, which is often studied in seismic diffraction theory. Secondly, the absolute value of the amplitude as function of the angle θ is *asymmetric* across the shadow boundary, the shadow branch being stronger than the illuminated branch. The reason for the asymmetry is that both shadow boundaries (II and IV) interact with each other (both for the exact formula (11.8) and the high-frequency approximation (11.12)). Kirchhoff's diffraction theory, by contrast, is effectively a boundary layer approximation (as Poincaré had shown in 1889) and predicts a symmetric absolute amplitude across the shadow boundary.

The significance of Sommerfeld's rigorous solution is based on various aspects. First of all, it is the first *exact* solution in history for a diffraction problem, that, although an idealization, is physically feasible. As such, it served as a source of inspiration for exact solutions of other diffraction problems obtained by Sommerfeld's students and successors (such as Carslaw, MacDonald and Bromwich, see Bouwkamp (1954)). Also, it serves as a reference for more general, but approximate solutions. For this contribution, Sommerfeld can be considered to be the founder of the rigorous mathematical theory of diffraction.

The second significance of Sommerfeld's solution is that it shows that the wavefield can be expressed as the superposition of geometrical optics waves and diffracted waves. To appreciate this, let us recall that Fresnel and Kirchhoff, basing themselves on Huygens' principle, attempted to explain the total wavefield in one integral expression which combined the geometrical optics waves and diffraction. By contrast, Thomas Young argued that the formation of a diffracted wave is of local character and that it takes place in the vicinity of the boundary shadow behind the edge of obstruction. In fact, this view can be traced back to Francesco Grimaldi (referred to by Sommerfeld as the "*father of diffraction*"), who described the edge as a luminous line and source of diffraction. It was Gian Antonio Maggi (1856–1937, student of Kirchhoff), who in 1888 first showed that Kirchhoff's diffraction integral may be reduced to a sum of a geometrical optics term and a diffraction term (Sommerfeld seems to have been unaware of this). With Sommerfeld's rigorous solution, for the first time Grimaldi's and Young's view came back to the forefront. In fact, ever since Sommerfeld, diffracted waves have been treated as separate from the other components of the total wavefield. This rebirth of the *boundary diffracted wave* opened the way for a *boundary layer theory* of diffraction as well as the concept of the *diffracted ray*.

Some key quotes from Sommerfeld (Mathematische Theorie der Diffraktion, 1896) are:

- on the shadow zone: "For us, who must presuppose the wavelength as a finite, albeit small quantity, the term "shadow" becomes illusory and the word "shadow boundary" means for us only a line along which one approximation formula loses its validity and is replaced by another.";
- on diffracted rays: "In the language of geometric optics one would interpret the formula in the following way: rays propagate from the edge in the direction of the radius vector throughout the entire region, exactly as if the edge were a luminous point.";
- on diffraction into the shadow zone: "From the standpoint of geometric optics, the strange distribution of light in this shadow region is an exceptional state, whose appearance is called diffraction. We can, if we wish, accept this name and say: in the geometric shadow there is diffracted light".

As mentioned above, Sommerfeld's solution applies to the idealized, but nevertheless physically realizable, case of the perfect conductor/reflector. Many applications deal with imperfect reflectors that both reflect and transmit energy. These cases do not pose a fundamental limitation to Sommerfeld's solution, as was argued by Raman and Krishnan (1927) and later experimentally verified by Savornin (1939) (as related in Sunil Kumar and Ranganath (1991)).

Diffracted Rays—Grimaldi and Young Back to the Forefront

After Sommerfeld, the diffraction scientific community was clearly ready for the concept of the *diffracted ray*. The term had first been used by Alexey Kalashnikov in 1912, who also devised a photographic method to examine diffraction phenomena at very large diffraction angles, thereby obtaining results which are consistent with

Sommerfeld's theory. The natural question following Sommerfeld's exact solution was whether under more general conditions the wavefield could also be decomposed into geometrical optics and diffracted components. This was investigated first by Maggi (mentioned above), Hermann Rudolf Eugen Maey (1869–1946, student of Sommerfeld) and later further developed by Wojciech Rubinowicz (1889–1974, assistant of Sommerfeld—his first name is given as Adalbert in German literature). The general conclusion of these investigations was that the decomposition is indeed possible, but is rigorous only in very special cases.

Rubinowicz's contributions are twofold: the transformation of Kirchhoff's diffraction *surface integral* over the aperture in a screen into a *line integral* over the boundary of the aperture and a *stationary phase approximation* of the remaining line integral to isolate the single diffracted ray. In summary and in formula, in accordance with Sommerfeld, the total wavefield is expressed as $U = U^{(g)} + U^{(d)}$, where the geometrical optics wave $U^{(g)}$ is given by

$$\begin{array}{lll} U^{(g)} = \frac{\exp(ikr)}{r} & \text{illuminated zone} \\ U^{(g)} = 0 & \text{shadow zone} \end{array}, \tag{11.14}$$

and the diffracted wave $U^{(d)}$ is expressed by a line integral over the edge ∂A

$$U^{(d)} = \frac{1}{4\pi} \oint_{\partial A} \frac{\exp ik(\mathbf{r}+\mathbf{s})}{\mathbf{rs}} \frac{\cos(\mathbf{n},\mathbf{s})}{1+\cos(\mathbf{n},\mathbf{r})} \sin(\mathbf{n},\mathbf{l})dl, \tag{11.15}$$

referred to as the *Maggi-Rubinowicz representation* of Kirchhoff diffraction integral (all symbols have the same meaning as in (11.6) and Fig. 11.23, l denotes the arc length along the edge, $\mathbf{l}$ denotes the unit tangent vector along the edge). The line diffraction integral (11.15) describes the superposition of spherical waves arising (diffracting) at the boundary of the aperture and may be considered as the mathematical formulation of Young's diffraction theory. The factor $\cos(\mathbf{n},\mathbf{s})/(1+\cos(\mathbf{s},\mathbf{r}))$ represents the inclination factor, which is singular (infinite) at the shadow boundary, where r is antiparallel to s. This ensures that the resulting integral for the diffracted wave $U^{(d)}$ has a (finite) discontinuity at the shadow boundary, again to compensate for the discontinuity in the geometrical wave $U^{(g)}$ and to ensure smoothness of the total wave U (Rubinowicz, 1917; Born and Wolf, 1999).

In historical context, it should be noted that Maggi (1888) was the first to realize that Kirchhoff's surface integral over the aperture does not depend on the surface spanning the aperture, but only on its boundary, and therefore can be substituted by a line integral over the boundary. Unfortunately, this seems to have gone unnoticed and undergone the same fate as Maggi's observation that Kirchhoff's diffraction integral may be reduced to a sum of a geometrical optics term and a diffraction term.

The second contribution of Rubinowicz, this one unprecedented, was to apply a stationary phase argument to isolate points along the aperture edge which add up most to the composite diffraction signal. By the principle of stationary phase these are the points where the phase of the integrand in (11.15) is stationary, $(d/dl)k(r+s) = 0$, or for which

$$\cos(\mathbf{r},\mathbf{l}) = -\cos(\mathbf{s},\mathbf{l}). \tag{11.16}$$

The diffracted wave at stationary points along the aperture edge is (Rubinowicz, 1924; the double prime r'' denotes the second derivative with respect to arclength l):

$$U^{(d)} = \frac{1}{2\sqrt{2\pi k}} \frac{1}{\sqrt{r''+s''}} \frac{\exp \mathrm{i}[k(r+s)+\pi/4]}{rs} \frac{\cos(\mathbf{n},\mathbf{s})}{1+\cos(\mathbf{s},\mathbf{r})} \sin(\mathbf{r},\mathbf{l}) \tag{11.17}$$

We note that the Kirchhoff-Rubinowicz expression (11.17) has a singularity at the shadow boundary $\mathbf{r} = -\mathbf{s}$. Also, for the half plane geometry it is symmetric around the shadow boundary (as can be seen by taking $\mathbf{n} = (1,0,0)$, $\mathbf{r} = (1,0,0)$ and $\mathbf{s} = (\cos\theta,\ \sin\theta,\ 0)$, so that $\frac{\cos(\mathbf{n},\mathbf{s})}{1+\cos(\mathbf{s},\mathbf{r})} = \frac{\cos\theta}{1+\cos\theta}$ which is symmetric in θ). We also note that Miyamoto and Wolf (1962) show that the Rubinowicz construction is valid not just for plane and spherical incoming waves, but for any wave.

Rubinowicz was aware that his first stationary phase expression had a singularity at the shadow boundary, and proceeded to derive an alternative, complementary, expression which is valid within the boundary layer around the shadow boundary:

$$\begin{aligned} U^{(d)} &= \sigma \frac{\mathrm{e}^{-\mathrm{i}(\pi/4+kR)}}{\sqrt{2}R} \int_{v_1}^{v_2} \mathrm{e}^{\mathrm{i}\pi/2v^2} \mathrm{d}v, \\ \text{where } v_1 &= +\infty \text{ and } v_2 = \sqrt{\tfrac{2k}{\pi}(\rho + r - R)}, \end{aligned} \tag{11.18}$$

where R is the distance from observation point to source, r the distance from source to edge, ρ from edge to observation point, $\sigma = -1$ in the shadow zone, $\sigma = +1$ in the illuminated zone (Rubinowicz 1924, 1957 Eq. 4.17ab). In this expression (11.18), the Fresnel integral is back again to take care of the regularity of the diffracted wavefield, and therefore the smoothness of the total wavefield, as further demonstrated by Pauli (1937). Only the explicit factor σ accounts for the discontinuity in the diffracted wavefield.

Figure 11.25 summarizes the transition from Kirchhoff's surface integral via Rubinowicz's line integral towards the single diffracted ray.

The condition for the stationarity of diffraction rays (11.16) appears to be a fundamental one. We recognize a formulation of *Snell's reflection law*: the components of incoming and reflecting rays along the reflector are equal. In the case of diffraction on an edge, this means that the angle of the incoming ray on the edge is equal to the angle of the diffracting ray. However, it should be noted that although we know the angles, this does not imply that the diffracted rays are all in the same plane. In other words, the stationarity condition defines a "reflection cone" (in the formulation of Rubinowicz), which is now known as "*Keller's diffraction cone*" (see Fig. 11.26). The same geometrical property was also shown later to follow from a generalized form of *Fermat's theorem* (see next section on Joseph Keller). An experimental verification of the cone of diffracted rays had already been given by Maey (1893) for the case of oblique incidence to the edge.

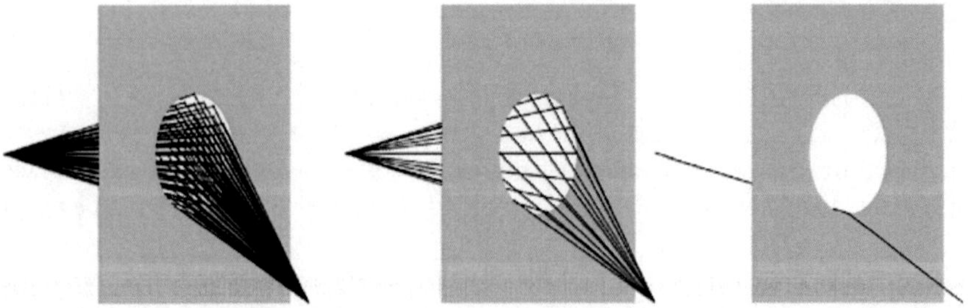

Fig. 11.25 Towards the diffracted ray. Left: Kirchhoff's surface integral over the whole aperture; middle: Young-Rubinowicz's line integration over aperture rim; right: Rubinowicz's stationary phase approximation for a single diffracted ray

Fig. 11.26 Edge diffracted rays, forming a "reflection cone" (Rubinowicz) or "Keller's diffraction cone". The angle of the incident ray with the edge is equal to the angle of the diffracted rays with the edge, but the diffracted rays are confined to a cone with the edge as its axis. [with permission from ©The Optical Society]

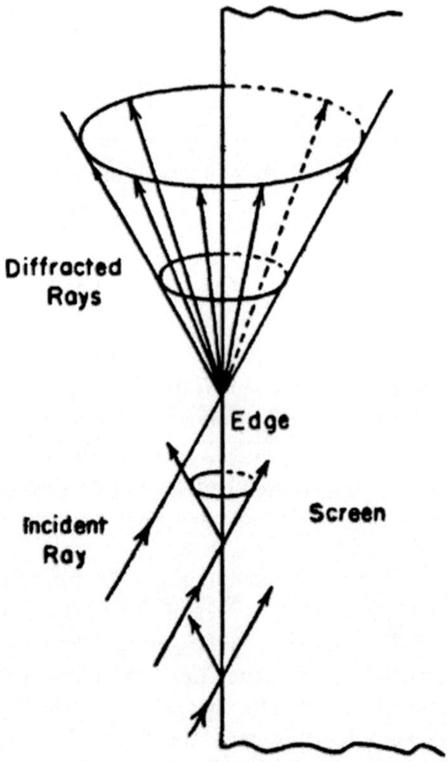

We have now arrived in the mid twentieth century. To summarize the achievements in diffraction theory up to this point: Kirchhoff's solution for diffraction through an aperture was criticized and its validity was determined by Poincaré (shadow boundary, far field, high frequency). Rubinowicz first showed how the surface integral of Kirchhoff over the aperture can be replaced by a line integral over the edge of the aperture and secondly how this integral can be replaced by the contributions from isolated points along the aperture edge. This provided the basis for Keller's Geometrical Theory of Diffraction. A parallel development was the rigorous diffraction theory of Sommerfeld, who derived the exact formula for the idealized case of diffraction on a perfectly reflecting infinitely thin screen. The Sommerfeld solution rigorously demonstrated that the total wavefield can be decomposed into the geometric optical wave and the diffracted wave, thereby putting a firm foundation under Young's concept of the boundary diffracted wave.

As a matter of fact, Rubinowicz can be credited for showing that Fresnel's view on diffraction—arising from a surface integral over the unobstructed part of the wavefront—and Young's view—as a separate boundary wave—are fully equivalent (Hecht, 2001). We could even argue that, after all, there was nothing to fight over, if not for the historical fact that Fresnel and Young showed, as we have seen, an unprecedented collegial respect to each other.

Huygens and Newton would have been proud.

In 1901, when Guglielmo Marconi (1874–1937) established wireless communication across the Atlantic Ocean, the ability of telegraphic waves to travel over great distances over the horizon of the curved surface of the Earth was met with great surprise. Two explanations were offered: surface diffraction and atmospheric reflection. In 1908, Poincaré offered a simple way of explaining the exponential decay of *surface diffracted waves* as function of propagation distance. In the situation shown in Fig. 11.27, an antenna OC would emit radiation along the tangent OF to the Earth. As the waves propagate along the surface, they emit rays tangent at each point (OF, HG, KL), each time losing a constant fraction $\alpha < 1$ of the remaining energy. Hence the intensity after n iterations is α^n, so that it decreases exponentially with distance. Unfortunately, several attempts to match the field data with the theory of surface diffraction eventually failed. In 1919, the English mathematician George Neville Watson (1886–1965) developed a mathematical theory of atmospheric reflection that is able to correctly predict the amplitude behavior of telegraphic waves (see the detailed account by Darrigol, 2012). Nevertheless, the attempts to explain the effect by diffraction have provided useful insight in the properties of surface diffraction. It goes without saying that in both cases Huygens' principle is instrumental. As always it is looming in the background.

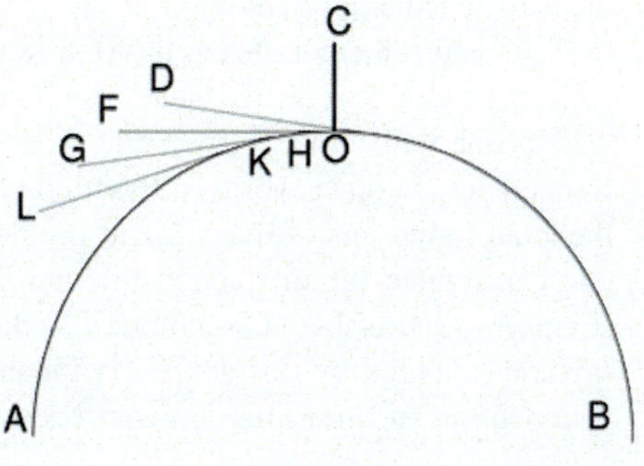

Fig. 11.27 Surface diffracted rays and Poincaré's explanation of Hertzian waves (redrawn after Darrigol, 2012)

Keller's Geometrical Theory of Diffraction

With the American mathematician Joseph Keller (1923–2016), student of Richard Courant (1888–1972, German-American mathematician, himself assistant of German mathematician David Hilbert (1862–1943)) at New York University and later professor at the same university and Stanford University, we are closing in on modern theory of diffraction. Keller is most of all celebrated because of the Geometrical Theory of Diffraction and his paper of 1962 is the most cited in modern diffraction literature. This combines ideas and developments of Sommerfeld and Rubinowicz, and outlines a systematical theory of the geometrical properties of diffraction Keller (1953, 1956, 1962, 1978), Keller, Lewis and Seckler (1957), Buchal and Keller (1960). Geometrical Theory of Diffraction is an extension of geometrical optics accounting for diffraction.

To appreciate the scope of this extension, let us recall the basic principles of geometrical optics or ray theory. Ray theory describes wave propagation by means of rays and wavefronts. Energy flows along the rays without exchange of energy across the rays (diffusion) taking place. The phase of the wavefield along a ray is proportional to its travel time, the amplitude is inverse proportional to the width of a narrow tube of neighboring rays. At smooth boundaries, the rays are reflected and transmitted with corresponding amplitudes. The geometry of the ray trajectories is given by Snell's law, or equivalently, by Fermat's principle, which states that the ray trajectory between two points has the shortest (or stationary) propagation time among all curves between the points. A big advantage of the ray method is that elementary waves can be treated independently and therefore easily interpreted at the observation plane. Also ray theory allows one to describe the wavefield locally at pre-selected locations. Most importantly, ray tracing is a computationally very fast technique for wave modeling.

However, ray theory breaks down in a number of important cases. Whenever the ray field is incident on a non-smooth boundary, for instance with an edge, the reflected and transmitted fields will split and a shadow zone is generated, where no energy penetrates. Whenever a ray field grazes a curved surface it will generate a shadow zone between the surface and the last ray tangent to it. Whenever the ray field folds over itself, the rays become tangent to a caustic, the ray tubes become infinitely narrow and the amplitude infinitely strong, while a shadow zone is generated beyond the caustic (caustic shadow).

The local character of rays allowed Keller to set up the same theory for diffracted rays, in analogy to reflected and transmitted rays. In this respect, Keller's construction was in line with Young's idea that diffraction is an edge effect, for waves at high frequencies. In other words, diffraction is a local phenomenon, determined only by the local properties of the incident ray field and the diffracting object. As a result, one can reduce the diffraction problem by complex objects to simpler canonical solutions, such as that of Sommerfeld in the half plane. The amplitudes of a diffracted ray are governed by diffraction coefficients, which play a role analogous to those of reflection and transmission coefficients. Diffraction by an edge is characterized by an edge-diffracted ray, whereas diffraction from a vertex is characterized by a vertex-diffracted ray.

One of the main advantages of generalization ray theory to diffraction, is again the computational speed. Tracing geometrical and diffracted rays is typically orders of magnitude faster than the evaluation of the corresponding Kirchhoff diffraction integrals.

A key ingredient of the Geometrical Theory of Diffraction is the generalization of Fermat's principle: the trajectory of a ray between two points which is diffracted underway by a diffracting obstacle has the shortest (or stationary) travel time among all curves between the two points via the diffracting obstacle. By this generalization, the ray trajectory of edge, vertex and surface diffracted rays can be constructed and hence the generalized Fermat's principle accounts for the kinematic definition of diffracted rays.

For the ray between two points which is diffracted on an edge, *Fermat's principle for edge diffraction* states that it has stationary travel time among all curves between the two points with one point on the edge. As explained by Keller (1962) this is particularly simple to prove when the edge is straight and the incident and diffracted ray lie in the same homogeneous medium. In fact, one can rotate the plane containing the start point and the edge around the edge until it contains the end point, to see that before and after rotation Snell's law of reflection is satisfied. Thus the incident and diffracted rays must make equal angles with the edge. By rotating the end point of the diffracted ray around the edge, a one-parameter family of edge-diffracted rays is generated which is confined to a cone with its axis along the edge; this is the cone already recognized by Rubinowicz (previous section). In case the incident ray is normal on the edge, the diffracted wave is *cylindrical,* in case of oblique incidence it is *conical.*

For the ray between two points which is diffracted at a vertex or tip (for instance the corner of an edge), *Fermat's principle for vertex diffraction* is particularly simple and states that the vertex-diffracted ray is the curve which has stationary travel time among all curves between the two points passing through the vertex. By rotating the end point of the diffracted ray around the vertex, a two-parameter family of of vertex-diffracted rays is generated which leave the vertex in all directions. The vertex diffracted wave is therefore *spherical.* Quote from Keller (1978): "*it's not much of a law, but at least it's democratic*".

In the third variant, Keller formulated *Fermat's principle of surface diffraction* for the ray between two points diffracted on a curved surface: it is the curve with stationary travel time among all curves between the two points having an arc on the boundary surface.

With the kinematical properties of diffracted ray defined by the generalized Fermat's principle, Keller (1973) arrived at the following classification of rays:

Geometrical optics	Geometrical theory of diffraction
Direct ray	Edge diffracted ray
Reflected ray	Vertex/tip diffracted ray

(continued)

(continued)

Geometrical optics	Geometrical theory of diffraction
Refracted ray	Surface diffracted ray
	Complex ray

All classes of rays obey Fermat's principle in one way or another. Light rays are always diffracted in such a way that the total path from the source via the diffracting boundary to the observation point is stationary.

The direct ray in a homogeneous medium is evidently a straight line. The rays reflected and transmitted on an interface between media obey Snell's law: the component along the interface of the incident and reflected/transmitted rays are equal and the incident, reflected and transmitted ray are all co-planar with the interface normal (see Fig. 11.28). Quote from Keller (1978): *"In 1637 Descartes also published Snell's law, but it is not clear how he obtained it—perhaps by reading Snell's paper."*

The complex ray occupies a special class in the system: it is a complex-valued curve which satisfies the ray equations and provides a general ray solution capable of describing edge-diffraction as well as caustic diffraction; however, a limitation is that the medium needs to be defined in analytic or piecewise analytic functions.

Based on the Geometrical Theory of Diffraction, the kinematic properties of diffraction are relatively simple and well understood. The dynamical properties are more involved. For the dynamical description of edge diffraction (diffraction coefficient), Keller (1962) resorted to a peculiar procedure, namely matching the ray theory amplitude with Sommerfeld's exact solution for diffraction of a planar scalar wave by a half-plane. We consider a semi-infinite screen with a straight edge and incident rays normal to the edge. The edge diffracted field u_e is then expressed as

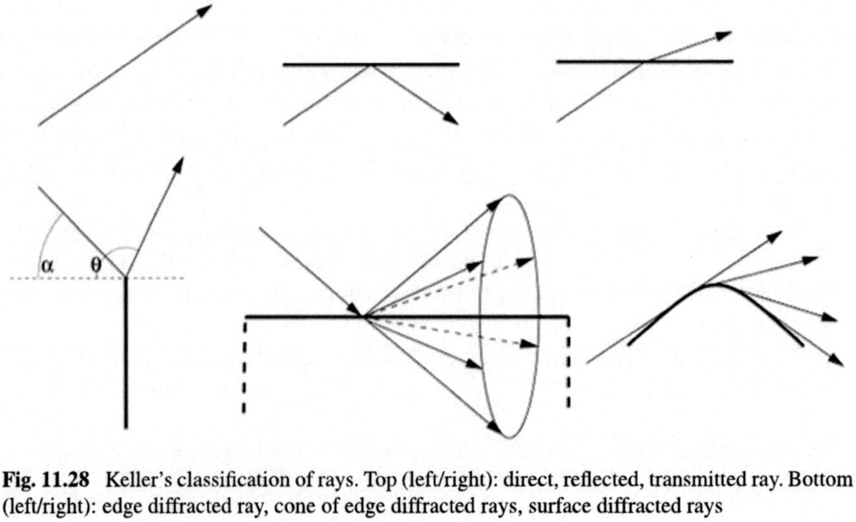

Fig. 11.28 Keller's classification of rays. Top (left/right): direct, reflected, transmitted ray. Bottom (left/right): edge diffracted ray, cone of edge diffracted rays, surface diffracted rays

$$u_e = DA_i r^{-1/2} \exp ikr, \tag{11.19}$$

where A_i is the incident amplitude, r the distance from the edge, $k = 2\pi/\lambda$ and D the diffraction coefficient. If the far-field/high-frequency assumption is made ($kr \to \infty$) then (11.19) agrees with Sommerfeld's solution *provided* that D has the following form

$$D = \frac{-e^{i\pi/4}}{2(2\pi k)^{1/2}}\left[\frac{1}{\cos(\theta-\alpha)/2} \pm \frac{1}{\sin(\theta+\alpha)/2}\right] \tag{11.20}$$

where α and θ are the angles between the incident and diffracted rays and the normal to the screen, respectively (see Fig. 11.29); the upper sign refers to the boundary condition on the screen $u = 0$, the lower sign to $\partial u/\partial n = 0$. For oblique incidence at angle β with the edge, the coefficient D is scaled by a factor $1/\sin\beta$.

Keller's edge diffraction coefficient (11.20) has opposite polarity across the shadow boundaries ($\theta = \alpha + \pi$ for the direct wave and $\theta = -\alpha$ for the reflected wave). Also, the amplitude is asymmetric around both shadow boundaries. The asymmetry is a characteristic property of Sommerfeld's solution, which is caused by the interaction between both diffracted waves (for direct and reflected wave). The polarity reversal

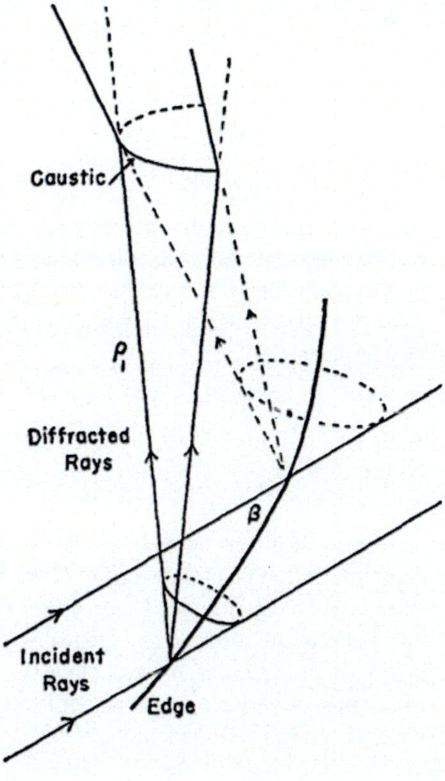

Fig. 11.29 Curved edge diffraction. The curvature of the diffracting edge causes the diffraction cone to move and rotate along, so that a caustic is generated (where edge diffracted rays touch each other). [with permission from ©The Optical Society]

around the shadow boundaries and amplitude asymmetry are features observed and studied in diffraction theory. Another feature of Keller's edge diffraction coefficient (11.20) (also inherited from Sommerfeld's solution) is the singularity at both shadow boundaries: $D \to \infty$ for both $\theta \to -\alpha, \alpha + \pi$. As we will see in the following sections, several authors after Keller attempted to address this singularity.

In the class of edge diffracted rays Keller considered both diffraction on a straight edge and on a curved edge. For the straight edge the expression is relatively simple (Eq. (11.20)). For diffraction on a curved edge the expressions become more involved, because depending on the curvature of the edge the diffraction cones may bend along and generate a caustic (Fig. 11.29). At the caustic the curved edge diffracted wave has an infinite amplitude. Curved edge diffraction has important applications in the seismic characterization of certain geological structures (such as saltdome-sediment interfaces).

Keller (1962) noted that the diffraction coefficients depend on the wave number k. In the high frequency limit all diffraction coefficients tend to zero, the diffracted wavefield disappears and the geometrical optics fields remains. However, the asymptotic behavior is different for edge-, tip- and surface diffraction. Dimensional considerations show that edge-diffraction coefficients are proportional to $k^{-1/2}$, tip/vertex-diffraction coefficients to k^{-1} and surface diffraction coefficients to e^{-k}. Surface diffracted waves are therefore weaker than tip diffracted waves which are in turn weaker than edge diffracted waves (note that the amplitude of surface diffracted waves also decreases exponentially with the distance along the surface, as first shown by Poincaré in 1908 and noted in the previous section). Amplitude considerations are important to classify diffraction phenomena with respect to their significance. In a different context, Sunil Kumar and Ranganath (1991) note that the shape of diffracting objects can have different impact on the diffracted amplitudes. Raman and Krishnan (1926) were probably the first to note that the diffraction amplitude behind a sphere and a disc are very different. In particular, the classical Poisson spot behind a circular disk (Fig. 11.18) has a much larger amplitude than a comparable bright spot behind a sphere. As noted by Keller, a ray representation for the surface diffraction phenomenon known as a creeping wave was introduced by Franz and Deppermann (1952).

Since Keller

As the present age approaches, the task of the historian becomes increasingly complex. Different applications of diffraction theory fan out in different branches of science and technology, and there is not always a complete cross-reference between publications. As a result the "genealogy of ideas" becomes more and more difficult to unravel. Optics, electromagnetics and seismology are major applications of diffraction. A very extensive and at that time up to date bibliography of diffraction papers is given by Bouwkamp (1954).

Let us recapitulate. For diffraction by an aperture in an opaque screen, Kirchhoff proposed formulas based on the assumption that the incident wave is not disturbed by the aperture and that the backside of the screen is totally dark. Poincaré was one of those who pointed to the intrinsic contradictions of these assumptions, but also showed that the Kirchhoff approach is valid in the combined high-frequency, far-field and boundary-layer limits, which makes it in fact a very useful approximation which is applicable in many situations. In a separate development, Sommerfeld derived an exact expression for diffraction on a perfectly reflecting half plane, which carried no assumptions at all, but was difficult to generalize beyond a very small number of special situations. A high-frequency asymptotic expansion of Sommerfeld's formula showed that the main wavefield can be decomposed into a reflected and diffracted component, and this gave rise to notion of the *diffracted wave*. Rubinowicz started from the Kirchhoff approximation, and first showed that the surface integral over the aperture can be reduced to a line integral over the aperture edge (without any further approximation by itself), and second that a stationary phase approximation over the aperture edge allows to isolate individual *diffracted rays*. This naturally led to Keller's Geometrical Theory of Diffraction which entails a fairly complete description of the geometrical properties of rays diffracted by edges, tips and curved surfaces, all based on generalizations of Fermat's principle. The kinematic part of the diffraction theory was therefore more or less completely solved: GTD allows to compute the trajectory and travel time of diffracted rays for a very general class of models.

The dynamic part of diffraction theory was still unclear and incomplete: Rubinowicz's formula (11.17) for the stationary phase diffracted ray had a singularity (infinite amplitude) at the shadow boundary, and so did Keller's solution (11.20) (taken from the high-frequency limit of Sommerfeld's exact solution)—both formulas result in valid amplitudes only away from the shadow boundary. A correct feature predicted by both formulas is the polarity reversal across the shadow boundary; Keller in addition correctly predicted an amplitude asymmetry, which Kirchhoff/Rubinowicz failed to predict. Rubinowicz later gave the diffraction formula (11.18) which is regular (finite) at the shadow boundary and valid within the boundary layer—the polarity reversal across the shadow boundary is explicitly given by a sign function. A common feature of diffraction formulas that are regular at the shadow boundary is that they contain the Fresnel integrals, which are necessary for smoothing out singularities.

In order to resolve the problem of the shadow boundary singularity of the edge diffracted wave, Keller and co-workers applied boundary-layer approximations. Even before Keller's groundbreaking paper on the Geometrical Theory of Diffraction appeared (Keller, 1962), Buchal and Keller (1960) already studied both edge and caustic diffraction and obtained separate asymptotic expansions for diffraction in various regions. They also resolved the *ad-hoc* nature of Keller's diffraction coefficient, which was obtained by comparison of the asymptotic method with the exact solution of a canonical problem. However, this still required separate expansions in separate situations and as a whole rather complicated computations. A first proposal to combine these separate expansions into a uniform theory was presented by Lewis and Boersma (1969), and generalized by Ahluwalia, Lewis and Boersma (1968). This led to a Uniform Asymptotic Theory of Edge Diffraction (UATD), which aimed at a uniform asymptotic (high-frequency) expansion of the diffracted wavefield equally valid in all cases. The historian casually notes that some of these theories were already generalized before they were even published. Also, diffraction literature became populated with theories with resounding abbreviations: GTD (Geometrical Theory of Diffraction), UATD (Uniform Asymptotic Theory of Diffraction), UGTD (Uniform Geometric Theory of Diffraction), RTD (Rigorous Theory of Diffraction), ATD (Analytic Theory of Diffraction), MTD (Mathematical Theory of Diffraction), PTD (Physical Theory of Diffraction) and more.

Building on Keller's GTD, Kouyoumjian and Pathak (1974) developed a uniform theory for edge diffraction for electromagnetic waves obliquely incident on a curved edge formed by perfectly conducting surfaces (UGTD). Returning to Keller's ad-hoc approach of the diffraction coefficient obtained by matching asymptotic theory with exact canonical solutions, they arrived at expressions for edge diffraction which are valid in regions near the shadow boundaries as well as outside the boundary layers. As a matter of fact, the uniform expressions for edge diffraction simplify to Keller's expression away from the shadow boundaries. The Fresnel integrals again play a pivotal role in these expressions, to ensure that the total wavefield is smooth and regular throughout—and that Huygens' principle is taken care of. Kouyoumjian and Pathak took care to arrive at compact expressions, which have the same form for different types of edge illumination, the only difference being the argument of the associated Fresnel integral. By addressing singularities in the geometrical theory of diffraction, such as the discontinuity across the shadow boundary, the extended solution of Kouyoumjian and Pathak is a significant step towards the unified theory of diffraction.

We conclude with a quote from Born and Wolf (Sect. 8.1): *Of all approximate methods for diffraction the theory of Huygens-Fresnel is by far the most powerful and is adequate for the treatment of the majority of problems encountered in instrumental optics.*

Chapter 12
Huygens and Geophysics

And perpendicular now, and now transverse. Pierce the dark soil and as they pierce and pass, Make bare the secrets of the Earth's deep heart.
— *Percy Bysshe Shelley, Prometheus Unbound, Act 4, Scene 1*

Remote Sensing

At a number of places in the preceding chapters we have already alluded to Huygens and his importance for remote sensing. Remote sensing refers to methods for obtaining information about a remote object without having direct access to it. Remote sensing can be done in an active and passive mode. In active remote sensing, the observer intentionally sends out a signal to the object, detects the backscattered signal and tries to derive information from it about the object. This information can consist of the object's geometric shape, mass, density distribution, luminosity or other physical parameters. In passive remote sensing, the observer sits and waits until a useful signal is emitted from the object, either directly from the object itself, or scattered from another distant source not related to the observer.

As we have indicated in the other chapters, Huygens was the first to distinguish between active and passive remote sensing and apply both of them in astronomy. In his *Traité de la Lumière* he outlined an active remote sensing method that would find the distance between the Earth and the Moon, which unfortunately did not work with the technology of his time. In the same chapter, he outlined a more successful way to measure the speed of light through passive remote sensing, using observations of the nearest moon of Jupiter. These two applications of remote sensing are both related to the physical properties of outer space and belong to the field of astronomy. Geophysics is another field where remote sensing bears fruit, or even where it is the only tool available to us to obtain information.

T. J. Moser and E. A. Robinson, *Walking with Christiaan Huygens*, History of Physics,
https://doi.org/10.1007/978-3-031-46158-3_12

Geophysics is the branch of physics which studies physical processes and properties of the interior of the Earth and its immediate environment. It uses physics and mathematics to obtain quantitative information of the subsurface geology. Geophysics as a separate branch of science did not emerge before the end of the nineteenth century, but many of its notions and problems were already known since antiquity. Since the dawn of mankind people have suffered from the destructive power of earthquakes and volcanic eruptions and as a result seismology and volcanology have developed as subsciences of geophysics. Topics of geophysical study include the force of Earth gravity, heat flow, seismic activity, electromagnetism, radioactivity, and fluid dynamics. Applications include mitigation of natural hazards and environmental protection as well as the exploration for natural resources, such as petroleum and minerals. Archaeogeophysics uses physical methods to reveal archaeological artifacts, such as pharaonic tombs, hidden in the shallow subsurface.

Tools at the disposal of the geophysicist include electrical, gravimetric, magnetic and seismic methods—the latter being most significant and most widely used. In seismic methods, Huygens' principle for wave propagation plays a pivotal role, and combined with the notions of active and passive remote sensing, everything comes together nicely.

Huygens is of course known for his many investigations and discoveries in astronomy: the rings of Saturn, Saturn's moon Titan, the Orion Nebula, the surface of Mars and its rotation and so forth. He also had a keen interest in the study of our Earth. In his book *Discours de la Cause de la Pesanteur* (Discourse on the cause of weight, published in 1690), Huygens studied the Earth's gravity. He concluded that the Earth is not a perfect sphere, but rather an ellipsoid which is flattened at the poles. As a reason for this fact, he argued that all matter on Earth is subject to a composite of two forces: gravity and the centrifugal force because of the Earth's rotation. Since gravity is a radially symmetric central force, but the centrifugal force is perpendicular to the Earth's rotation axis and its magnitude inversely proportional to the distance to the axis, the resultant does not point to the Earth's center, but is deviated. Huygens speculated that the interior of the Earth is fluid so the surface assumes a shape such that forces are in balance. This shape could not be a sphere but had to be elongated at the equator. Huygens estimated the flattening at a conservative 1/578. Evidence of the flattening of the Earth was provided in 1735 by measurements in Lapland, by Pierre Louis de Maupertuis, and in Peru, by Charles Marie de La Condamine, taken at the order of King Louis XV of France. In 1670 Robert Hooke argued that all planets must be ellipsoids with some degree of flattening. Newton had a great interest in gravity and obtained a value for the Earth flattening of 1/230. Modern insights have provided more precise numbers of the flattening (at 1/298) and refinements of the ellipsoidal shape due to tectonics.

Earthquakes

On 18 September 1692 Huygens had the opportunity to observe an earthquake which occurred in the Ardennes region in the Southern Netherlands with a magnitude 6.2 on the Richter scale, used in seismology to measure and classify earthquakes to their strength (also quoted in the book "Huygens and Hofwijck" by Van der Leer and Boers, 2022):

> Sitting at the Hofwijck near Voorburg at 2.30 pm, reading a book, I felt the Earth's motion suddenly and not without panic. The Earth was visibly shaken and quivering, so that the paintings, in the upper chamber, which covered the walls with gilded leather dashed to pieces. The pavement on which I was standing was raised and it sank again, and this was several times during the period of about 10 or 12 seconds. I suspected for some time that the arsenal of Dunkirk, which was full of gunpowder, was set ablaze, since it was expected that this city would be besieged by our army every day, and set on fire by the explosion of cannon balls. But scarcely at such a distance [170 km] could the pressure wave from such an explosion have traveled through the air, nor could its sound be heard. Two days later we learned that the Dunkirkers were still alive, and that the earthquake was indeed real. It terrified all in both Amsterdam and Antwerp. It looked like a wave from Amsterdam to the south had advanced, and thus everything had been shaken. It was said that towers were shaken and some of them sounded spontaneously. If the Earth rises up and sinks due to the passage of certain waves, it must be hollow below, or rely on the presence of water in such spaces, through which such waves would then travel. But from where their motions? More likely is the possibility that these waves travel in an underground gas layer, which will not burn should it penetrate the surface rock layers and leak into the air. Can anything be deduced about the depth of such a cavity filled with vapors?

In these last sentences, Huygens clearly identifies the questions of remote sensing, and passive remote sensing in this case: can we derive information from the subsurface from observations on the surface? Later, the geophysics of the twentieth century would prove him right with regard to the propagation of waves in the Earth (although mistaken in his speculation about caverns and vapors). The modern seismologist sits and waits for earthquakes; the more the better—setting aside here the issue of damages and casualties. The more earthquakes, the more data about the Earth's interior become available. For example, in the context of global seismology we have established that the Earth consists of a crust, a mantle and a core. The crust-mantle boundary is about 5–75 km deep and is called the Moho discontinuity after the Croatian seismologist Andrija Mohorovičić (1857–1936) who discovered it in 1909 from longitudinal (primary or P-) seismic waves. The core-mantle boundary consists of a sharp compositional contrast that generates a strong seismic reflection response. In 1926, the British seismologist Sir Harold Jeffreys (1891–1889) discovered, by analyzing both P- and S- (secondary or transverse) waves, that the core is liquid, at least at its outer boundary. In 1936, the Danish seismologist Inge Lehmann discovered by further analysis of both P- and S-waves that it has a solid inner core. At a local scale, seismicity induced by the production of oil and gas or the heat flow from geothermal reservoirs, typically at 0–4 km depth from the surface, are all monitored by passive remote sensing.

In active seismic remote sensing, or seismic exploration, a controlled seismic source such as dynamite is used to send seismic waves into the Earth's subsurface. Their backscattering from geological discontinuities is then recorded by a network of recorders. The seismic recordings are then processed and converted into a geological model. However, this is easier said than done.

Seismic Inversion

Converting geophysical data, and seismic data in particular, into a reliable model of the subsurface Earth is a formidable task which belongs to the field of inverse problem theory. The subsurface of the Earth is unknown. The only information we have are seismic wave recordings at the surface (and to some extent in boreholes). The single most important question in geophysics is how to invert surface seismic recordings into a reliable Earth model. This leads to inverse problem theory. A mathematical problem is called an inverse problem if one wants to infer the cause underlying the effect from an observed or desired effect of a system. Inverse problems arise in areas such as medical imaging, machine learning, non-destructive testing, radar detection, inverse scattering and earthquake detection.

Inverse problems are notoriously difficult or sometimes impossible to solve. The opposite of an inverse problem is a direct problem (sometimes also called a forward problem), where one wants to derive the effect of the system starting from the known cause. Often the inverse problem is solved by trial and error, by repeating the solution of a direct problem many times while changing the underlying model, until the data generated in this way optimally fit the observed data. The main challenge of the geophysical inverse problem is that the data can be incomplete (they cover only part of the Earth's surface), inaccurate (they are contaminated by noise) and even worse, are often inconsistent (there is no model which actually can explain them). The geophysical inverse problem is often ill-posed, in the sense given by the French mathematician Jacques Hadamard (1865–1963, already introduced in Chap. 7, see Fig. 7.13). Hadamard stated that a problem is "well-posed" if.

1. a solution exists,
2. it is unique and
3. it depends continuously on the input data.

For geophysical inverse problems, these conditions are more often violated than satisfied, making them ill-posed. Existence of an Earth model from a given class of models that fits the data is not guaranteed, on the other hand there may be (infinitely) many models that fit them to a certain degree, and model variations can have discontinuous effects on the data fit. For numerical treatment often requires regularization of the problem in the form of model smoothing, but this can involve a subjective choice by the geoscientist, who must be aware of the pitfalls of over- or underparametrization of the model.

Solutions to the inverse problem often decompose the Earth model into a smooth background model representing the large-scale global physical variations and a perturbation on top of it, which represents the small-scale local variations and relevant geological details. *Seismic tomography* is often used to obtain a reliable smooth background model. This can be constrained by additional data, for instance from shallow boreholes or from non-seismic methods (gravity, electromagnetic properties and others). The reconstruction of the local perturbation field on top of the smooth global background model is the topic of *seismic imaging,* also called *seismic migration.*

Seismic Imaging and Huygens' Principle

Huygens' principle is ubiquitous in seismic imaging, as a simple literature search can testify—as an example, we can mention the conceptual paper by Hagedoorn and Diephuis (2001). As we will show, there is hardly a better way than seismic imaging to illustrate Huygens' principle and Kirchhoff's completion of it.

Let us recall that Huygens devised his wave principle to explain the propagation of light. He was inspired by the propagation of water waves (which could be seen) and the propagation of sound waves (which could be heard). For light the two main fundamental problems were: what is the nature of light (particles or waves?), and if we assume the latter, in which medium do light waves propagate (what oscillates?). Huygens' assumption of a luminiferous ether held sway until the work of James Clerk Maxwell (1831–1879), but was finally discredited as unnecessary by Einstein's special theory of relativity.

Geophysics is primarily concerned with classical concepts. Apart from some well logging tools and some concepts in seismic wave propagation, quantum mechanics does not play a major role in geophysics. Similarly, relativity does not play an important role, apart again from some sophisticated concepts. Yet, even when based on classical physics the geophysical inverse problem is formidable enough.

In seismology we clearly do have a medium, so there is no problem with an ether. Also, the wave character of seismic events is well established. So we do not need to be bothered by the two questions ("waves or particles?" and "is there an ether?") which dominated the theory of light propagation. There are no photons having their own eternal and universal reference frame, just local material wavelets traveling at speeds much less than that of light. By contrast, seismic waves can be much more complicated than light waves, because of the character of the medium.

The medium is considerably more complicated than the media studied in other branches of (wave) physics. Recall from Chap. 7 (Eq. 7.1) that the three-dimensional wave equation reads.

$$\nabla^2 u - \frac{1}{c^2}\frac{\partial^2 u}{\partial t^2} = 0,\ \nabla^2 = \frac{\partial^2}{\partial x^2} + \frac{\partial^2}{\partial y^2} + \frac{\partial^2}{\partial z^2}\,, \tag{12.1}$$

where u is the wave function, c the medium velocity, x, y, z are spatial coordinates and t the propagation time. In geophysics, the difficulty lies in the medium velocity c, which can be a complicated function of space, so that very quickly simple solutions become unavailable and one has to resort to advanced solution techniques and sophisticated computer hardware.

The medium is *inhomogeneous*, varying with depth as well as laterally. The straight-line assumption which is often sufficient to describe light in a constant medium no longer applies. Rays are curved and wavefronts are no longer spherical but are deformed into more complicated shapes. Depending on the degree of inhomogeneity of the medium, the ray field may fold over itself, so that *caustics* appear.

The medium has *discontinuities* in the form of geological interfaces. These act as reflectors, more precisely as half-mirrors, where the incoming waves are both reflected back and transmitted into the layer below. The same process happens at other reflectors in the same medium, creating multiple reflected and transmitted waves. The complete wavefield takes a tree structure where each branch corresponds to a specific ray history or sequence of visited reflectors. The interfaces themselves, being a result of geological processes, can also be of considerable complexity. They can have edges and tips, which distort the smooth reflection response and generate additional diffracted waves. They can have local rugosity. Faults and fractures add to their complexity. Despite their intricate nature, obtaining accurate information about them is crucial in the exploration of Earth resources (such as hydrocarbons, minerals and geothermal energy).

If the medium has a dense set of reflectors separating multiple fine microlayers, it is said to be *anisotropic* when observed with seismic wavelengths much longer than the layer thickness (Helbig, 1994). The wave propagation in an medium of fine layering is more accurately described by anisotropy than by isotropic reflection/transmission, and the waves emitted from a point source become elliptical (non-spherical) rather than spherical. Geology is almost always anisotropic.

All these effects pose an important challenge to the modeling of wave propagation in a known model of the subsurface Earth. And all these effects have already been predicted by Huygens and can be described by his—Huygens'—principle.

Examples

Let us illustrate with some examples. Consider Fig. 12.1. Here we have a solid reflector at some depth in the Earth's subsurface. The depth axis is denoted by z and points downwards. The exploding reflector model is a simple model to simulate a single reflection on the reflector measured at the surface by sources and receivers which are located at the same position (at zero offset). The numerical simulation is done by assuming a dense set of elementary source points at the reflector, all exploding at once. For simplicity we assume that the medium is known and simple: spatially constant, isotropic, and acoustic (so that shear waves do not play a role).

The reflector is smooth but curved and exhibiting a geological syncline (a downward bump). After some propagation time the elementary waves originating from the elementary source points form a wavefront which is constructed by taking their envelope. At a later time (Fig. 12.2), the wavefront is propagated further towards the surface. Due to the syncline it has developed a caustic, where the wavefield folds over itself. Upon arrival at the surface, arrival times are observed and recorded (Fig. 12.3, where the time axis is denoted t and points upwards). The caustic has manifested itself in a triplication of the arrival time curve: in the central part there are three wave arrivals. On the flatter sides there is one arrival. Triplication, or even multiplication, of wave arrivals is a very common feature in seismology, which points to subsurface model variations.

Huygens' younger contemporary Gottfried Wilhelm Leibniz said: "When a truth is necessary, the reason for it can be found by analysis, that is, by resolving it into simpler ideas and truths until the primary ones are reached. Let us use this idea for *inversion* of the recorded wave arrival times". See Fig. 12.4. Suppose we have the recordings but no information of where they come from. We will backpropagate the waves into the subsurface. Again for simplicity we assume that the medium is simple and constant. For each individual arrival we can draw a circle with a radius proportional to the arrival time of the wave. All circles obtained in this way must be tangent to the original reflector. Therefore the envelope of the circles must be the reconstruction of its shape and depth. We see in Fig. 12.4 (bottom figure, fat

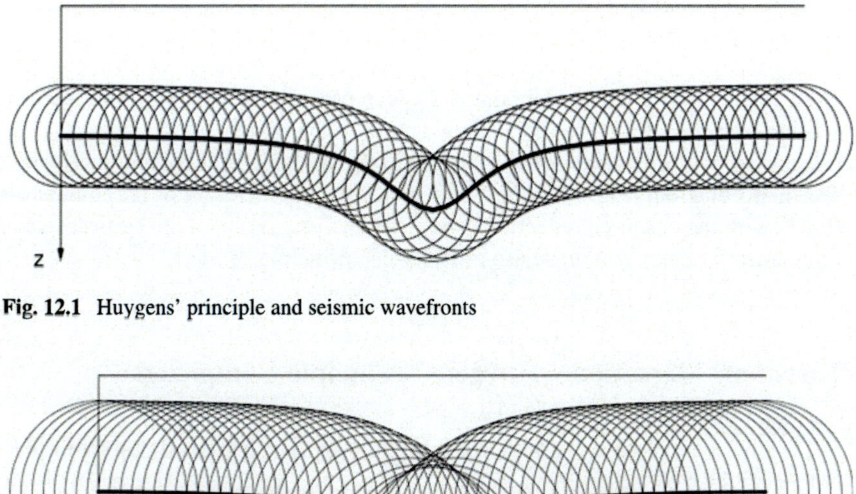

Fig. 12.1 Huygens' principle and seismic wavefronts

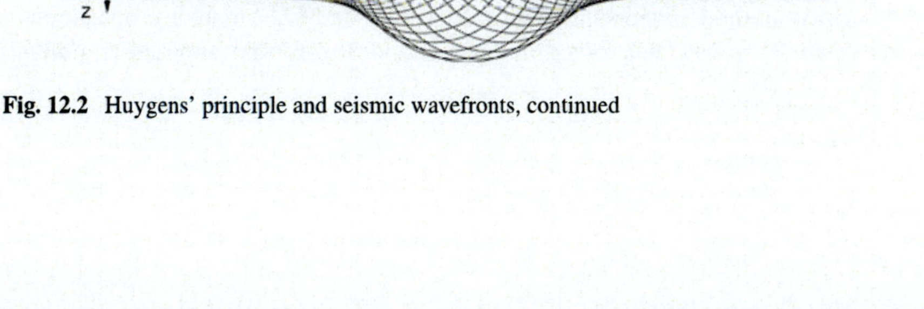

Fig. 12.2 Huygens' principle and seismic wavefronts, continued

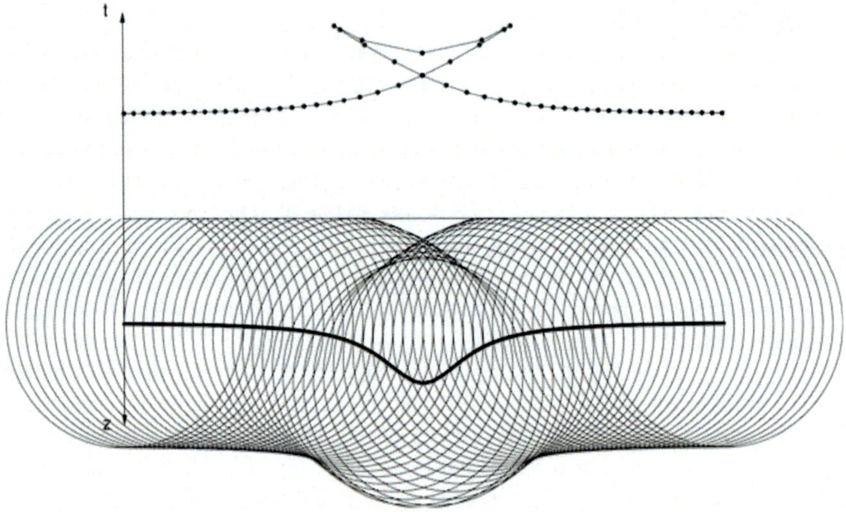

Fig. 12.3 Huygens' principle and seismic wavefronts, continued

line) that the syncline (downward bump) is nicely reconstructed from the triplicated arrival times. *Seismic migration* is the name given to this process, which migrates the seismic wave arrivals in the time domain to the correct locations in the depth domain. To paraphrase our earlier statements in this book: the fundamental role of Huygens' principle in imaging gives Huygens a strong claim as the father of modern geophysics.

In 1921 at Belle Isle, Oklahoma, J. C. Karcher was the first person to record a seismic reflection line. For the Viola event he threw out each reflection time on a circular arc with its center on the surface at the midpoint of shot and receiver. As seen in his diagram (Fig. 12.5) the Viola interface is the envelope of the circular arcs that fit with the observed reflection time. This analysis is one of the earliest seismic migration examples that made use of Huygens' principle.

Kirchhoff Migration—Huygens' Principle Completed

Let us now connect a wavelet to the wavefronts—apparently a small step at this point, but a giant leap for science and mankind.

As noted in previous chapters, Huygens' principle in its original form was still incomplete. This had to do with its qualitative, rather than quantitative character. It did not make any difference between forward and backward propagation. Therefore in the forward modeling displays (Figs. 12.1, 12.2 and 12.3) there is propagation up- and down. This was later solved by Fresnel and Stokes who imposed an obliquity

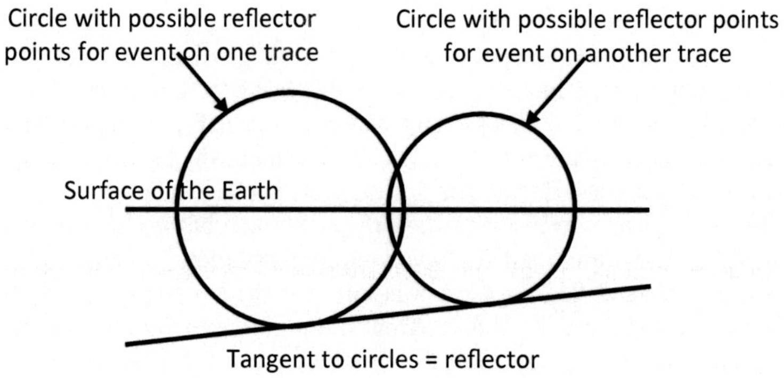

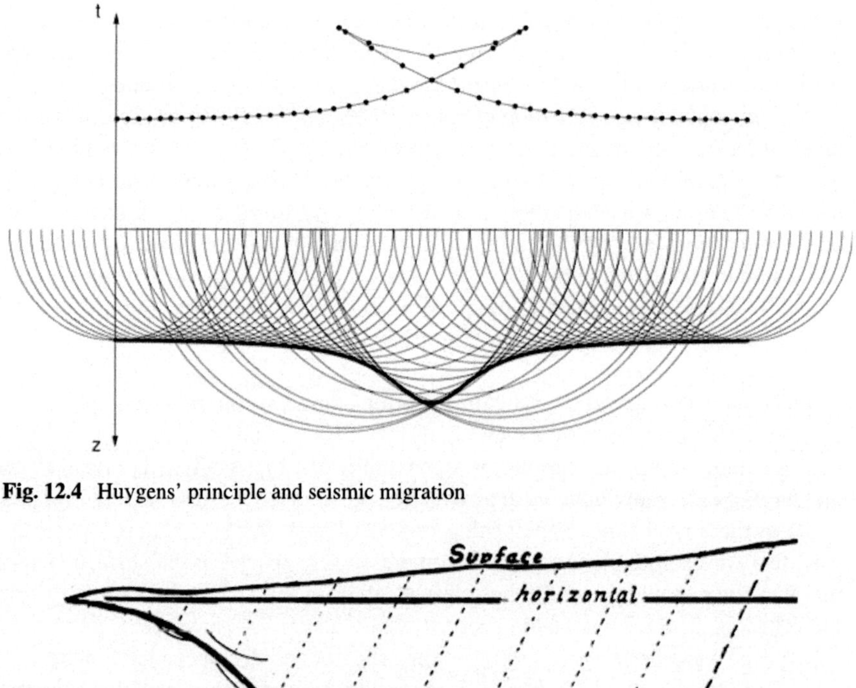

Fig. 12.4 Huygens' principle and seismic migration

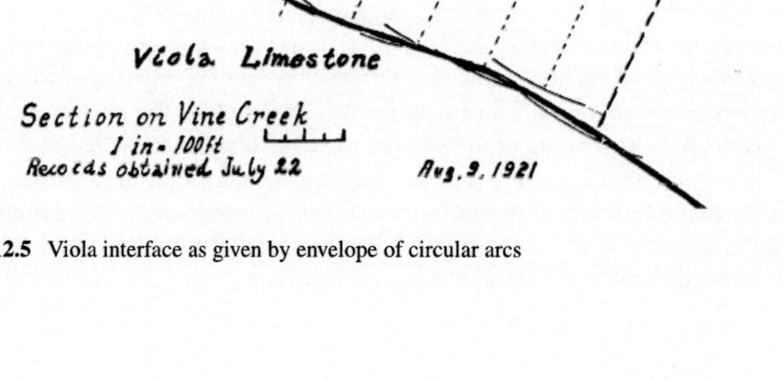

Fig. 12.5 Viola interface as given by envelope of circular arcs

factor on the waves which tapered out the undesired propagation. The obliquity factor was finally formalized and put on a physical basis by Kirchhoff—see Chap. 7.

More importantly, interference was still missing in the original formulation of Huygens' principle. The backpropagated wavefronts in Fig. 12.4 have their envelope at the true reflector. However, there is no explanation why the waves would interfere constructively at the reflector, and destructively above and below.

To recapitulate: we must know how the amplitudes of the elementary waves vary with direction. It turns out that this variation of amplitude with direction of propagation is complicated. This problem was solved in the nineteenth century by Fresnel, Green, Helmholtz and Kirchhoff. Their more advanced formulation of Huygens' principle states that we can obtain the wavefield at any point by first considering each point on any closed surface (it may be taken as a wavefront for convenience) as a source of secondary waves and then superposing the effects of these secondary waves at the point in question. However, these secondary waves have different amplitudes in different directions. If we take dependence of amplitude and phase into account, then the wavefront traveling backward turns out to have zero amplitude.

The principle of *superposition* of waves states that when two or more waves are incident on the same point, the total displacement at that point is equal to the vector sum of the displacements of the individual waves. This is a direct consequence of the *linearity* of the wave equation. If a crest of a wave meets a crest of another wave of the same frequency at the same point, then the magnitude of the displacement is the sum of the individual magnitudes; this is known as *constructive interference*. If a crest of one wave meets a trough of another wave, then the magnitude of the displacements is equal to the difference in the individual magnitudes; this is known as *destructive interference*.

Our simple example of the syncline reflector shows how this works. See Fig. 12.6. Here we attach a wavelet to the wave arrivals of Fig. 12.3 and start a process of wave superposition. Again for simplicity, we assume that the medium is constant and known. For each individual wave arrival we construct a spherical wave with radius equal to the arrival time. Superposing, in each panel, more and more elementary spherical waves with centers more and more closely spaced at the surface, we see that they start to add up along the original reflector and start to cancel away from the reflector. In the limit of the process, when all arrivals have been treated, the superposed wave perfectly interferes at the reflector and disappears elsewhere.

This process is named *Kirchhoff migration*, in reference to his complete mathematical formulation of Huygens' principle. Kirchhoff himself did not consider seismic imaging and the term was introduced into geophysics in the 1970s.

Back to our assumptions on the medium (constant and known). What happens if the medium is not accurately known? In this case the backpropagated waves will fail to interfere constructively and the image becomes defocused. The modern geophysicist has a comprehensive set of tools to improve the focusing by updating the background; seismic tomography is one of them. In the case that the medium is not homogeneous or not even isotropic or acoustic, these tools can account for inhomogeneities, anisotropy and shear arrivals. The Huygens-Fresnel-Kirchhoff principle gives the recipe how to construct waves in such media.

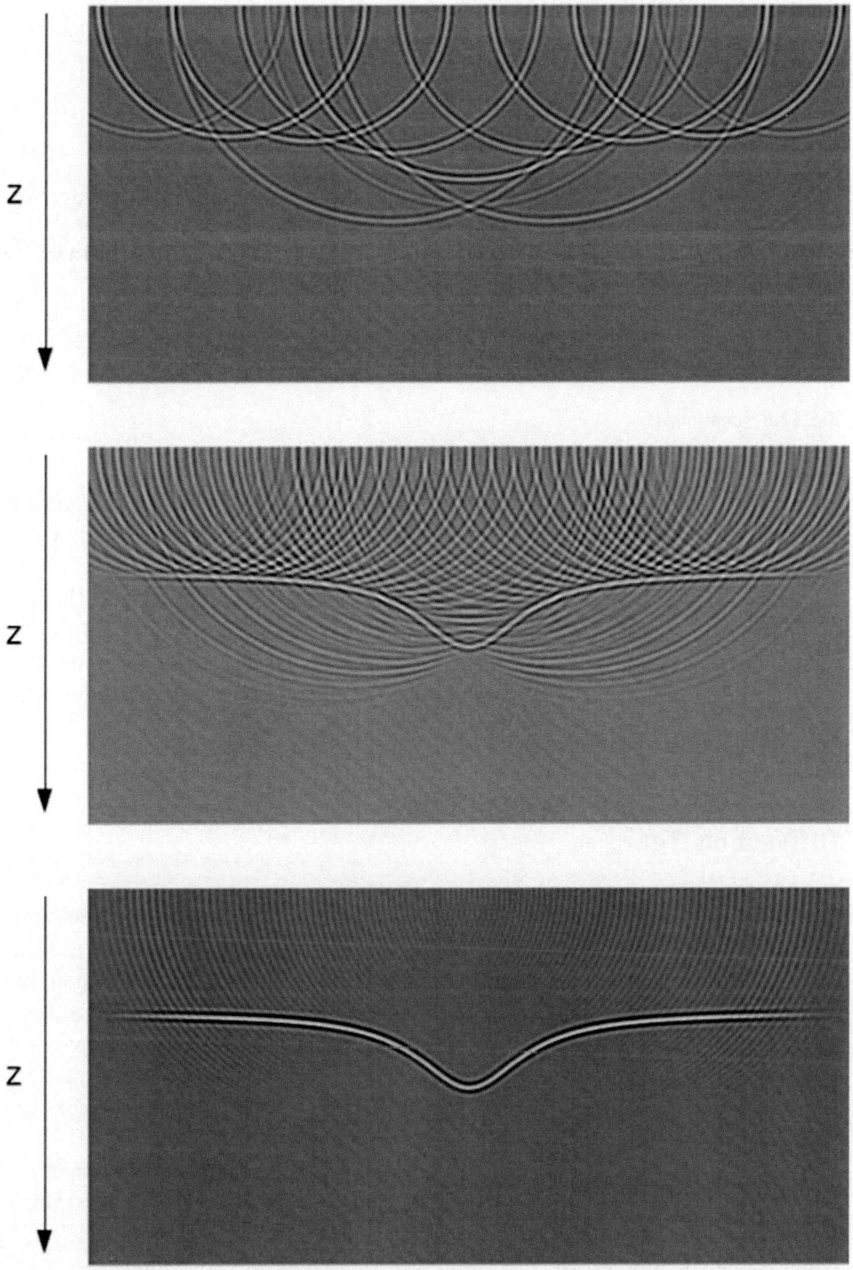

Fig. 12.6 Huygens' principle, interference and Kirchhoff migration

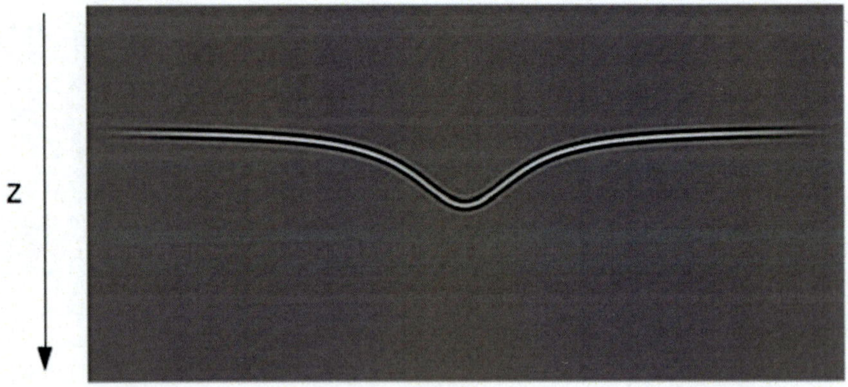

Fig. 12.6 (continued)

Figure 12.7 shows a more realistic example over a complicated faulted geology. The input data are in time. The time axis is now pointing downward to allow interpretation of data before imaging. The data contain many reflected wave arrivals, recognizable by the continuous curved events running from left to right. Using the procedure described above and assuming an accurate smooth (isotropic and acoustic) medium, we propagate the wave arrivals back in the subsurface and superpose them. The resulting image in the depth domain (with the z axis pointing down again) provides an accurate and well-focused image of the geology.

Diffraction Again

Enter diffraction. As pointed out in the previous Chap. 11, Huygens' principle and its later completion by Fresnel and Kirchhoff is very well suited for the modeling of diffracted waves. In the context of seismic imaging we consider as an operational definition for diffraction any wave which does not satisfy Snell's reflection law. In *diffraction imaging,* the deviation from Snell's law is taken as a way to separate out diffraction energy from the main wavefield. This allows the geophysicist to obtain separate images (diffraction images) with diffracting features isolated. This has some important advantages compared to the full wave images: the diffraction image has an increased resolution for small scale structural details and since it ignores Snell's law it has better illumination properties. Applications of diffraction imaging are effective over a wide range of geologic objectives, such as faults, fractures, karsts, stratigraphic edges, channels, fluid escape pipes, volcanic pipes and sand injectites.

Fig. 12.7 Input data in time domain (top) and depth image (bottom)

As an example Fig. 12.8 shows the diffraction image over the same data as Fig. 12.7. Diffraction events in the input data of Fig. 12.7 are manifested by the hyperbolas which are visible. In the diffraction of Fig. 12.8 all reflecting interfaces have been effectively removed, and a wealth of additional structural detail recovered.

Together with the full wave image of Fig. 12.7 this provides useful information for the geologic interpreter.

Figure 12.9 offers a zoom on these images, with diffraction hyperbolas highlighted. The reflection image provides the main structural interfaces. The diffraction image provides the structural skeleton of the subsurface geology.

Who is not tempted to make a comparison with Huygens' Laterna Magica (Fig. 11.14)?

Fig. 12.8 Diffraction image

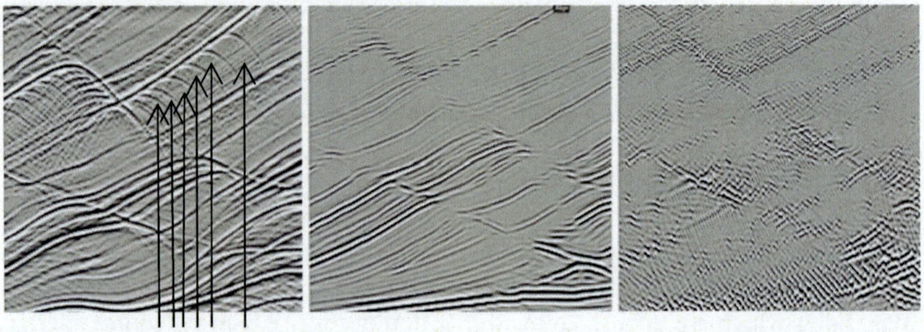

Fig. 12.9 Zoom on input data with some diffraction hyperbolas indicated by arrows (left); reflection image (middle) and diffraction image (right)

Epilogue

Traditionally, the word "theoretical" is seen as the opposite of "empirical". Theoretical refers to the mathematical and physical foundations of science. Empirical refers to conclusions drawn from the analysis of data. In the field of physics today, there is an abundance of observational data, and computational power is virtually unlimited. Experimentation was almost unknown in antiquity. Aristotle classified the theoretical sciences into three categories: physics, mathematics, and theology. Aristotle's physics is the study of nature, in a wide sense, encompassing biology, chemistry, geology, and meteorology. Remarkably, the discipline of physics, as formulated by Aristotle, had little to do with experimentation or quantitative measurement. Aristotle's physics uses a priori arguments to explain physical phenomena. The Scholastic tradition continued this practice throughout the Middle Ages. In the early 1600s, using experimental facts and very simple argumentation, Galileo showed that Aristotle's description of motion is wrong. Galileo's correct description paved the way to modern science.

The Renaissance marked the onset of an experimental tradition in Western Europe, accompanied by a concurrent tradition of applied mathematics. A prevailing notion held that the value of experimental physics lay in quantifying physical principles. The significance of mathematics for experimental physics faced scrutiny, with some scientists criticizing its excessive use, alleging an inappropriate reliance on abstract systems. Nevertheless, the general consensus maintained that both experimentation and mathematics were indispensable for acquiring knowledge. Understanding the physical world necessitated resorting to experiment, as knowledge from first principles was deemed insufficient. Reason advocated for a middle ground, combining experiment with quantitative measurement to provide a continuous validation of theoretical frameworks.

In the seventeenth century in the Netherlands, Christiaan Huygens made his entrance in the scientific world. He found new methods for grinding and polishing lenses, making telescopes more powerful. His work on perfecting timepieces to determine accurate longitude at sea yielded the pendulum clock. Huygens' combination

T. J. Moser and E. A. Robinson, *Walking with Christiaan Huygens*, History of Physics,
https://doi.org/10.1007/978-3-031-46158-3

of mathematics, mechanics, and optics within the realm of astronomy yielded his design and construction of the telescope he used to discover Saturn's largest moon. Beyond these practical instruments, Huygens established the wave theory of light.

In parallel developments in the Netherlands, Antoni van Leeuwenhoek pioneered the creation and use of the microscope. By 1688, William of Orange ascended to the English throne, fostering increased scientific collaboration between England and the Netherlands. Newton, in his seminal work 'Opticks' (1704), brilliantly combined experimentation with mathematics. Across the Continent, especially in Holland, scientists further refined the experimental Newtonian method. At the University of Leiden in the early 18th century, figures like Willem 's-Gravesande, Herman Boerhaave, and Pieter van Musschenbroek, following Newton's example, organized experiments that underscored the pivotal role of mathematics in experimental physics. 's-Gravesande's work particularly emphasized this, and he, along with other Dutch physicists, effectively redefined physics by synthesizing both theoretical and empirical methods. At the time, Western Europe, with a notable focus on England, France, and the Netherlands, emerged as the cradle of scientific development, with Huygens finding himself in the midst of this intellectual ferment.

Putting together a book on Huygens is both an intellectually rewarding and responsible undertaking. Delving into his scientific achievements in fields like astronomy, optics, and mathematics provides an opportunity for in-depth research and understanding. Furthermore, the pleasure of studying original texts by Huygens and his contemporaries cannot be overstated. A wealth of material is still waiting to be explored. The choice of topics in this book is admittedly a personal one. Many topics have not been covered. To name a few: Huygens had a keen interest in planar curves (algebraic and transcendental) and in algebraic equations (solutions for the cubic and quartic equations by Cardano and Ferrari had been published in 1545, the impossibility of a general solution of higher order equations was finally proved in the early 19th century). He spent quite some time on notes and correspondence regarding the catenary, the line hanging between two endpoints under its own weight. While we have only briefly mentioned his work on probability, we have not touched upon his contributions to pneumatics, nor on Huygens' technical drawings, which are often considered genuine works of art. In 1657, Huygens demonstrated his skill by providing a proof of Pythagoras' theorem. However, its complexity may have limited its historical significance. Huygens' philosophical and religious views expressed in his book the Cosmotheoros, in correspondence and in scattered manuscripts deserve attention.

In the Netherlands numerous schools, research laboratories, and students' associations have been named after Christiaan Huygens. Monuments to Huygens can be found in Rotterdam, Delft, Leiden, Haarlem, and Voorburg, at the Hofwijck Estate. A students' association in Utrecht named after Huygens, which was founded in 1887 and counts a few Nobel prize winners among its members, has as its motto "Docendo Discimus" ("By teaching we learn"). Underlying our exploration of Huygens' contributions and legacy, this premise has been the main motivation for our book.

We conclude with a dedication to Huygens from the preface to the Dutch translation (from Latin) of the *Cosmotheoros* by Pieter Rabus (1660-1702), a Dutch poet and translator: "Nog streelt dat Hemelwerk mijn zinnen, En 't komt my telkens weêr te binnen, Hoe Hugens [sic] my van d' Aarde om hoog, Opvoerde boven 't zwerk der Wolken, En Dampen, na de Starrevolken, door kreits op kreits, door boog op boog" ("Still, that Celestial Work delights my senses, And it comes to my mind again and again, How Huygens lifted me from the Earth upwards, Above the sky of Clouds and Vapors, Towards the Starry Realms, through circle upon circle, through arch upon arch").

References

Aarts, J. 2015. *Christiaan Huygens: Het Slingeruurwerk, een studie,* Epsilon Uitgaven 80.

Ahluwalia, D.S., R.M. Lewis, and J. Boersma. 1968. Uniform asymptotic theory of diffraction by a plane screen. *SIAM Journal on Applied Mathematics* 16: 783–807.

Aki, K., P.G. Richards. 1980. *Quantitative Seismology, Theory and Methods.* W.H. Freeman and Company.

Aldersey-Williams, H. 2020. *Dutch Light, Christiaan Huygens and the Making of Science*, London, Picador.

Andriesse, C.D. 2005. *Huygens: The Man Behind the Principle*, Cambridge University Press. (Dutch: 1993. *Titan kan niet slapen*, Uitgeverij Contact, Amsterdam.)

Anonymous. 1676. Démonstration touchant le mouvement de la lumière trouvé par M. Roemer de l'Académie des sciences, *Journal des Sçavans.*

Baker, B.B., E.T. Copson. 1939. *The Mathematical Theory of Huygens' Principle*, Oxford at the Clarendon Press.

Bartholin, E. 1669. *Experimenta Crystalli Islandici disdiaclastici, quibus mira et insolita refractio detegitur*, Copenhagen, Daniel Paulli.

Bernoulli, J. 1697. Curvatura radii in diaphanis nonuniformibus (The curvature of a ray in nonuniform media). *Acta Eruditorum*, 206–211.

Born, M., E. Wolf. 1999. *Principles of optics*, Cambridge University Press.

Bouwkamp, C.J. 1954. Diffraction theory. *Reports on Progress in Physics* 17: 35–100.

Brekhovskikh, L.M., and Yu.P. Lysanov. 2003. *Fundamentals of Ocean Acoustics.* Springer.

Brown, K. 2018. *Reflections on Relativity.*

Buchal, R.N., and J.B. Keller. 1960. Boundary layer problems in diffraction theory. *Comm. Pure Appl. Math.* 13: 85–114.

Buchwald, J.Z., and C.-P. Yeang. 2016. Kirchhoff's theory for optical diffraction, its predecessor and subsequent development: The resilience of an inconsistent theory. *Archive for History of Exact Sciences* 70: 463–511.

Buchwald, J.Z. 1989. *The Rise of the Wave Theory of Light.* The University of Chicago Press.

Cauchy, A.L. 1821. *Cours d'Analyse (Cours in Analysis).* Paris: L'Imprimerie Royale.

Cauchy, A.L. 1823. *Résumé des leçons sur le calcul infinitésimal (Summary of Lessons on Infinitesimal Calculus).* Paris: L'Imprimerie Royale.

Cauchy, A.L. 1830. Sur la dispersion de la lumière. *Bull. Des Sc. Math.* 14: 9.

Cauchy A.L. 1826–28. *Leçons sur les applications du calcul infinitésimal à la géométrie (Lessons on the Applications of Infinitesimal Calculus on Geometry)*, Paris, L'Imprimerie Royale.

Cauchy, A.L. 1868. *Sept Leçons de physique générale.* Deuxième Leçon, Bureau du Journal Les Mondes, Paris.

Červený, V. 2001. *Seismic Ray Theory.* Cambridge University Press.

T. J. Moser and E. A. Robinson, *Walking with Christiaan Huygens*, History of Physics, https://doi.org/10.1007/978-3-031-46158-3

Clausius, R. 1851. On the moving force of heat, and the laws regarding the nature of heat itself which are deducible therefrom. *Philosophical Magazine* 4(2): 1–21, 102–119.

Copernicus, N. 1543. *De revolutionibus orbium coelestium.* Nuremberg.

Crommelin, C.A., J.H. Plantenga, and J.A. Volgraff. 1929. Christiaan Huygens. 1629–14 April—1929. *Zijn geboortedag, 300 jaar geleden, herdacht.* Uitgever. H.J. Paris. Amsterdam

D'Alembert, J.-le-R. 1743. *Traité de dynamique.* Paris, David l'Aîné.

D'Alembert, J.-le-R. 1744. *Traité de l'Équilibre et du Mouvement des Fluides.* Paris, David l'Aîné.

D'Alembert, J.-le-R. 1745. *Théorie Générale des Vents.* David l'Aîné, Paris.

Darrigol, O. 2012. Poincaré's Light, *Séminaire Poincaré* XVI, 1–43.

Del Monte, G. 1600. *Perspectivae libri VI.* Pesaro, Girolamo Concordia.

Deppermann, K., and W. Franz. 1954. Theorie der Beugung an der Kugel unter Berücksichtigung der Kriechwelle. *Annalen Der Physik* 6: 253–264.

De Saint-Vincent, G. 1647. *Opus geometricum quadraturae circuli et sectionum coni decem libris comprehensum.* Antwerp. Joannes et Jacobus Meursius

Descartes, R. 1937. *Œuvres et lettres.* Édition d'André Bridoux. Parution en Mai. Bibliothèque de la Pléiade, n 40.

Diderot, D., and J.-le-R. d'Alembert. 1751–1772. *Encyclopédie ou dictionnaire raisonné des sciences, des arts et des métiers,* Paris.

Doyle, A.I.C. 1986. *The Complete Illustrated Sherlock Holmes.* Omega Books.

Duistermaat, J.J. 1990. Huygens' principle for linear partial differential equations. *Huygens' Principle 1690–1990, Theory and Applications (The Hague and Scheveningen).* In: eds H. Blok, H. A. Ferwerda and H. K. Kuiken.

Einstein, A. 1905. Zur Elektrodynamik bewegter Körper. *Annalen Der Physik* 17: 891–921.

Einstein, A. 1940. Considerations concerning the fundaments of theoretical physics. *Science* 91: 2369.

Ellicott, J. 1739a. An account of the influence which two pendulum clocks were observed to have upon each other. *Philosophical Transactions* 41:453.

Ellicott, J. 1739b. Further observations and experiments concerning the two clocks above mentioned. *Philosophical Transactions,* 41:453.

Ellis, W. 1873. On sympathetic influence between clocks. *Monthly Notices of the Royal Astronomical Society* 33 (8): 480–484.

Feynman, R.P. 1948. Space-time approach to non-relativistic quantum mechanics. *Reviews of Modern Physics* 20: 367.

Feynman, R.P. 1985. *QED: The strange theory of light and matter.* Princeton University Press.

Feynman, R.P., R.B. Leighton, M. Sands. 1989. *The Feynman Lectures on Physics.* Addison-Wesley.

Franz, W. 1949. Zur Theorie der Beugung. *Zeitschrift Für Physik* 125: 563–596.

Franz, W., and K. Deppermann. 1952. Theorie der Beugung am Zylinder unter Berücksichtigung der Kriechwelle. *Annalen Der Physik* 6: 361–373.

Fresnel, A.-J. 1866. *Œuvres Complètes d'Augustin Fresnel.* Paris: Imprimerie Impériale.

Galilei, G. 1638. *Discorsi e dimostrazioni matematiche intorno a due nuove scienze (Discourses and Mathematical Demonstrations Relating to Two New Sciences),* Leiden, Elsevier.

Green, G. 1828. *An Essay on the Application of Mathematical Analysis to the Theories of Electricity and Magnetism.* Nottingham.

Gregory, J. 1667. *Vera Circuli et Hyperbolae Quadratura.* Padua.

Grimaldi, F. 1665. *Physico-mathesis de lumine, coloribus et iride aliisque adnexis* (in Latin). Girolamo Bernia: Johann Zieger.

Günther, P. 1988. *Huygens principle and hyperbolic equations.* Academic Press.

Hadamard J. 1923. *Lectures on Cauchy's Problem.* Yale.

Hadamard, J. 1924. Le Principe de Huygens. *Bull. de la Soc. Math. de France,* 52:610–40.

Hagedoorn, J.G. 2001. In: G. Diephuis. The seismic transmission volume. *Geophysical Prospecting* 49:697–707.

Hecht, E. 2001. *Optics.* Addison-Wesley.

Helbig, K. 1994. *Foundations of Anisotropy for Exploration Seismics*. Elsevier Science & Technology Books.
Helmholtz, H. von 1859. *Theorie der Luftschwingungen in Röhren mit offenen Enden: Journal für die reine und angewandte Mathematik (Crelle's Journal)* 57:1–72.
Hess, H.-J. 1980. Bücher aus dem Besitz von Christiaan Huygens (1629–1695) in der Niedersächsischen Landesbibliothek Hannover. *Studia Leibnitiana* 12: 1–51.
Hevelius, J. 1647. *Selenographia, sive Lunae description*. Gedani. Hünefeld.
Hooke, R. 1665. *Micrographia: or Some Physiological Descriptions of Minute Bodies Made by Magnifying Glasses. With Observations and Inquiries Thereupon*, Royal Society.
Hordijk, B. and R. Memel. 2021. Salomon Coster, the clockmaker of Christiaan Huygens. The production and development of the first pendulum clocks in the period 1657–September 1658. *Antiquarian Horology* 42(3):323–344.
Huygens, Chr. 1888–1950. *Œuvres complètes de Christiaan Huygens (Complete Works of Christiaan Huygens)*, 22 volumes, Martinus-Nijhoff. The Hague.
Icke, V. 2013. *De principes van Huygens*. Historische uitgeverij.
Ikelle, L.T. 2012. On elastic-electromagnetic mathematical equivalences. *Geophysical Journal International* 189 (3): 1771–1780.
Kalashnikov, A. 1912. The Gouy-Sommerfeld diffraction. *Journal of the Russian Physical-Chemical Society* 44: 137–144.
Keller, J.B. 1962. Geometrical theory of diffraction. *Journal of the Optical Society of America* 52: 116–130.
Keller, J.B. 1978. Rays, waves and asymptotics. *Bulletin of the American Mathematical Society* 8: 727–750.
Keller, J.B., R.M. Lewis, and B.D. Seckler. 1957. Diffraction by an aperture II. *Journal of Applied Physics* 28: 570–579.
Keller, J.B. 1953. The geometrical theory of diffraction, *Proceedings of the Symposium on Microwave Optics*. 2(1):207–210. McGill University.
Keller, J.B. 1956. Diffraction by an aperture I, New York University Institute of Mathematical Sciences. *Research Report* EM-92:1–61.
Kirchhoff, G. 1882. Zur Theorie der Lichtstrahlen, *Annalen der Physik*, 18:663–695. In: K. Klem-Musatov, H.C. Hoeber, T.J. Moser, and M.A. Pelissier. 2016a. Classical and Modern Diffraction Theory. SEG Geophysics Reprint.
Kirchhoff, G. 1891. *Vorlesungen über mathematische Physik, zweiter Band, Mathematische Optik*, Teubner. Leipzig.
Koehler, P.J. 2015. Huygens' headache. *Cephalalgia* 35 (13): 1215–1219.
Kouyoumjian, R.G., and P.H. Pathak. 1974. A uniform geometrical theory of diffraction for an edge in a perfectly conducting surface. *Proceedings of the IEEE* 62: 1448–1461.
Van der Leer, K., and H. Boers. 2022. *Huygens and Hofwijck: The Inventive World of Constantijn and Christiaan Huygens*. Amsterdam University Press.
Lewis, R.M., and J. Boersma. 1969. Uniform asymptotic theory of edge diffraction. *Journal of Mathematical Physics* 10: 2291–2305.
Maesumi, M. 1992. Parabolic mirrors, elliptic and hyperbolic lenses. *American Mathematical Monthly* 6: 558–560.
Maey, E. 1893. Ueber die Beugung des Lichtes an einem geraden, scharfen Schirmrande. *Annalen der Physik*, 285, 69–104.
Maggi, G.A. 1888. Sulla propagazione libra e perturbata delle onde luminose in un mezzo isotropo. *Annali Di Matematica Pura Ed Applicata* 16: 21–48.
Marchand, E.W., and E. Wolf. 1966. Consistent formulation of Kirchhoff's diffraction theory. *J Journal of the Optical Society of America* 56: 1712–1721.
Maxwell, J.C. 1859. *On the stability of the motion of Saturn's rings*. London: Macmillan and Co.
Maxwell, J.C. 1868. On governors. *Proceedings of the Royal Society of London* 16: 270–283.
Maxwell, J.C. 1880. On a possible mode of detecting a motion of the solar system through the luminiferous either. In a Letter to Mr. D. P. Todd. *Nature* 21: 314–315.

Maxwell, J.C. 1861–1862. On physical lines of force. *Philosophical Magazine* 21, 23 (4) I, II, III, IV.
Miyamoto, K., and E. Wolf. 1962. Generalization of the Maggi-Rubinowicz theory of the boundary diffraction wave—part I and II. *Journal of the Optical Society of America* 52: 615–637.
Necker, L.A. 1832. Observations on some remarkable optical phænomena seen in Switzerland; and on an optical phænomenon which occurs on viewing a figure of a crystal or geometrical solid. *The London and Edinburgh Philosophical Magazine and Journal of Science.*
Neumann, P.M. 1998. Reflections on Reflection in a Spherical Mirror, *American Mathematical Monthly*, 105 (6): 523–528.
Newton, I. 1729. *The Mathematical Principles of Natural Philosophy*, translated by Andrew Motte with a preface by Roger Cotes. To which are added, The Laws of the Moon's Motion, according to Gravity. By John Machin. London: Benjamin Motte.
Pascal, B. 1998, 2000. *Œuvres complètes Tome I et II*. Édition de Michel Le Guern. Bibliothèque de la Pléiade.
Pauli, W. 1938. On asymptotic series for functions in the theory of diffraction of light. *Physical Review* 54: 924–931.
Pemberton, H. 1728. *A View on Sir Isaac Newton's Philosophy*. London. S. Palmer
Pietrow, A.G.M. 2023. Did Christiaan Huygens need glasses? A study of Huygens' telescope equations and tables notes and records. *Royal Society Journal of the History of Science.*
Poe E.A. 1848. Eureka: a prose poem. In: *Edgar Allan Poe Fiction and Poetry*, Complete and Unabridged, 2006, Barnes and Noble, New York.
Poincaré, H. 1889. Leçons sur la Théorie Mathématique de la Lumière, *Cours de Physique Mathématique,* Rédigées par J. Blondin, G. Carré, Éditeur.
Poor, V.C. 1925. The Huygens governor. *American Mathematical Monthly* 32: 115–121.
Raman, C.V., and K.S. Krishnan. 1927. The diffraction of light by a metallic screen. *Proceedings of the Royal Society of London* A116: 254–267.
Rayleigh Lord, F.R.S. 1897. On the passage of waves through apertures in plane screens and allied problems. *Philosophical Magazine* 43: 259–272.
Riesz, M. 1949. L'integrale de Riemann-Liouville et le problème de Cauchy. *Acta Mathematica* 81: 1–223.
Rouse Ball, W.W. 1908. *A short account of the history of mathematics*, 4th ed. New York: Dover.
Rubinowicz, A. 1917. Die Beugungswelle in der Kirchhoffschen Theorie der Beugungserscheinungen. *Annalen Der Physik* 358: 257–278.
Rubinowicz, A. 1924. Zur Kirchhoffschen Beugungstheorie. *Annalen Der Physik* 378: 339–364.
Rubinowicz, A. 1965. Darstellung der Sommerfeldschen Beugungswelle in einer Gestalt, die Beiträge der einzelnen Elemente der beugende Kante zur gesamten Beugungswelle erkennen last. *Acta Physica Polonica* 28: 841–860.
Rubinowicz, A. 1957, Die Beugungswelle in der Kirchhoffschen Theorie der Beugung. *Polnische Akademie der Wissenschaften Physikalische Monographien.* Warschau: Panstwowe. Wydawnictwo Naukowe.
Savornin, J. 1939. Étude de la diffraction éloignée. *Annales De Physique* 11: 129–255.
Schelkunoff, S.A. 1951. Kirchhoff's formula its vector analogue, and other field equivalence theorems. *Communications in Pure and Applied Mathematics* 4: 43–59.
Schouten, J.A. 1920. Die relative und absolute Bewegung bei Huygens. *Jahresbericht der Deutschen Mathematiker-Vereinigung.*
Snellius, W. 1621. *Cyclometricus.* Leiden. Elsevier.
Sommerfeld, A. 1896. Mathematische theorie der diffraktion. *Mathematische Annalen* 47: 317–374.
Sommerfeld, A. 1949. *Vorlesungen über theoretische Physik*, vol. IV. Optik: Deutsch Verlag.
Sonar, T. 2018. *The History of the Priority Diſpute between Newton and Leibniz*. Birkhäuser.
Stein, J. 1941. Christiaan Huygens en de Jezuieten. *Philosophische en Theologische Faculteiten der Nederlandsche Jezuieten* 4 (1): 166–191.
Stokes, G.G. 1849. On the dynamical theory of diffraction. *Transactions of the Cambridge Philosophical Society* 8: 1–62.

Stokes, G.G. 1864. On the discontinuity of arbitrary constants which appear in divergent developments. *Transactions of the Cambridge Philosophical Society* 10: 106–128.

Sunil Kumar, P.B., and G.S. Ranganath. 1991. Geometrical theory of diffraction, *Pramana. Journal of Physics* 37: 457–488.

Taylor, B. 1713. De motu nervi tensi. *Philosophical Transactions*, 28:337.

Taylor, B. 1715. *Methodus Incrementorum Directa et Inversa.* London. William Innys.

Tiscareno, M.S. 2020. Planetary rings. In (eds.): Oswalt, T.D., French, L.M. and Kalas, P. *Planets, Stars and Stellar Systems.* Dordrecht. Springer.

Trorey, A.W. 1970. A simple theory for seismic diffractions. *Geophysics* 35:762–784.

Whewell, W. 1834. On the connexion of the physical sciences by mrs. Somerville. *Quarterly Review* 51: 54–68.

Wilcox, C.H. 1956. A generalization of theorems of Rellich and Atkinson. *Proceedings of the American Mathematical Society* 7: 271–276.

Wolf, E. 1990. The life and work of Christiaan Huygens. In: (eds.) H. Blok, H. A. Ferwerda and H. K. Kuiken. *Huygens' Principle 1690–1990, Theory and Applications (The Hague and Scheveningen).*

Yoder, J.G. 1988. *Unrolling time: Christiaan Huygens and the mathematization of nature.* Press: Cambridge University.

Young, T. 1807. *A Course of Lectures on Natural Philosophy and the Mechanical Arts.* London.